EUROPEAN PERCEPTIONS OF TERRA AUSTRALIS

European Perceptions of *Terra Australis*

Edited by

ANNE M. SCOTT
University of Western Australia

ALFRED HIATT
Queen Mary, University of London

CLAIRE MCILROY
University of Western Australia

and

CHRISTOPHER WORTHAM
University of Western Australia

Routledge
Taylor & Francis Group
LONDON AND NEW YORK

First published 2011 by Ashgate Publishing

2 Park Square, Milton Park, Abingdon, Oxon OX14 4RN
711 Third Avenue, New York, NY 10017, USA

Routledge is an imprint of the Taylor & Francis Group, an informa business

First issued in paperback 2017

British Library Cataloguing in Publication Data
European perceptions of Terra Australis.
1. Australia--Discovery and exploration--European.
2. Australia--Discovery and exploration--Early works to
1800. 3. Australia--Foreign public opinion, European.
4. Public opinion--Europe--History. 5. Europeans--
Australia--Attitudes.
I. Scott, Anne M. [et. al.].
919.4'042-dc22

Library of Congress Cataloging-in-Publication Data
Scott, Anne M.
European perceptions of Terra Australis / Anne M. Scott... [et. al.].
p. cm.
Includes bibliographical references and index.
ISBN 978-1-4094-2605-9 (hardcover) -- ISBN 978-1-4094-3941-7 (ebook)
1. Cartography--Australia--History. 2. Geography--Australia--History. 3. Geographical perception--Europe--History. 4. Australia--In literature 5. Australia--Public opinion. 6. Public opinion--Europe. 7. Australia--Discovery and exploration--European. 8. Australia--Description and travel. I. Title.

GA1681.S36 2012
919.4--dc23

2011026572

ISBN 13: 978-1-4094-2605-9 (hbk)
ISBN 13: 978-1-138-11097-7 (pbk)

Contents

List of Figures

Notes on the Contributors

Mercedes Maroto Camino, Lancaster University

Mercedes Camino is Professor of Hispanic Studies at Lancaster University. Professor Camino's 2008 publication *Exploring the Explorers: Spaniards in Oceania 1519–1794* (Manchester University Press) extends the research done for an earlier book, *Producing the Pacific: Maps and Narratives of Pacific Exploration 1567–1606* (Rodopi, 2005), to include Spanish voyages to the Pacific during the eighteenth century. Her current research concentrates on the use of women and children as points of identification in films dealing with the Spanish *maquis* of the 1940s. Her fifth book, *Film, Memory and the Legacy of the Spanish Civil War* (2011) has been published by Palgrave.

Norman Etherington, University of Western Australia

After he completed his BA and PhD at Yale, Norman Etherington lectured for 20 years at the University of Adelaide before taking up a chair of history at the University of Western Australia. In addition to being Emeritus Professor of History, he is a Fellow of the Academy of Social Sciences in Australia, the Royal Historical Society and the Royal Geographical Society. He also holds an honorary appointment to the History Department at the University of South Africa. His most recent books include *Missions and Empire* (Oxford University Press, 2005) and *Mapping Colonial Conquest: Australia and South Africa* (University of Western Australia Press, 2007).

Jean Fornasiero, University of Adelaide

Jean Fornasiero is Professor of French Studies at the University of Adelaide. Her principal area of expertise is the history of ideas in nineteenth-century France. She is a member of the ARC project, 'The Baudin Legacy' and, with John West-Sooby, she is preparing a critical edition of François Péron's *Memoir on the English Settlements of New Holland*. She is also co-author with John West-Sooby and Peter Monteath of *Encountering Terra Australis: The Australian Voyages of Nicolas Baudin and Matthew Flinders* (Wakefield Press, 2004) (Frank Broeze Memorial Prize, 2005; revised edition, 2010). Recent edited volumes include *Littoral*, an art catalogue on the marine works of Charles-Alexandre Lesueur (with Vivonne Thwaites; 5 Star Print, 2010) and *Explorations and Encounters in French* (with Colette Mrowa-Hopkins; University of Adelaide Press, 2010).

Alfred Hiatt, Queen Mary, University of London

Alfred Hiatt is a Reader in English Literature at Queen Mary, University of London. His research interests centre on historical writing and spatial representation. His first book, *The Making of Medieval Forgeries*, was published in 2004 (British Library/University of Toronto Press). He has a particular interest in map history, and the results of this can be seen in a study of the representation of unknown, particularly antipodal, land, in classical, medieval and early modern literature and cartography, entitled *Terra Incognita: Mapping the Antipodes before 1600* (British Library/University of Chicago Press, 2008). He is currently working on a new map-related project, under the working title of 'Dislocations: Reading Medieval Maps'. Related interests include fourteenth- and fifteenth-century humanism, geographical and travel writing, and the role of maps and geography in the history of European colonization in the medieval and early modern periods.

Bill Leadbetter, Edith Cowan University

Bill Leadbetter is currently Dean of Studies for the Anglican Diocese of Perth. He trained as an Ancient Historian at Macquarie University and taught at Macquarie University, the University of Sydney and Edith Cowan University. He has published widely in the history of the Later Roman Empire and taken part in archaeological surveys of Roman sites in Turkey. His study of the Roman Empire at the beginning of the fourth century, entitled *Galerius and the Will of Diocletian*, was published by Routledge in October 2009.

Michael McCarthy, Western Australian Museum

Michael McCarthy, a member of the Western Australian Museum's Department of Maritime Archaeology holds a doctorate in maritime archaeology and a masters in Western Australian history. Additionally he holds an adjunct professorship at the University of Notre Dame Australia in Fremantle. He was the Museum's Inspector of Wrecks, is responsible for the Western Australian Museum's HMAS *Sydney*/HSK *Kormoran* programme and was official observer during the examination of the two wrecks in 2008. He was also responsible for the Department's Outreach and its Indigenous studies (Strangers on the Shore and indigenous maritime depictions) programmes. Dr McCarthy has led excavations at the Dutch East India ship *Zuytdorp* (1711), the iron-hulled SS *Xantho* (1872), the Broome flying boats from the Second World War and other endangered sites. In 2001 he led the successful searches for *L'Uranie* (1820), the vessel that transported the French explorers Rose and Louis de Freycinet, in the Falkland Islands and William Dampier's famous *Roebuck* (1701) at Ascension Island. These works have led to numerous museum exhibits, including a 'Steamships to Suffragettes' exhibit, and most recently he co-curated the Museum's 'Voyages of Grand Discovery' and 'Journeys of Enlightenment' exhibits. He is editor of two compilations and has produced two books, and over 200 articles, reports, interpretive works and websites.

Claire McIlroy, University of Western Australia

Claire McIlroy is a Research Associate for International Liaison in the Australian Research Council Centre of Excellence for the History of Emotions based at the University of Western Australia. She is an Honorary Research Fellow in the School of Social and Cultural Studies at the University of Western Australia, and author of *The English Prose Treatises of Richard Rolle* (Boydell & Brewer, 2004), now working on the perception of solitude and/or solitary living in the writings of the English mystics and related texts. She is also the Conference Manager for CARMEN (Co-operative for the Advancement of Research through a Medieval European Network), a member of the editorial board for the Brepols Publishers Series, 'Early European Research, 1200–1650', and a member of the editorial committee for *Parergon*, the journal of the Australian and New Zealand Association for Medieval and Early Modern Studies.

Katrina O'Loughlin, University of Western Australia

Having recently completed her doctorate at the University of Melbourne (2009), Katrina O'Loughlin has joined the University of Western Australia as an Honorary Research Fellow. Her research interests include eighteenth-century writing, cultural history, travel and cultural difference. She has published work on women's travel, writing and subjectivity, and taught in the School of Culture and Communication at the University of Melbourne. Current projects include the development of her doctoral research for publication, and a proposed edition of two female-authored Russian travel narratives of the 1730s. She is also pursuing new research in the area of eighteenth-century cosmopolitanisms, with a focus on correspondence, cultural exchange and intellectual sociability in the second half of the century.

Leigh T.I. Penman, Goldsmiths, University of London and University of Oxford

Leigh Penman graduated in Arts and Law from the University of Melbourne, before completing a doctorate concerning seventeenth-century intellectual and religious history, undertaken at Melbourne in conjunction with the former Max-Planck-Institut für Geschichte in Göttingen, Germany. At present he is a Research Associate at Goldsmiths, University of London, and an Associate Member of the History Faculty of the University of Oxford. He is the author of the forthcoming *Unanticipated Millenniums: Chiliastic Thought in Post-Reformation Lutheranism, 1600–1630*, in addition to more than a dozen articles on various areas of cultural and intellectual history, ranging from the seventeenth century to the present.

W.A.R. (Bill) Richardson, Flinders University

Bill Richardson retired as an Associate Professor (Spanish and Portuguese) at Flinders University in 1987; since then he has been a Visiting Scholar there, in the Department of Languages, School of Humanities, Faculty of Education, Humanities and Law. His research interests lie in the area of the history of cartography (especially fourteenth to seventeenth centuries), enigmatic place-names on early maps and charts, and in manuscript and printed sailing directions. He has published some 40 articles, mainly in specialist academic journals in Australia, the United States, the UK, and Portugal, and his book *Was Australia Charted before 1606? The Jave la Grande Inscriptions* was published by the National Library of Australia in 2006 (repr. 2008). He has lectured in Australia, New Zealand, the United States, the UK, France, Germany, Spain and Portugal and is an honorary member of the Australian and New Zealand Map Society and member of the Advisory Council of the International Map Collectors' Society.

Margaret Sankey, University of Sydney

Margaret Sankey is Professor Emerita in the Department of French Studies at the University of Sydney. Her research focus is on cultural and intellectual history and her specialist period spans the French seventeenth and eighteenth centuries. She has published extensively on the work of the writer Cyrano de Bergerac. For some years now, she has been working on early French notions of *Terra Australis*, publishing in 2006 a critical edition of the Abbé Jean Paulmier's *Mémoires pour l'établissement d'une mission dans les Terres australes* (Honoré Champion). She has also published in the area of early French exploration of Australia and is at present heading a team of ARC-funded researchers working on the Baudin expedition in 1800–1804 to Australia.

Anne M. Scott, University of Western Australia

Anne M. Scott is an an Honorary Research Fellow in English and Cultural Studies at the University of Western Australia. Her field of research is in fourteenth-century English literature, and her monograph entitled *'Piers Plowman' and the Poor* (Four Courts Press) was published in 2004. She has published several essays on fourteenth- and fifteenth-century English literature, co-edited a volume of essays entitled *Renaissance Poetry and Drama in Context* (Cambridge Scholars, 2008), and work is currently in progress for a book on the iconography and representations of poverty in medieval English literature and art. She is editor of *Parergon*, the journal of the Australian and New Zealand Association for Medieval and Early Modern Studies, now available as part of the Project Muse database.

John West-Sooby, University of Adelaide

John West-Sooby is Associate Professor of French at the University of Adelaide. His principal area of expertise is nineteenth-century French literature. He is a member of the ARC project, 'The Baudin Legacy', and, with Jean Fornasiero, he is preparing a critical edition of François Péron's *Memoir on the English Settlements of New Holland*. He is also co-author with Jean Fornasiero and Peter Monteath of *Encountering Terra Australis: The Australian Voyages of Nicolas Baudin and Matthew Flinders* (Wakefield Press, 2004) (Frank Broeze Memorial Prize, 2005; revised edition, 2010). He has edited several volumes of essays, the most recent of which is *Nowhere is Perfect: French and Francophone Utopias/*

Dystopias (University of Delaware Press, 2008). An edited volume on the theme *Discovery and Empire: The French in the South Seas* is due for publication in 2011/2012.

Christopher Wortham, University of Western Australia

Christopher Wortham is an Emeritus Professor and Senior Honorary Research Fellow in English and Cultural Studies at the University of Western Australia. He is also Professor of Theatre Studies and English Literature at the University of Notre Dame Australia in Fremantle. His main research area and interests are sixteenth- and seventeenth-century literature. He has published articles on Marlowe's *Dr Faustus*, Shakespeare's *King Lear* and *Antony and Cleopatra*, Milton's *Paradise Lost* and Marvell's Cromwell poems. Editions published include the morality play, *Everyman*, Marlowe's *Dr Faustus* and Marvell's Poetry, *On Shakespeare: Readers, Audiences, Players*, edited with R.S. White and C. Edelman (University of Western Australia Press, 1998) and *Renaissance in the South*, edited with Tom Burvill (University of Queensland Press, 1998) He is also the former editor of the journal *Parergon*.

Acknowledgements

The papers which have become the chapters of this book were originally presented at a symposium held at the University of Western Australia in June 2009, entitled 'Perceptions of *Terra Australis*'. This symposium was developed under the Australian Research Council Network for Early European Research (NEER), within its research theme: 'Early European-Australasian Connections'. The editors acknowledge the initial generous sponsorship of NEER, and the ongoing support of the Perth Medieval and Renaissance Group, and the Centre for Medieval and Early Modern Studies (CMEMS) at the University of Western Australia, in the production of this volume. Particular thanks are due to Kate Attwood for help in preparing the manuscript and index, and to Brett Hirsch for help in preparing the images. The editors also thank Peter Woods and the Woodside Valley Foundation for their long-standing sponsorship of research, some of which underpins chapters in this volume.

Chapter 1
Perceptions

Anne M. Scott

To invite a group of scholars from different disciplines to write about 'perceptions' of *Terra Australis* is to give them an opportunity to define the intellectual responses to *Terra Australis* that have been made across centuries of thought and exploration, to record the personal perceptions of cartographers, geographers, explorers and travellers, and to contextualize these perceptions in their historical moment and cultural environment. The chapters in this collection bring a new dimension to the already wide field of publications dealing with cartography, exploration and the history of discovery in the southern land, *Terra Australis*.[1]

The term 'perception' is richly evocative, for perception is essentially an individual activity. Driven by the mind, it is nourished through the senses, and given expression in as many ways as there are people. To perceive engages the individual in some form of judgement about what is perceived, so a perception becomes an assessment, an appraisal – always a personal activity and therefore, by definition, subjective. But, in being subjective, perceptions transmitted to others become immensely rich, evoking, as they do, the wealth of experience, knowledge and understanding that inform the person making the 'perception'. This quickly becomes apparent in reading definitions of the term. Isaac Watts says: 'Perception is that Act of the Mind (or as some Philosophers call it, rather a Passion or Impression) whereby the Mind becomes conscious of any Thing, as when I feel Hunger, Thirst, or Cold, or Heat; when I see a Horse, a Tree, or a Man; when I hear a human Voice, or Thunder' (*Logick* I. i, 1725). In this definition, perception is uniquely individual. Seen in this light, the term has proved to be an immensely stimulating one around which to construct an essay collection. It allows for two kinds of individual perception – that of the scholarly author, and that of the historical subject on which the scholar has conducted research. Non-

---

1 The wealth of literature is to be found referenced in the bibliography to this volume. Particular mention must be made of two publications which were not available when the chapters in this volume were being written. The first is Douglas Mark Goldie, *The Idea of the Antipodes: Place, People and Voices* (New York, 2010). The second is Avan Judd Stallard, 'Origins of the Idea of the Antipodes: Errors, Assumptions, and a Bare Few Facts', *Terrae Incognitae*, 42 (2010): 34–52.

prescriptive, the term also allows for a truly interdisciplinary approach to the topic, for perceptions can be expressed through works of imaginative art and literature, works of scientific observation, personal records and even through national policy-making, all of which present themselves as fruitful fields of scholarly enquiry.

The second major element of the project's title, *Terra Australis*, is equally evocative. Never confined to the specific continental landmass that is now known as Australia, the term originates in the speculative thought of antiquity about the material world and man's geographical place in it. To trace the use of the term *Terra Australis* through the perceptions of philosophers, poets, artists, writers, geographers, cartographers, explorers, scientists and circumnavigators is to gain insights into a wide variety of cultural and historical experience. The chapters of this volume collectively explore such insights. The epithet 'European' is added to the title in acknowledgement of the historical reality that the concept expressed by the two words *Terra Australis* is essentially European. As will become evident, the south was defined by the ancients in terms of the then known European world, and in the course of the chapters, searching questions are asked about the growing understanding over the centuries in Europe of what might be a *Terra Australis*, what might be the nature of its peoples, and what might be the responsibility of Europeans towards those peoples.

The unique contribution of this volume is to unite two areas that have traditionally remained separate in scholarly discussions: *Terra Australis* as an intellectual concept, and the early exploration and conceptualization of Australia. There has been relatively little sustained scholarly work done on the concept of *Terra Australis* itself, despite the fact that it was one of the most widespread concepts in European geography in the sixteenth and seventeenth centuries. At the same time the intellectual background to European voyages of discovery and exploration also remains understudied. It is this intellectual background that acquires definition through the different perspectives that emerge from the chapters in the volume.

The second and the final chapters demonstrate the essential European quality of the term *Terra Australis*. Alfred Hiatt explains how the classical and medieval ideas of the antipodes exerted an influence, not only upon the imagination of the sixteenth- and seventeenth-century explorers, cartographers and writers, but upon decisions they and their paymasters took, and attitudes they formed towards new lands and peoples. Drawing on a variety of sources, cartographic and literary, medieval and early modern, Hiatt suggests that the appeal of *Terra Australis* lay in its capacity to act both as a mirror and as an outlet for political and cultural ambitions, for dreams of reform – and for mockery of self and others.

Leigh Penman takes the event of the *Batavia* shipwreck, and traces its treatment in fiction and newspaper reportage as providing a vehicle for Europeans' fear of the land – Western Australia – and its indigenous inhabitants. The reactions described are those of European visitors, the shipwrecked Dutch and, three centuries after the event, Australians of European descent. In the 1890s, the *Batavia* wreck became a rallying point for nationalist sentiment in the lead-up to Federation. In the years following the Second World War, the search for the physical remains of the *Batavia* wreck was employed as something of a metaphor for the nation's attempt to find its place in the new post-war international scene. And although, during the bicentennial years and afterward, the wreck and its aftermath became the basis for projections of a multicultural reading of Australia's past, and a potential rallying point for future efforts towards indigenous reconciliation, the European nature of the *Batavia* icon was fundamental to all interpretations.

The earliest perceptions explored in this volume are those of the intellectual world of imperial Rome and of Hellenistic philosophy and science. The south remained a source of both mystery and speculation to Roman and Greek geographers, philosophers, poets and historians. Was it a space for similarity or antithesis to the north, or for something between the two? In his contribution, Bill Leadbetter demonstrates how much the theorists of antiquity built their concepts of the makeup of the globe on perceptions of symmetry 'used to generate fictive worlds that existed almost parallel to the empirical world of experience'. These perceptions led some to develop the zone theory that placed the southern lands in an inaccessible place of the utmost heat, an idea that, as Christopher Wortham demonstrates, exerted long-standing cultural influence on later thinkers and writers. The Romans and Greeks were not confined to intellectual and theoretical views of geography; actual exploration took place by land and sea, which demonstrated the reality of land much further south than had been imagined. Yet while authors like Pomponius Mela, Strabo, Pliny the Elder and the third century AD novelist Heliodorus all knew something of the mysterious African south, they were more inclined to fantasize about it than describe it; and the discoveries of the Alexandrian school of pragmatic geographers were overlooked by Roman intellectuals who 'either missed this work or preferred the versions of Virgil and Cicero' (Leadbetter). This left the way open for medieval and early modern perpetuation of a mythology of the southern lands and southern peoples. Although the experience of voyagers contradicted some of these myths, such as that which held that the antipodeans lived in a world upside down (Hiatt), nevertheless, the thought processes that had peopled the southern lands with monsters and with creatures beyond salvation lived on in the fear and suspicion with which the peoples of the southern lands were regarded (Wortham, Penman, Hiatt, McCarthy).

Wortham traces the way seemingly elemental fear of peoples from the south, expressed tragically in Shakespeare's *Othello*, are based on a deep-rooted perception of the composition of the world that had been portrayed since early medieval times in the *mappaemundi*, and continued to be developed in early modern cartography. Wortham's analysis of the *mappaemundi* in conjunction with his reading of Shakespeare's *Othello* makes a convincing case for the way medieval and early modern writers and thinkers equated the lands of the south with areas where excess and unbridled immorality are the norm. Relating his thoughts to the zone theory described by Leadbetter, in which the torrid zone south of Europe is presented as uninhabitable by human beings, Wortham finds further evidence to support medieval and early modern perceptions of the south as unfit for the civilized and inhabited by physically and morally monstrous people.

Moving from the world of literature to that of personal observation, Michael McCarthy further develops the idea that perceptions are influenced by culture. He compares the unpublished version of Dampier's journals with that published by James Knapton, and concludes that the much-quoted account of Indigenous Australians as 'the miserablest people in the world',[2] is at odds with the more neutral words originally penned by Dampier in his unpublished journal. It would appear that the publisher's entrenched European perceptions that the peoples of the southern lands must be less human than Europeans flatly contradicted the more humane and tolerant account penned by the circumnavigator himself. McCarthy adduces other instances to support his thesis that field drafts and accounts prepared solely for personal reasons and for limited private readership convey perceptions that can often be at variance with those eventually brought to public view. McCarthy's study of material written by Rose and Louis de Freycinet as well as by Dampier finds evidence of subsequent rewriting and significant alteration of material, according to the perceptions of those who had a particular agenda to promote.

Approaching their subject through the lens of 'perceptions', several authors find remarkable evidence of underlying agendas that have come to light in their researches into voyages of exploration and enlightenment. Exploration has frequently been treated as a form of curiosity-driven research; yet national rivalries, ambitions of empire, the prospect of self-aggrandizement, in addition to what might appear more laudable ambitions of furthering scientific and geographical knowledge, are revealed in the chapters that discuss the perceptions of explorers and travellers. Perhaps the most startling of these agendas is exposed in John West-Sooby and Jean Fornasiero's chapter which discusses the systematic rewriting of the records and emendation of nomenclature on the maps produced as a result of the Baudin expedition. Baudin's hydrographers produced detailed maps of the south and west coasts, where they also made a

[2] Quoted in this volume by Hiatt and Wortham, as well by McCarthy.

number of important discoveries. It is no surprise to note, then, that this voyage contributed significantly to the list of French names that dot the coasts of modern Australia – names that serve as a reminder of the contribution made by a succession of French mariners to the charting of *Terra Australis*. In the case of the Baudin expedition, however, the process of conferring names on places visited and discovered was complicated by a variety of factors, mostly political in nature. This chapter offers a striking analysis of the differences between the naming practices of Baudin and those of Freycinet and Péron, highlighting two quite different approaches and ultimately two very different perceptions of *Terra Australis*.

The Gregory Expedition of 1855–57 was motivated by simpler political intentions. Norman Etherington emphasizes the economic and strategic imperatives that drove the mid-Victorian exploration of Northern Australia by recalling the imperial context of the Gregory expedition – the last major expedition funded by the British Parliament. Even though the exploration of the coasts and interior of Australia was an extremely expensive business, the reasons that British and colonial governments elected to spend vast sums have been relatively neglected in Australian annals. Patrick White's *Voss* carries the notion of discovery without gain to its logical conclusion – portraying the expedition leader as so supremely indifferent to practical considerations that he merely shrugs his shoulders when his surveying instruments are lost. Etherington's study of the imperial context of nineteenth-century exploration reveals a less heroic, more hard-headed, economically driven approach.

While the Dampier, Freycinet, Baudin and Gregory expeditions all took place within the territory that had come to be designated as *Terra Australis*, maps and memoirs have proved powerful theoretical tools for promoting particular perceptions, as the chapters by Margaret Sankey, W.A.R. Richardson, Mercedes Camino and Katrina O'Loughlin demonstrate. Sankey recounts the strong case built upon conjecture and theory by the Abbé Paulmier who hoped that Pope Alexander VII would grant a mission to the *Terres Australes* in order to convert its inhabitants to Christianity. Paulmier never achieved this ambition, although, in order to persuade the pope that the *Terres Australes* did in fact exist, he had summarized and analysed the writings of previous European authors who had written on *Terra Australis*, as well as exploring a wide variety of cartographic representations. In querying why Paulmier never achieved his ambition, Sankey calls into question the reliability of his interpretation of the sources. Richardson, too, takes up and deals definitively with perceptions based on erroneous interpretations of cartographic features. He argues against some scholars who, rather than regarding the identity of *Jave la Grande* as a hypothesis to be tested, accept it as true first, and then try to prove it. His chapter, relying primarily on the place-name evidence, shows that *Jave la Grande* is not a landmass at all, but consists of Java and Sumbawa, to which were attached two very early, differently scaled, Portuguese sketch charts of southwest Java, and southern Vietnam, together with its offshore islands. The French, not surprisingly unable to identify them from their inscriptions and outlines, tentatively sited them as part of *Terra Australis Incognita*.

Mercedes Camino brings new material to bear on the influential reporting of Spanish explorations in the Pacific. The term Spanish Pacific has been mostly used to refer to the Philippines, which are often treated as an outpost of the Americas, but within this context, much interesting graphic and visual material produced during the early modern Spanish voyages to the Pacific remains unknown. Camino's chapter analyses the perceptions of Pedro Fernández de Quirós who represented the Pacific as a Terrestrial Paradise, using terms which feminize the land, as well as the peoples encountered. The influence of this perception is recognizable in subsequent and little known prints and bird's-eye views produced during seventeenth- and eighteenth-century Spanish voyages to the Pacific which catalogue the shift from a concept of discovery that rested largely on the enactment of rituals to an enlightened belief based on representation and classification of geographical and ethnographic data. Nevertheless, both the earlier and later interpretations gendered the relative positions between the voyagers and the indigenous land and peoples subject(ed) by a masculine eye/I.

O'Loughlin writes on what is perhaps the most personal of all the records explored in this volume. Like Rose de Freycinet, but a few years before her in 1791, Mary Ann Parker joined her husband Captain John Parker on a voyage to the fledgling colony at Port Jackson. *The Gorgon*, under Parker's command, accompanied the transport vessels of the Third Fleet, carrying critical supplies and manpower for the fragile community: her passengers included Philip Gidley King, the new Governor of Norfolk Island, and a fresh corps of soldiers for the convict settlement. On her return to England, the recently-widowed Parker published her small travel book by subscription, reluctantly, and 'solely for the benefit of her little flock'. Her tentative and rather self-deprecating *Voyage* provided not only the first account of the colony written by a woman, but the first published impressions of Port Jackson by a private individual.

O'Loughlin's chapter explores Parker's perceptions of the landscape and indigenous peoples of *Terra Australis* in the contexts of contemporary discourses of landscape aesthetics, and heated debates about the politics of slavery and sovereignty across Europe and her colonies. In particular, it explores the way Parker's narrative reflects the role of writing and travel in articulating forms of subjectivity and self-representation for women and Britons in the late modern period. She deliberately recasts the travel narrative as an alternative technology of historical cartography: the *Voyage* constitutes a powerful form of cultural description which is oriented toward the fertile location *Terra Australis*, but perhaps more reliably maps late eighteenth-century British preoccupations and subjectivities.

Conclusion

This volume makes a new contribution to our understanding of the concept *Terra Australis*. Because the project is interdisciplinary, it brings together the work of experts in a variety of fields, and provides a context in which each chapter can explore 'perceptions' in a way appropriate both to the discipline and to the subject of the research. The chapters bring new evidence to bear on the reactions of early travellers to what they found, and provide thought-provoking explanations for the intellectual basis of what was written, or interpreted in maps and images. The explorers of many nations find their way into the chapters: Portuguese, Dutch, French and British; and Indigenous Australians have a strong, if often incomprehensible, presence. But the book attempts neither to give a chronological account, nor to cover all the facets of early exploration of *Terra Australis*. Acknowledging its debt to existing scholarship, the volume sheds new light on familiar texts, personages and events by setting them alongside unexpected contexts and unfamiliar texts and people (the Abbé Jean Paulmier, Rose de Freycinet, Mary Ann Parker). It also offers a broad historical focus, from classical antiquity through to the twenty-first century. Above all, the chapters of this volume celebrate something that, for centuries, was a product of the intellect – *Terra Australis* – and provided an intellectual background for the early exploration of what eventually became known as Australia. Linking these two facets of the Great Southern Land in one volume will, we hope, lead to a deeper understanding of the intellectual and cultural forces that drove European expansion and colonization towards the formation of a global economic and political system.

Chapter 2
Terra Australis and the Idea of the Antipodes

Alfred Hiatt

On 4 May 1699 the *Antelope* set off from Bristol on a voyage to the South Seas. Initially successful, the ship ran into trouble as it neared the East Indies, and was 'driven by a violent Storm to the North-west of *Van Diemen*'s Land'. By the time the *Antelope* reached the latitude of 30 degrees 2 minutes south, 12 of the crew were dead and the rest weak with exhaustion. Land was sighted on 5 November, but strong winds drove the ship against a rock, causing it to split. The sole survivor of the wreck was the ship's surgeon, one Lemuel Gulliver who, after swimming for some hours, reached a shore where he collapsed exhausted. He awoke to find himself bound and surrounded by the native inhabitants, human creatures 'not six Inches high'.[1] Some 15 years later, traumatically expelled from the island of the Houyhnhnms, where he had been elevated to the status of 'gentle Yahoo', Gulliver sailed east in a canoe, his aim 'to reach the *South-West* coast of *New-Holland*'. His success in reaching that continent was only temporary, however. A group of natives, naked, hostile and firing arrows, forced his retreat, but he was even more appalled by the prospect of rescue by a passing Portuguese vessel: 'I set out in the Morning; choosing rather to trust my self among these *Barbarians*, than live with *European Yahoos*.'[2]

Looking at the four maps that illustrated Jonathan Swift's *Gulliver's Travels* on its publication in 1726 is a disconcerting experience.[3] Post-Abel Tasman,

---

1 Jonathan Swift, *Gulliver's Travels*, ed. Herbert Davis (Oxford, 1959), pp. 20–21.

2 *Gulliver's Travels*, pp. 284–5.

3 It is generally assumed that Swift himself did not compose the maps in the first edition of *Gulliver's Travels*, which was printed in five separate printing houses in London for Benjamin Motte. On this edition's imitation of the formal features and design of a travel book see John Ross, 'The Framing and Printing of the Motte Editions of *Gulliver's Travels*', *Bibliographical Society of Australia and New Zealand Bulletin*, 20 (1996): 5–19. Earlier twentieth-century critics castigated the mapmaker for geographical inaccuracy and inconsistency with Swift's descriptions: *Gulliver's Travels*, ed. Harold Williams (London, 1926), pp. lxxix–lxxx; Arthur E. Case, 'The Geography and Chronology of *Gulliver's Travels*', in his *Four Essays on* Gulliver's Travels (Gloucester, MA,

but pre-James Cook, these images serve to introduce us to a moment at which imagination and exploration shared a close kinship. The volume's first map shows the incomplete outline of Van Diemen's Land, south of Sumatra; but alongside these emerging contours of reality (as we know it) appear the equally plausible looking islands of Blefuscu and Lilliput (Figure 2.1). Similarly, on the map that illustrates Gulliver's fourth adventure, the south coast of New Holland appears with its markers of Dutch exploration, 'Edels Land', 'Nuyts Land' and 'Lewins Land', in the region that would become the Great Australian Bight (Figure 2.2). Such realism is, however, given a jolt by the island 'Houyhnhnms Land', along with the punctilious inscription 'Discovered AD 1711'.

The maps, and the precise descriptions that mimic the discourses of exploration, indicate that Swift wanted to invest Gulliver's narrative with an air of verisimilitude. Behind the hostile and naked inhabitants of New Holland, for example, seem to lie the natives described in William Dampier's *Voyage round the World* of 1697, though Swift may also have been drawing on accounts of early Dutch encounters with indigenous populations.[4] Such is the level of the novel's mock participation in exploration as ongoing process, in the coastline not mapped but being mapped, that Swift has Gulliver solemnly propose a correction to the maps and charts that 'place [New Holland] at least three Degrees more to the *East* than it really is'.[5] Yet *Gulliver's Travels* and its accompanying cartographical fictions also register the shadow of *Terra Australis Incognita*, the vast continental landmass that had been conjured into being by sixteenth-century mapmakers, and whose possible existence would only conclusively be disproved by Cook's first two voyages of 1768–71 and 1772–75. In its standard form, *Terra Australis* extended across 180 degrees of longitude. Its northernmost points were promontories at around 15 and 20 degrees of latitude south, in the vicinity of Java and New Guinea, respectively [see Figure 2.7]. From there the coastline descended steadily to Tierra del Fuego, and then on beneath Africa. The plausibly jagged coastline of *Terra Australis* was the product not simply of speculation, but of calculation, since the existence of a significant austral landmass was thought essential to balance the weight of land in the northern hemisphere. Its sparse toponyms emanated from an amalgam of sources, chief amongst them the book of Marco Polo, as well as sixteenth-century accounts of New World exploration. Slowly, from the early 1600s onwards, the credibility of

1958), pp. 50–68, who argues that Swift based his geography principally on Herman Moll's *A New and Correct Map of the Whole World* (1719). See more recently David Fausett, *Images of the Antipodes in the Eighteenth Century: A Study in Stereotyping* (Amsterdam, 1994), pp. 39–54.

4 Fausett, *Images of the Antipodes*, p. 46. William Dampier, *A New Voyage Round the World* (London, 1697), pp. 463–70: 'The Inhabitants of the Country are the miserablest People in the world' (p. 464).

5 *Gulliver's Travels*, p. 284.

Figure 2.1 Jonathan Swift, *Gulliver's Travels* (London, 1726), map illustrating Part I. By permission of The British Library

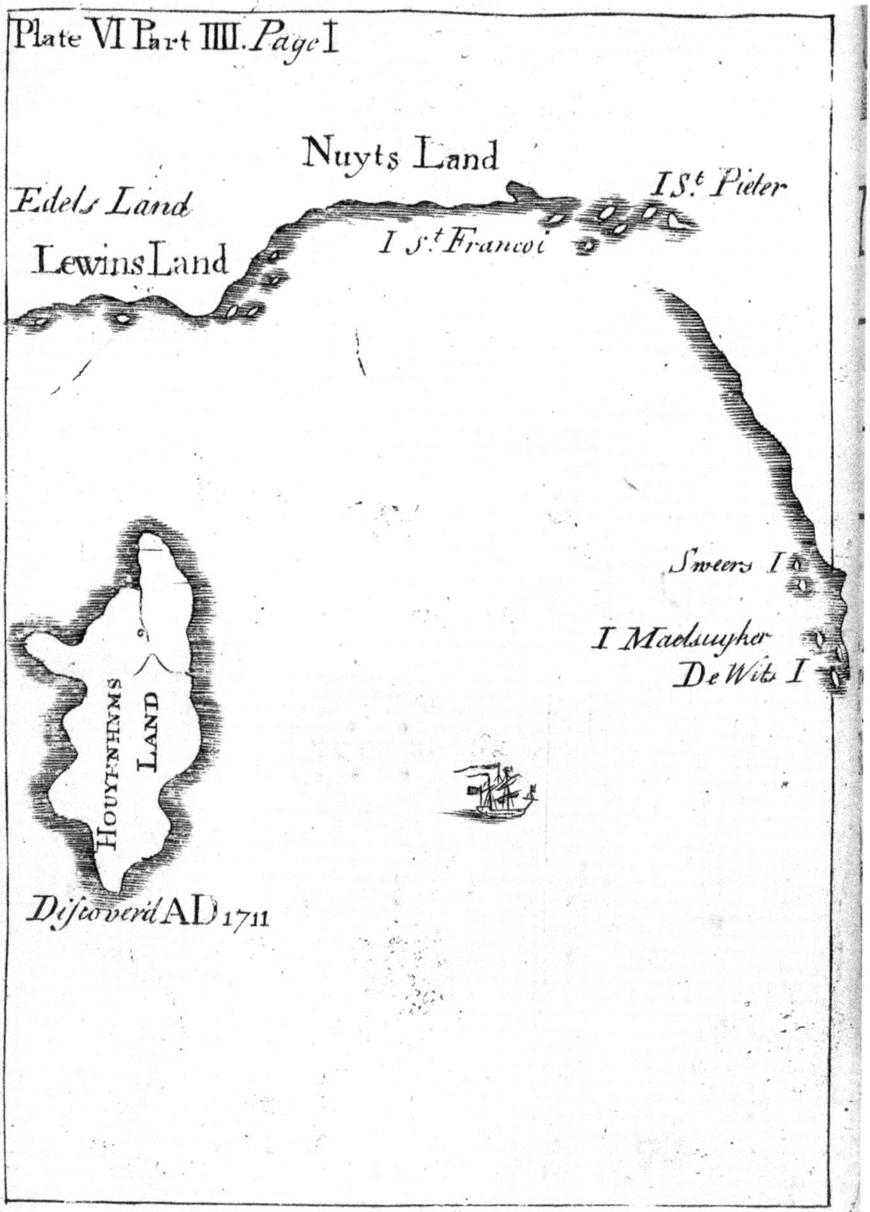

Figure 2.2 Jonathan Swift, *Gulliver's Travels* (London, 1726), map illustrating Part IV. By permission of The British Library

Terra Australis diminished, particularly as a result of Pacific exploration and the Dutch mapping of the coasts of what were to become Australia and New Zealand. And yet the spectral presence of the great southern land remained on certain maps at the end of the seventeenth century,[6] and it continued to draw influential eighteenth-century adherents, such as the hydrographer Alexander Dalrymple.[7]

Precisely because of their partially mapped status, the South Seas and the gaping canvas of *Terra Australis* could, in turn, become fiction's archipelago, a multitude of possible islands, with their possible encounters. In fact, though, as much as any aboriginal inhabitants, by 1711 on the shores of *Terra Australis* Gulliver was just as likely to encounter the succession of marooned Europeans who since Thomas More's *Utopia* had populated the genre of travel literature. Two centuries of nearly-but-not-quite discovery had led to a rich and to our eyes eccentric tradition, in full bloom by the second half of the seventeenth century. In 1668 the best-known English example of the genre appeared: Henry Neville's *The Isle of Pines, or, A late Discovery of a fourth Island near Terra Australis, Incognita* told the story of an Englishman, William Pine, shipwrecked on an island near *Terra Australis* with four women. The consequences are predictable. Pine dies after siring a modest 47 children, but 100 years later the island is populated by over 1,700 Pines, and riven by dissention, gambling and prostitution.[8] In 1676 the ex-monk and libertine Gabriel de Foigny had revealed in *La Terre Australe connue* a rather different fantasy of the southern land. Here *Terra Australis* emerged as a kind of earthly paradise, populated by hermaphrodites, who held property in common and to whom the word 'father' was unknown: 'To see these people,' reflects the narrator, Sadeur, 'you would think that Adam had not sinned in them, and they are what we would have been but for that fatal Fall.'[9] In the same year a two volume *History* issued from the pen of Denis Veiras, in

---

6 See for example Jean Baptiste Nolin's 'Globe Terrestre' or Pierre Mortier's 'Mappe-Monde Geo-Hydrographique', both of c. 1700, which show a speculative coastline connecting New Guinea with Van Diemen's Land, along with the parts of the coastline of the southern continent not conclusively disproved by exploration: Rodney W. Shirley, *The Mapping of the World: Early Printed World Maps 1472–1700*, 2nd edn (Riverside, 2001), pp. 598–601, 612–13.

7 Alexander Dalrymple, *An Historical Collection of the Several Voyages and Discoveries in the South Pacific Ocean* (2 vols, London, 1770–71), pp. i–xxx.

8 Neville initially published a pamphlet containing a narrative allegedly written by Pine himself, and recited by one of Pine's grandchildren to the crew of a Dutch ship that had come upon the island in 1667. He then expanded the narrative by publishing (initially separately, then along with the first pamphlet) an account of the island by one of the Dutchmen, Cornelius Van Sloetten: *The Isle of Pines, or, A late Discovery of a fourth Island near Terra Australis, Incognita by Henry Cornelius Van Sloetten* (London, 1668).

9 Gabriel de Foigny, *La Terre Australe connue* (1676), ed. Pierre Ronzeaud (Paris, 1990), p. 105: 'A voir ces gens, on diroit facilement qu'Adam n'a pas peché en eux, & qu'il sont ce que nous

which the inhabitants of the southern continent were revealed to be the sun-worshipping Sevarambians, an ancient society of Persian ancestry governed by Reason.[10] Just under two decades before Swift's satire the Dutch author Hendrik Smeeks described *The Mighty Kingdom of Krinke Kesmes*, in which a surgeon in the service of the Spanish discovers in *Terra Australis* a learned, pacific community living in quasi-Confucian wisdom and – as in so many examples of utopic literature – anxious to resist excessive European influence.[11]

What was the appeal of this essentially fictional, 'not yet discovered', southern land to the European imagination? No categorical answer can be given, because the nature of austral fiction was capacious, protean. But running through many examples of this sub-genre is the opposition between what is described as savagery and barbarism – the state of the Yahoos, or less graphically the natives of New Holland met by Gulliver – and the advanced, even superior civilizations encountered by the alienated European. The discourse is not simply satirical, but lends itself to philosophical modes: in the case of de Foigny, to Platonic dialogues between European and Australian on the nature of perfection; in the case of *Krinke Kesmes*, debates over the soul and body and the teachings of Descartes. Undoubtedly driving this capacity for philosophizing was the ancient antipodal trope of the world turned upside down: the world in which horses use men as cattle, or in which a pair of spectacles, hidden in Gulliver's pockets, are used not to aid reading but as a shield against a hail of Lilliputian arrows.

This strange location for philosophy – place of reflection, isolation and self-discovery, a blank space, and so a mirror for humanity – had a long history. The following pages outline the intellectual lineage of *Terra Australis* with particular attention to the concept of the antipodes. As I will suggest, European representations and perceptions of *Terra Australis* in the sixteenth and seventeenth centuries were in part a meditation upon classical and medieval ideas of the antipodes. But there were also critical differences between the antipodes as formulated by philosophers, poets and scientists from Plato to Albertus Magnus and Dante, and the *Terra Australis* sought and imagined by Gulliver and

---

aurions été sans cette cheute falle'; translation from *The Southern Land, Known*, ed. and trans. David Fausett (Syracuse, NY, 1993), p. 59.

10 Denis Veiras, *L'Histoire des Sévarambes* (2 vols, Paris, 1677); *The History of the Sevarites or Sevarambi: A Nation inhabiting part of the third Continent, Commonly called, Terrae Australes Incognitae* (2 vols, London, 1679).

11 Hendrik Smeeks, *The Mighty Kingdom of Krinke Kesmes (1708)*, trans. Robert H. Leek (Amsterdam, 1995). Later examples of fictional voyages to the southern land include Zaccaria Seriman's *Viaggi di Enrico Wanton ai terre incognite australi ed ai Regni delle scimie e de' cinocefali* (2 vols, Bern, 1764), and the anonymous *Description of New Athens in Terra Australis Incognita* (1720), reprinted in Gregory Claeys (ed.), *Utopias of the British Enlightenment* (Cambridge, 1994), pp. 27–53.

his fellow castaways. In particular, the idea of the antipodes consisted of a set of theoretical propositions about the lands and peoples on the other side of the earth which retained the possibility of contemplation without action. In the most common medieval formulation of the antipodes, it was considered impossible for Europeans to reach antipodal spaces, with the result that the only mode of comprehension was intellectual. As one medieval description put it, 'no-one ever crossed from that part to us, nor from us to them',[12] but it was nevertheless necessary to represent this hypothetical space in order to show the world in its entirety. The idea of travel to the antipodes – which gained credence in Europe from the thirteenth century, and finally attained some reality at the end of the fifteenth – crucially introduced the dynamic of encounter, real or imagined, between European and antipodean, with its attendant possibilities of assimilation or rejection. *Terra Australis* retained many of the tropes familiar from classical and medieval formulations of the antipodes: reversal of perspective, mirroring, but also reticence in the face of the unknown. At the same time, however, its striking image performed the function of bringing the unknown to the threshold of the Old World of Europe, Africa and Asia, and offering a contrast to the New World of the Americas and the East Indies. While to some eyes the shimmering presence of *Terra Australis* promised not only a continuation but an expansion of narratives of European exploration and colonization, a 'third world'[13] ripe for settlement and yet more commerce, to others it also contained a promise of correction and to some extent purgation of the excesses associated with the New World.[14] Simultaneously a learned invention and an incitement to exploration, a goad and a reproach to vanity and ambition: as its Swiftian remnants of Lilliput and Houyhnhnms Land were to testify so wittily, *Terra Australis* offered a wealth of representational possibilities as it straddled reality and fiction.

At the Earth's Antipodes

Classical Greek science devised three fundamental starting points for the theory of the antipodes. The first of these was the conception of the earth as spherical

---

12 London, BL, MS Cotton Julius D.VII, f. 46r: 'numquam enim aliquis ab illa parte ad nos uel econtra transmigrauit'. The statement is contained in a map from the 1250s attributed to John of Wallingford, a monk of St Albans and contemporary of Matthew Paris.

13 Distinct, that is, from the Old World (first) and the New World (second). The formulation is that of Gerard Mercator, as reported by his biographer Walter Ghim: *Vita Mercatoris*, ed. Hans-Heinrich Geske, *Duisburger Forschungen*, 6 (1962): 246–72, 260–61.

14 See especially Henri Lancelot Voisin de la Popelinière, *Les trois mondes*: *Les Trois Mondes de La Popelinière*, ed. Anne-Marie Beaulieu (Geneva, 1997).

and the calculation, credited to Eratosthenes of Cyrene (c. 285–194 BC), that the circumference of the earth was 252,000 stadia (about 45,000 km).[15] The second was the division of the earth from pole to pole, into climatic zones. That theory, associated particularly with the philosophers Parmenides of Elea (fifth century BC) and the Stoic Posidonius (c. 135–c. 51 BC), insisted on symmetrical relationships between places and peoples in different parts of the earth. In essence zonal theory posited five zones: two frigid, uninhabitable zones at either end of the earth, extending from the poles; a central zone of intolerable heat; and two temperate zones, one in each hemisphere, both habitable by humans (Figure 2.3).[16] There was no intrinsic hierarchy between zones: north and south, east and west were equal and reversible terms. The third starting point for theories of the antipodes was the willingness to posit lands beyond the Mediterranean basin. In Plato's *Phaedo*, Socrates had famously compared inhabitants of the Mediterranean to 'ants or frogs around a marsh', insisting that 'many others live elsewhere in many such places'.[17] This notion of multiple other worlds was given formal clarity around 150 BC by a commentator on Homer, Crates of Mallos. Crates was apparently the first to posit the fourfold division between the known world (the *oikoumene*), the *antoikoumene* beneath it to the south, the *perioikoumene* ('around from' it in the northern hemisphere), and the *antipodes* or *antikthones* (diametrically opposed, in the southern hemisphere).[18] Of several possible terms for people beyond the known world it was 'antipodes' (literally, those with their feet opposite) that eventually won out as the generic term for the other side of the earth, remaining with us still. But as Crates' scheme indicates, the classical and medieval concept of the antipodes was fundamentally pluralist: there was more than one world, and more than one antipodal space. The effect of such theorizing of the earth's shape and extent was double. On the one hand it delineated the boundaries of experience: the other side or sides of the earth were known and defined geometrically but not geographically or topographically. The geographer Strabo therefore excluded the antipodes from his *Geographia* precisely because they could only be described through

[15] Strabo, *Geography*, trans. H.L. Jones (Cambridge, MA, 1917), 2.5.7.

[16] Strabo, *Geography*, 2.2.2–2.3.2.

[17] Plato, *Phaedo*, ed. C.J. Rowe (Cambridge, 1993), 109B.

[18] For zonal theory, the quadripartite division of the globe and the views of Crates see Strabo, *Geography*, 1.1.7, 1.2.24, 2.5.10, Macrobius, *Commentarii in Somnium Scipionis*, ed. J. Willis (Leipzig, 1963), 2.9.1–2, and the Greek astronomer Geminos, *Introduction aux phénomènes*, ed. Germaine Aujac (Paris, 1975) 15.1–4, 16.1, 16.19–22; English translation and commentary in James Evans and J. Lennart Berggren, *Geminos's Introduction to the Phenomena: A Translation and Study of a Hellenistic Survey of Astronomy* (Princeton, NJ, 2006). On Crates see Hans Mette, *Sphairopoiia: Untersuchungen zur Kosmologie des Krates von Pergamon* (Munich, 1936), pp. 66–7.

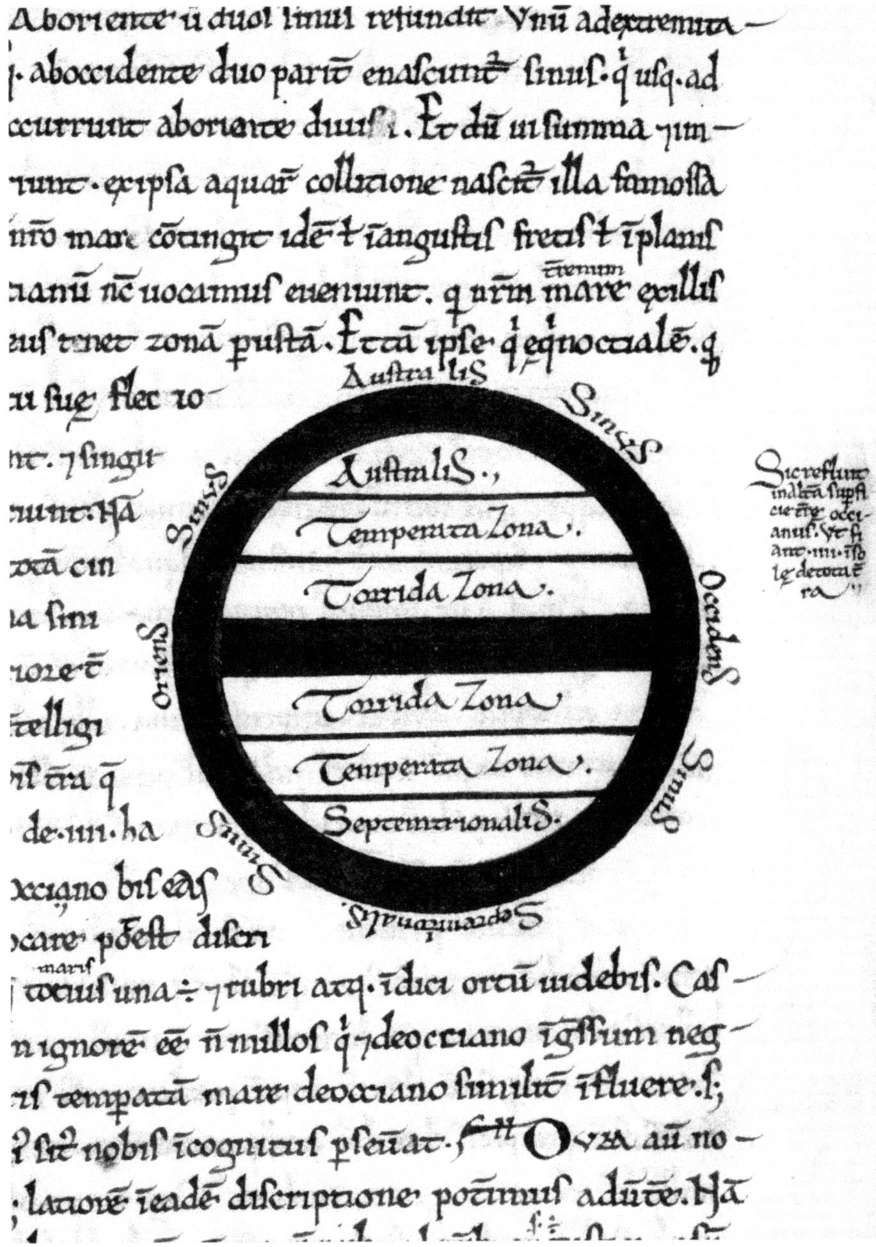

Figure 2.3 Zonal map of the world, in a copy of Macrobius' *Commentary on the Dream of Scipio*, c. 1100. South is at the top. London, British Library, Cotton MS Faustina C.1, f. 87v. By permission of The British Library

conjecture.[19] But that exclusion made the intellectual resonance of antipodal lands all the greater. Precisely because they were conceivable but not reachable, lands on the other side of the earth could function as a sign of the limitations of human knowledge, or as a check on ambition and power. And consequently the desire to break the boundaries of knowledge, to go *plus ultra*, became a sign of overreaching ambition to the point of madness, a consistent theme of the classical and medieval Alexander tradition.[20] Amongst the multiplicity of meaning, though, two motifs stand out in my mind for their lasting importance to medieval and early modern antipodal representation: the antipodes as retreat or recess; and the antipodes as world turned upside-down.

In Plato's *Republic*, Er, a dead soldier, returns to earth to narrate the passage of his soul to the celestial spheres. He tells of a system of divine penalties and rewards and the allotment of souls to bodies.[21] The 'myth of Er' in the *Republic* initiated a tradition of political fiction in which the known world was seen from beyond its limits. It was Er's vision of the heavens that Cicero surely had in mind when he wrote the 'Dream of Scipio', the final section of his own work *De re publica*. But there was a significant difference. Plato's soldier sees and describes the heavens, not the earth. Scipio, also a soldier, but Roman this time, is meant to see the celestial spheres and to hear their music, when his grandfather guides him above the earth. Scipio is not dead, however, merely dreaming, and his eyes return repeatedly to the earth, where he sees not just the Roman empire, but also the parts beyond it. The unfamiliar nature of the sight demands the explication provided by Scipio's grandfather. Referring to zonal division, he tells Scipio that what he can see are five belts surrounding the earth:

> e quibus duos maxime inter se diversos et caeli verticibus ipsis ex utraque parte subnixos obriguisse pruina vides, medium autem illum et maximum solis ardore torreri. duo sunt habitabiles, quorum australis ille, in quo qui insistunt adversa vobis urgent vestigia, nihil ad vestrum genus; hic autem alter subiectus aquiloni quem incolitis cerne quam tenui vos parte contingat.[22]

> of those the two furthest apart from each other, supported at each end by the very poles of heaven, you see stiffened with frost; that middle one however, the biggest,

19 Strabo, *Geography*, 2.5.13.

20 For example, Lucan, *De bello civili*, ed. D.R. Shackleton Bailey, 2nd edn (Stuttgart, 1997), 10.36–45; Walter of Châtillon, *Alexandreis*, ed. Marvin L. Colker (Padua, 1978), 9.569–70, 10.98–100.

21 Plato, *The Republic*, trans. Paul Shorey (London, 1935), 614B–621B.

22 Cicero, *De re publica*, ed. K. Ziegler (Leipzig, 1964), 6.21. Translations are mine unless otherwise indicated.

> you see burn with the heat of the sun. Two belts are habitable: those who dwell in the southern one press their feet against you, and have nothing to do with your people; and this other one lying to the north, which you inhabit – look what a meagre portion has anything to do with you!

The sight of antipodal peoples – those who press their feet against you – is the trigger for self-examination, for reflection on the transience and limitations of earthly glory. Scipio's reaction is immediate: 'the earth itself now seemed to me so small that I was ashamed of our empire, through which we possess just a single point of it' (iam vero ipsa terra ita mihi parva visa est, ut me imperii nostri quo quasi punctum eius attingimus paeniteret).[23] In panegyric for the emperor Augustus or the tradition of Alexander the Great the antipodes figured the ultimate temptation of political power, the desire to extend empire beyond the known world;[24] in Cicero's formulation, by contrast, the sight of the antipodes demands the restraint of Roman power within known limits.

Most obviously, and still today, the very notion of people who press their feet against ours, embedded in the term 'antipodes', carries an inherent reversal of perspective. Again Plato provided early expression of this idea: in his *Timaeus* it is asserted confidently that 'even were a man to travel round [the cosmos] in a circle he would often call the same part of it both "above" and "below", according as he stood now at one pole, now at the opposite'.[25] The world turned upside down also inherently lent itself to satire.[26] The briefest, yet most emblematic trace of this trope in classical literature is the fragment that remains of a letter composed by the late antique poet Tiberianus. All that is left of the letter, supposedly written by antipodeans, and carried north by the wind to Rome, is its opening line: 'We above send greetings to you below' (superi inferis salutem).[27] That fragment is enough to convey the satirist's shock tactic: our up is their down. Others, such as Seneca, adapted the trope to criticize moral deviance: he grumbled about Roman party animals who stay up all night and sleep during

23 Ibid., 6.16.

24 See Virgil, *Aeneid*, in *Opera*, ed. R.A.B. Mynors (Oxford, 1969), 6.794–7; the *Panegyricus Messallae*, in *Appendix Tibulliana*, ed. Hermann Tränkle (Berlin, 1990), 3.7.149–76; and Manilius, *Astronomica*, ed. G.P. Goold (Stuttgart, 1998), 1.384–6. For commentary see James S. Romm, *The Edges of the Earth in Ancient Thought* (Princeton, NJ, 1992), pp. 133–40; Gabriella Moretti, *Gli antipodi: avventure letterarie di un mito scientifico* (Parma, 1994), pp. 54–62.

25 Plato, *Timaeus, Critias, Cleitophon, Menexenus, Epistles*, trans. R.G. Bury (Cambridge, MA, 1929), 63A.

26 For treatment of the antipodes in the context of satire, particularly Menippean satire, see Moretti, *Gli antipodi*, pp. 31–54.

27 Silvia Mattiacci, *I carmi e frammenti di Tiberiano* (Florence, 1990), pp. 61, 201–6.

the day – 'antipodeans in the same city as us who … do not see the sun rise and do not see it fall' (in eadem urbe antipodes qui … nec orientem umquam solem viderunt nec occidentem).[28] But perhaps the most fertile satirical deployment of the antipodes came in the second century AD from Lucian of Samosata, in whose writing can be found the tone and spirit of the early modern faux travel narrative. Lucian's 'The Ship' contains the fantasy of a magic ring which enables its wearer to fly and to 'see how much of the earth is uninhabited and if people live head-downwards in the southern half of the world'. Most significant for his early modern descendents is Lucian's 'A True Story', in which a nautical voyage ends in 'the world opposite to the one which we inhabit', though Lucian's readers never learned what this opposite world contained.[29]

The classical tradition, then, contained at least two important seeds for future representations. One was the strong sense conveyed by Cicero's 'Dream of Scipio' that the conceptualization of the state was not complete without a vision of what lay beyond its borders, even if the function of that *terra incognita* was to return the gaze of the statesman to the known world, rather than to encourage expansionist fantasies. The other was the powerful trope of reversal thanks to which, in the hands of satirists, the antipodes could become a reflection of the known world. This reflection was far from flattering: the mirror of the antipodes functioned to lay bare folly, pomposity and vanity.

Beyond the Word of God

What was the impact of Christianity on this fertile, but also fleeting, idea of antipodal spaces? Two works of late antiquity, Augustine of Hippo's *De civitate dei* (*On the City of God*) and Macrobius' *Commentarii in Somnium Scipionis* (*Commentary on the Dream of Scipio*), shaped in profound ways the medieval reception of the classical antipodes. In a brief discussion of the matter in book 16 of *De civitate dei*, Augustine viewed the antipodes from the perspective of Christian history. Very obviously, the insistence of classical science that land on the other side of the earth could not be reached from the known world posed a problem for a religion that posited the descent of humanity from a single source, Adam and Eve, and a sole location, land in the vicinity of Eden. If the descendents of Adam and Noah had not reached the antipodes then there could be no human habitation there:

---

28 Seneca, *Epistulae morales*, ed. L.D. Reynolds (2 vols, Oxford, 1965), 122.3.

29 Lucian, 'The Ship or the Wishes', in *Opera* (8 vols, London, 1913–67), vol. 6, trans. K. Kilburn, p. 483; 'A True Story II', in *Opera*, vol. 1, trans. A.M. Harmon, p. 355.

> Quod uero et antipodas esse fabulantur, id est homines a contraria parte terrae, ubi sol oritur, quando occidit nobis, aduersa pedibus nostris calcare uestigia: nulla ratione credendum est ... Quoniam nullo modo scriptura ista mentitur, quae narratis praeteritis facit fidem eo, quod eius praedicta conplentur, nimisque absurdum est, ut dicatur aliquos homines ex hac in illam partem, Oceani immensitate traeiecta, nauigare ac peruenire potuisse, ut etiam illic ex uno illo primo homine genus institueretur humanum. Quapropter inter illos tunc hominum populos, qui per septuaginta duas gentes et totidem linguas colliguntur fuisse diuisi, quaeramus, si possumus inuenire, illam in terris peregrinantem ciuitatem Dei.

> And that there are supposed to be antipodeans, that is men at the opposite part of the earth, where the sun rises when it falls for us, who tread their footsteps opposite to our feet: there is no reason for belief in them ... Since scripture itself in no way lies (the credibility of its historical narratives is attested by the fulfilment of its predictions), it would be too absurd to say that certain men could have sailed and penetrated from this part to that, across the immensity of Ocean, so that there too from that one first man the human race might have been founded. On account of this let us see if we can find that city of God on pilgrimage on the earth amongst those races of humanity who are inferred to have been divided into seventy two peoples and the same number of tongues.[30]

The question of the antipodes in Augustine's thought was connected to, but carefully distinguished from, the question of the existence of monstrous races, which he considers in the previous chapter of *De civitate dei*. Augustine did not conceive of antipodeans as monsters, but as hypothetical men who, following the literal meaning of 'antipodes', 'tread their footsteps opposite to our feet'. Whereas monstrous beings seem for Augustine and other Christian thinkers to be positioned at the threshold of human and animal worlds, the 'fable' of the antipodeans falsely expanded the spatial extent of humanity. Augustine's sense of the authority of scripture on this point probably derives from the statements at the end of the synoptic Gospels, and the beginning of Acts, in which Christ appears to the disciples and enjoins them to evangelize 'all peoples' (Matthew 28:19), explaining that they should go throughout the world (Mark 16:15) or to the 'ends of the earth' (Acts 1:8) to do so, and promising the remission of sins for all (Luke 24:47). Acts 17:26, in which Paul instructs the Athenians that God made every race on the face of the earth from one man, provided incontrovertible

---

30 Augustine of Hippo, *De civitate dei*, ed. B. Dombart and A. Kalb, Corpus Christianorum, Series Latina (2 vols, Turnhout, 1955), 16.9.

support for the thesis of human monogenesis. Consequently, while Augustine was prepared to accept the notion of the earth as spherical, he insisted that it must be a-symmetrical in terms of habitation. The opposite side of the earth may be covered in water, he pointed out, with an allusion to Genesis 1:10 and God's 'gathering of the waters'. If uncovered by water, the land must be uninhabited. Either way, antipodeans were positioned outside of Christian belief.

Just a few years later, however, a very different, yet in its way equally influential, view of the antipodes was expressed. Writing in Rome around 430 AD, not a Christian, but like Augustine imbued with classical literature and philosophy, Macrobius Ambrosius Theodosius was concerned to interpret and thereby preserve the meanings of texts such as Cicero's *De re publica*.[31] Cicero's conception of the antipodes, as expressed in the 'Dream of Scipio', was consequently something Macrobius wanted not only to explain, but also to defend in his commentary. The antipodes were, Macrobius thought, unknowable: it was impossible to reach the other side of the earth, and while one could speculate that antipodeans enjoyed similar conditions and even topography (a southern-hemisphere Mediterranean was, he thought, probable) such features should not be represented on maps.[32] Nevertheless, and in contrast to Augustine's turn away from the representation of the antipodes, Macrobius supplied Cicero's text with explicatory images. He included a diagram designed to show rainfall on the earth (with the antipodes staying dry), zonal diagrams and a world map. From at least the ninth century the standard division of the earth on medieval maps of the known world was between the three *partes* of Europe, Asia and Africa, surrounded by an encircling ocean. But Macrobius' map, which survives in over 150 copies in manuscripts from the tenth through to the fifteenth century, served a different purpose (Figure 2.4). According to Macrobius' instructions, the map should show the flow of ocean from the equator to the poles, as well as illustrating key aspects of geography such as the location of the Mediterranean, Red, Indian and Caspian seas.[33] It was an elaboration of a zonal diagram, and hence retained representation of the five zones and equatorial ocean. As a consequence, the map showed the known world (often with just one or two toponyms to represent the areas familiar to a European audience) in apposition to the unknown temperate zone of the *antoikoi* or antipodeans – those dwellers in the southern temperate zone. The popularity of Macrobius' work ensured the diffusion of this map throughout the monastic libraries of much of western

[31] On Macrobius see Jacques Flamant, *Macrobe et le néo-Platonisme latin, à la fin du IVe siècle* (Leiden, 1977); Alan Cameron, 'Paganism and Literature in Late Fourth-Century Rome', in *Christianisme et formes littéraires de l'antiquité tardive en Occident* (Geneva, 1977), pp. 1–30.

[32] Macrobius, *Commentarii*, 2.9.7–8.

[33] Ibid., 2.9.7.

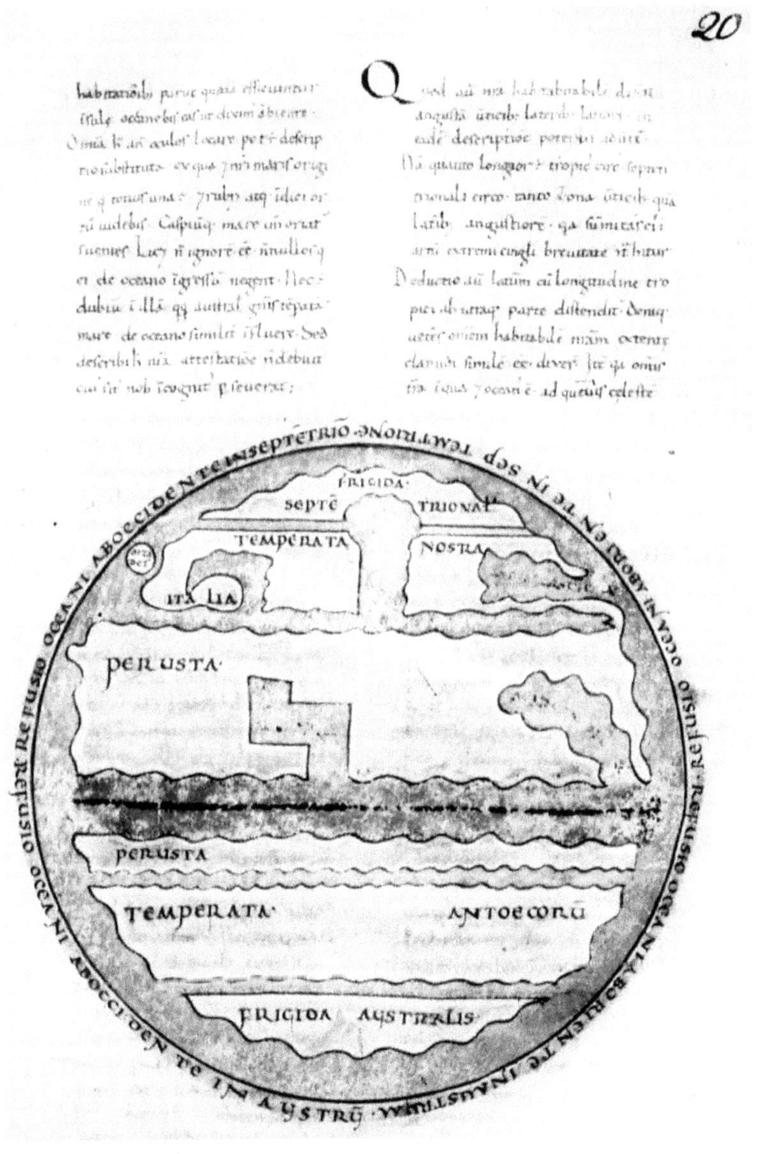

Figure 2.4 World map in Macrobius' *Commentary on the Dream of Scipio*, tenth or eleventh century. Italian. North is at the top. A large torrid zone ('perusta') separates the temperate zone of the northern hemisphere ('temperata nostra', our temperate zone) from its equivalent in the south ('temperata antoecorum', the temperate zone of the *antoikoi*). Bamberg, Staatsbibliothek Msc Class. 38, f. 20r (photo: Gerald Raab)

Europe,[34] and it helped to preserve not only an image of antipodal lands, but also the viability of the notion that, contrary to the views of Augustine, there may be men on the other side of the earth, their feet pressed against ours.

While we have examples of strong denials of the notion of the antipodes from late antiquity onwards,[35] it is equally the case that the clear and authoritative discussions of antipodal theory by Macrobius and another fifth-century author, Martianus Capella, remained of particular interest and utility to later medieval authors.[36] Encyclopaedic compilations such as Isidore of Seville's *Etymologiae* conveyed in their turn the multiplicity of classical antipodal thought. In book nine of his summa of classical learning, Isidore followed Augustine in rejecting the notion of antipodeans on the other side of the earth; in book 11 he described a monstrous race called the 'antipodes' whose eight-toed feet were turned in the opposite direction to their bodies; and in book 14 he stated that the 'antipodes' were said 'in fable' to live in the extremities of a 'fourth part of the world, across Ocean deeper in the south, unknown to us due to the heat of the sun' (Extra tres autem partes orbis quarta pars trans Oceanum interior est in meridie, quae solis ardore incognita nobis est; in cuius finibus Antipodes fabulose inhabitare produntur).[37] The confusing nature of Isidore's references to the antipodes derives from his use of diverse classical and early Christian sources, and his organization of information into frequently overlapping categories such as those of geography and ethnography. As these references show, the very uncertainty of the antipodes, their contested position between ratiocination, assertion and fable, meant that the question of the antipodes could never receive a definitive answer, and that debate on the topic would continue through the Middle Ages.

---

34 See further Alfred Hiatt, 'The Map of Macrobius before 1100', *Imago Mundi*, 59 (2007): 149–76.

35 Lactantius famously ridiculed the idea of the antipodes: *Opera omnia*, ed. Samuel Brandt and Georg Laubmann (2 vols, Vienna, 1890), vol. 1: *Divinae Institutiones*, 3.23. For the allegations made against Virgil of Salzburg c. 748 that he taught that there are 'another world, and other men beneath the earth', see *Die Briefe des heiligen Bonifatius und Lullus*, ed. Michael Tangl (Berlin, 1916), pp. 178–9. Manegold of Lautenbach's eleventh-century *Liber contra Wolfelmum*, ed. Wilfried Hartmann (Weimar, 1972), denounced the Neoplatonism of his contemporary Wolfelmus, including his belief in the antipodes.

36 Martianus Capella, *De nuptiis Mercurii et Philologiae*, ed. James Willis (Leipzig, 1983), 6.588–608. In the ninth century John Scottus Eriugena and his follower Remigius of Auxerre showed considerable interest in glossing the sections of Martianus describing the antipodes: *Iohannis Scotti Annotationes in Marcianum*, ed. Cora E. Lutz (Cambridge, MA, 1939), pp. 298–9. In the eleventh, if his opponent, Manegold, is to be believed, the Abbot of Brauweiler, Wolfelmus of Cologne, was a proponent of Macrobius' explanation of the antipodes.

37 Isidore of Seville, *Etymologiarum sive originum libri xx*, ed. W.M. Lindsay (2 vols, Oxford, 1911), vol. 1, 9.2.133–4; vol. 2, 11.3.24–5, 14.5.17.

Introspection and Expansion: The Later Middle Ages

In the twelfth and early thirteenth centuries at least two distinct modes of apprehending the antipodes can be discerned. One, fleetingly preserved in romance and learned anecdote, seems to emerge from a conflation of the antipodes with the underworld. In their chronicles and reminiscences Gerald of Wales, Walter Map and Gervase of Tilbury all told stories of people – usually unlettered – who stumble upon an underworld kingdom.[38] Gervase, for instance, tells of a Derbyshire swineherd who follows a sow into a cave, retrieves it from the lord of the land, and returns to 'our hemisphere'.[39] The antipodal motif was a favourite of Arthurian romance: amongst Arthur's baronage in Chrétien de Troyes' *Erec et Enide* is a certain dwarf Bilis, 'king of the Antipodes' (rois d'Antipodés), while in Étienne de Rouen's poem *Normannicus Draco* (1169–70) following his departure to Avalon, Arthur himself receives the lordship of the antipodeans, where he 'rules the lower hemisphere ... the other part of the world' (emisperium regit inferius ... Altera pars mundi).[40] In all these instances the antipodes derive potency from their position beyond the threshold of knowledge, yet somehow still within the world.

Judging by the surviving literature, however, the dominant tradition up to the mid thirteenth century remained the learned, scientific one. Here the emphasis of classical antipodal theory on unreachability encouraged the idea of the antipodes as something that could only be apprehended through the intellect, and that therefore functioned as a sign of intellection itself. The context of the emergence of universities, and the development of scholasticism more broadly, in twelfth- and thirteenth-century Europe is significant: the practice of formal debates on natural historical matters ensured repeated consideration of the question of the antipodes.[41] Scholastic debates in the thirteenth century included topics such as 'whether the zone that is between the tropic of Capricorn and the Antarctic circle is temperate'; 'whether inhabitants of the southern temperate zone are mortal, or immortal'; and 'whether the earthly paradise may

---

38 Gerald of Wales, *Itinerarium Kambriae*, ed. James F. Dimock, Rolls Series, 21 (London, 1868), 1.8, pp. 75–7; Walter Map, *De nugis curialium*, ed. and trans. M.R. James, rev. C.N.L. Brooke and R.A.B. Mynors (Oxford, 1983), Dist. 1, ch. 11, pp. 26–31; Gervase of Tilbury, *Otia imperialia*, ed. and trans. S.E. Banks and J.W. Binns (Oxford, 2002), pp. 2.12, pp. 334–7; 3.45, pp. 642–4.

39 *Otia imperialia*, 3.45, pp. 642–4.

40 *Le Dragon Normand et autres poèmes d'Étienne de Rouen*, ed. Henri Omont (Rouen, 1884), 2. 1165–8 (pp. 112–13).

41 Cf. Jacques Verger, 'Patterns', in Hilde de Ridder-Symoens (ed.), *A History of the University in Europe* (Cambridge, 1992), vol. 1, pp. 35–74; esp. 37–45.

be the site of those we call antipodeans'.[42] Learned authorities could be adduced on both sides of these questions, and in many ways the point was not so much the conclusion as the exercise of rational argument itself. The abstract nature of the antipodes comes through such topics clearly, and the traces of both Macrobius and Augustine seem to be discernable in them. It was in this tradition of scholastic debate that, at the end of the thirteenth century, the Catalan mystic and polymath Ramon Lull contrasted the intellect with the imagination, and used the antipodes as an example of a concept that only the intellect could grasp.

> Sicut imaginatio hominis, quae imaginatur, quod antipodes inferius cadant; et sicut gustus infirmus, qui dulcem saporem pomi, quod infirmus comedit, attingere non potest. Sed humanus intellectus sensitiuam et imaginatiuam transcendit, quando intelligit, quod hoc, quod uidetur imaginationi de antipodibus, uidelicet quod inferius cadant, esset supra ascendere. Item, quando intelligit hoc, quod sensitiuae infirmae uidetur esse amarum, in pomo dulci esse dulce.[43]

> As the imagination of a person, which envisages that the antipodeans fall down below, so is the diseased taste of a sick person, which cannot experience the sweet flavour of an apple. But the human intellect transcends the felt and the envisaged, when it comprehends that instead of appearing to fall downwards, the antipodes should seem to the imagination to rise upwards. The same, when it comprehends that what in a sweet apple seems to be bitter to the infirm taste, is in fact sweet.

Intellect here is a corrective to assumption, instinct, impression: the position of the antipodes is unexpected, yet rational – as true as the sweet taste of a sweet apple, whatever the palate may tell.

Dante's original contribution to the representation of antipodal spaces seems to have emerged precisely from this context of debate. The author of a treatise 'On the site and form of water and land' (*De situ et forma aque et terre*), in which the possibility of land in the southern hemisphere is rejected, in his *Commedia* Dante adapted cosmological material, including the matter of the antipodes, to poetic ends. Towards the end of *Inferno* (34) the poet's guide, Virgil, explains that when Lucifer fell, the earth, in terror, shifted elevation from the southern to the northern hemisphere:

> Da questa parte cadde giú dal cielo;
> e la terra, che pria di qua si sporse,

[42] Lynn Thorndike, *The Sphere of Sacrobosco and Its Commentators* (Chicago, IL, 1949).

[43] Ramon Lull, *Declaratio Raimundi*, ed. Theodor Pindl-Büchel, in *Raimundi Lulli Opera Latina 76–81*, Corpus Christianorum Continuatio Mediaevalis, 79 (Turnhout, 1989), p. 260.

per paura di lui fé del mar velo,
e venne a l'emisperio nostro; ...

On this side he fell down from Heaven; and the earth, which before stood out here, for fear of him made a veil of the sea and came to our hemisphere.[44]

Lucifer's descent literally turned the world upside-down; or rather, inside out. For, Virgil goes on, 'perhaps in order to escape from [Lucifer]' the mountain of purgatory 'left here the empty space and rushed upwards' (e forse/ per fuggir lui lasciò qui loco vòto/ quella ch'appar di qua, e sú ricorse).[45] The known world is revealed as formerly antipodal, having only entered the northern hemisphere in flight. Opposite to Jerusalem is the mountain of Purgatory, and at its summit the earthly paradise – located, in the scholastic formulation, 'in the site of those we call antipodeans'.[46] By replacing the geometry of classical science with a Christian topography, Dante transformed the space of the other side of the earth into a place simultaneously connected to the real and beyond it.

Even as Dante wrote, however, the tectonic plates of antipodal tradition were shifting away from the idea of the antipodes as a space only comprehensible by intellectual, rather than physical journeying. Authors had begun to imagine travel to or near antipodal places not as an otherworldly or wholly hypothetical experience, but as an extension of the known world. Scholars such as Roger Bacon and Albertus Magnus had bucked tradition by arguing that it would be possible to lead expeditions, including armies, into the torrid zone and beyond, and in the case of Roger speculating that the earthly paradise may have been located on the Tropic of Capricorn.[47] Marco Polo's account of the east included the description of lands where the Pole Star was not visible.[48] The compiler of *Mandeville's Travels* in the mid fourteenth century asserted the possibility of

44 Dante Alighieri, *La Commedia secondo l'antica vulgata*, ed. Giorgio Petrocchi (4 vols, Milan, 1966–67), vol. 2, *Inferno*, 34.121–4; translation from Dante Alighieri, *The Divine Comedy, vol. 1: Inferno*, trans. Charles S. Singleton (2 vols, London, 1970), vol. 1, p. 369.

45 *Inferno*, 34.124–6; Singleton, *Inferno*, 1, p. 369.

46 'Distinctio divisionis terre et paradisi delitiarum', summarized by Patrick Gautier Dalché, 'Le paradis aux antipodes? Une *Distinctio divisionis terre et paradisi delitiarum* (XIVe siècle)', in Dominique Barthélemy and Jean-Marie Martin (eds), *'Liber largitorius': Études d'histoire médiévale offertes à Pierre Toubert par ses élèves* (Geneva, 2003), pp. 615–37, 620–25.

47 Roger Bacon, *The 'Opus Maius' of Roger Bacon*, ed. John Henry Bridges (2 vols, Oxford, 1897), vol. 1, pp. 290–310, esp. 307–8. Albertus Magnus, *De natura loci*, ed. Paul Hossfield, *Opera Omnia*, vol. 2 (Aschendorf, 1980), 1.6–7, pp. 9–10, 14.

48 Marco Polo, *Le divisament dou monde*, ed. Gabriella Ronchi (Milan, 1982), chapter 166 (p. 543).

travel in the southern hemisphere, and even circumnavigation of the earth.[49] In the fifteenth century the visual image of the world changed markedly, and with it the representation of land beyond the known world. Claudius Ptolemy's *Geographia*, written in Alexandria in the second century AD, but barely known in the West until its translation into Latin in the first decade of the fifteenth century, encouraged the representation of unknown land at the ends of the known world, not separated by tracts of intemperate land, or expanses of ocean.[50] Ptolemy's description of the world left the Indian Ocean land-locked, with a bridge of land connecting Africa and Asia, but equally important was the *Geographia*'s acknowledgement of stretches of *terra incognita* in the south of Africa, in the east and north of Asia, and in northern Europe (Figure 2.5).[51] Contemporaneously, Portuguese exploration of Africa began to affirm the possibility of expansion beyond the borders known to classical antiquity. The notion of unbreachable distance between the known world and the antipodes came under pressure, as the extent and depth of knowledge of parts of Africa, Asia and even northern Europe increased, and Ptolemy's cartography began to offer a new way of assimilating such information. Map and globemakers increasingly included the antarctic regions in their world images, sometimes confusingly as an extension of Africa.[52] The possibility of habitation at the torrid zone and the relatively small size of ocean were now matters of serious debate and, by the end of the century, confident assertion.[53] Several factors, in short, contributed to the reformulation of the antipodes in the late fifteenth and early

---

49 *Mandeville's Travels*, ed. M.C. Seymour (Oxford, 1967), p. 132.

50 As I have argued in *Terra Incognita: Mapping the Antipodes before 1600* (London, 2008), pp. 145–83. On the fifteenth-century reception of Ptolemy's *Geographia* see Patrick Gautier Dalché, *La Géographie de Ptolémée en Occident (IVe–XVIe siècle)* (Turnhout, 2009).

51 Claudius Ptolemy, *Ptolemy's Geography: An Annotated Translation of the Theoretical Chapters*, trans. J. Lennart Berggren and Alexander Jones (Princeton, NJ, 2000), 7.5, pp. 108–11.

52 See for example the 1448 world map of Andreas Walsperger; the fifteenth-century copy of Johann of Udine's c. 1344 world map; and the map in a 1447 manuscript of Guglielmo Capello's commentary on the 'Dittamondo': for reproductions see, respectively, Marcel Destombes, *Mappemondes A.D. 1200–1500* (Amsterdam, 1964) pp. 212–14, no. 52.10; Scott D. Westrem, 'Against Gog and Magog', in Sylvia Tomasch and Sealy Gilles (eds), *Text and Territory: Geographical Imagination in the European Middle Ages* (Philadelphia, PA, 1998), pp. 54–75, 63; Marica Milanesi, 'Il commento al "Dittamondo" di Guglielmo Capello (1435–37)', in Marco Bertozzi (ed.), *Alla corte degli Estensi: Filosofia, arte e cultura a Ferrara nei secoli XV e XVI* (Ferrara, 1994), pp. 365–88, 375.

53 See for example the views of the French cardinal Guillaume Fillastre at the beginning of the fifteenth century: Patrick Gautier Dalché, 'L'œuvre géographique du cardinal Fillastre', *Archives d'histoire doctrinale et littéraire du Moyen Âge*, 59 (1992): 319–83, 357–8; and the notes of Christopher Columbus at the century's end: *Raccolta di documenti e studi pubblicati dalla R. Commissione Colombiana*, ed. Cesare De Lollis (Rome, 1894), 1, vol. 2, p. 291.

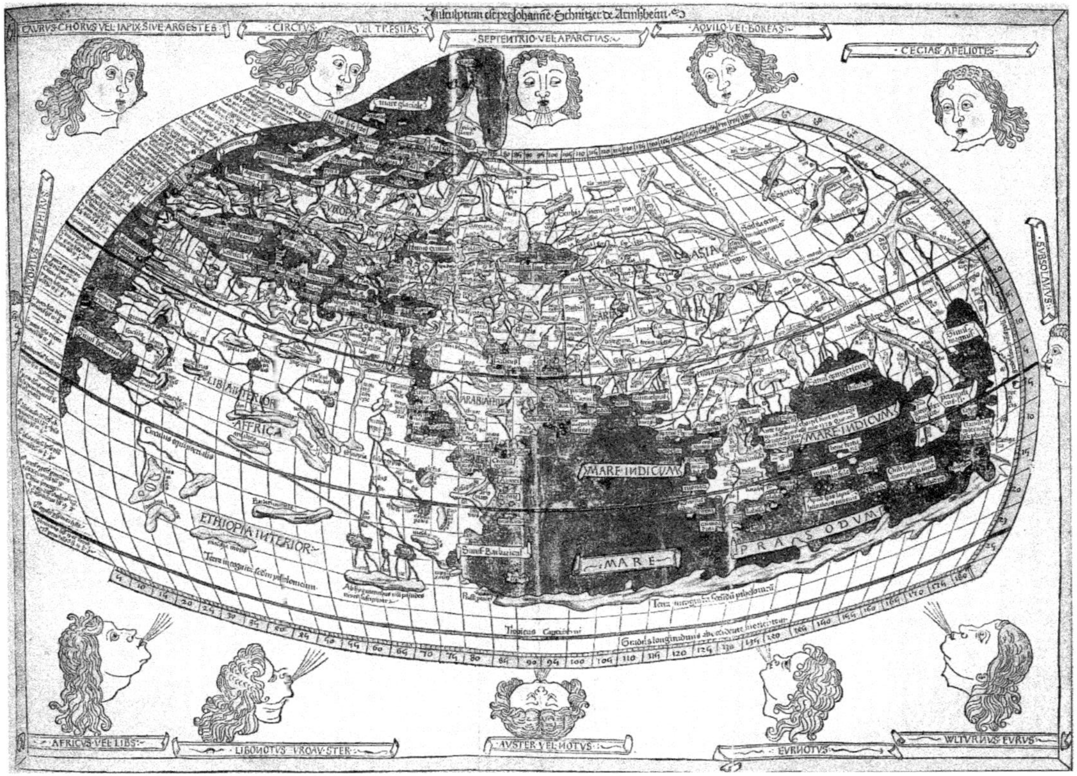

Figure 2.5 World map in Claudius Ptolemy, *Geographia* (Ulm, 1482). London, British Library, IC9306. By permission of The British Library

sixteenth centuries. But it was the discovery of the Americas, the rounding of Africa, and exploration in the Pacific that conspired, along with mathematical theory, to construct *Terra Australis*.

Terra Australis: Propaganda, Satire, Fantasy

Terra Australis was a continuation of classical and medieval traditions of land on the other side of the earth, and at the same time something new. Convinced that, in order to balance the earth, the ratio of land to sea must be the same in the southern hemisphere as it was in the north, sixteenth-century mathematician-cartographers such as Johann Schöner, Oronce Fine and Gerard Mercator calculated a giant southern continent. Unlike the temperate zone of the *antoikoi* on Macrobian maps, the southern land could be represented in coastal outline if not in much interior detail. Fine seems to have been the first to christen this continent 'Terra Australis', the southern land, as he put it in his map of 1531, 'recenter inuenta, sed nondum plene cognita' (recently discovered, but not yet fully known) (Figure 2.6).[54] Later, under the influence of Mercator, the name 'Magellanica' was coined for the southern land, testament to the belief that it had been discovered by Ferdinand Magellan on his voyage of 1519–21. As this title suggests, *Terra Australis* bore a particular relationship to the Americas. One of the few genuine parts of its vast coastline was formed by Tierra del Fuego, beneath the straits of Magellan. An early confusion with Brazil led to the toponym 'Brasilie Regio' appearing on the globes of the Nuremburg mathematician, Schöner, and 'Brasielie Regio' on Fine's 1531 bi-polar projection.[55]

Other aspects of *Terra Australis* were unmistakably extensions of Asia. The book of Marco Polo, viewed by sixteenth-century readers as an authoritative travel narrative, furnished the Land of Java, and its regions of Maletur, Beach and Locach. The authority of these toponyms was confirmed in the eyes of sixteenth-century cartographers such as Mercator by the account of the Italian Lodovico di Varthema's travels in Java.[56] In the lavish maps and atlases produced

54 Oronce Fine, 'Nova, et integra universi orbis descriptio', 1531. On Fine see Tom Conley, *The Self-Made Map: Cartographic Writing in Early Modern France* (Minneapolis, MN, 1996), pp. 88–134; Frank Lestringant and Monique Pelletier, 'Maps and Descriptions of the World in Sixteenth-Century France', in David Woodward (ed.), *The History of Cartography* (Chicago, IL, 2007), vol. 3.2, pp. 1463–79, esp. 1464–7.

55 W.A.R. Richardson, 'Mercator's Southern Continent: Its Origins, Influence and Gradual Demise', *Terrae Incognitae*, 25 (1993): 67–98, 82.

56 A caption on the southern continent on Mercator's terrestrial globe of 1541 stated that 'he who reads book 3, chapters 11 and 12 of the Venetian M. Polo, and compares it with book 6,

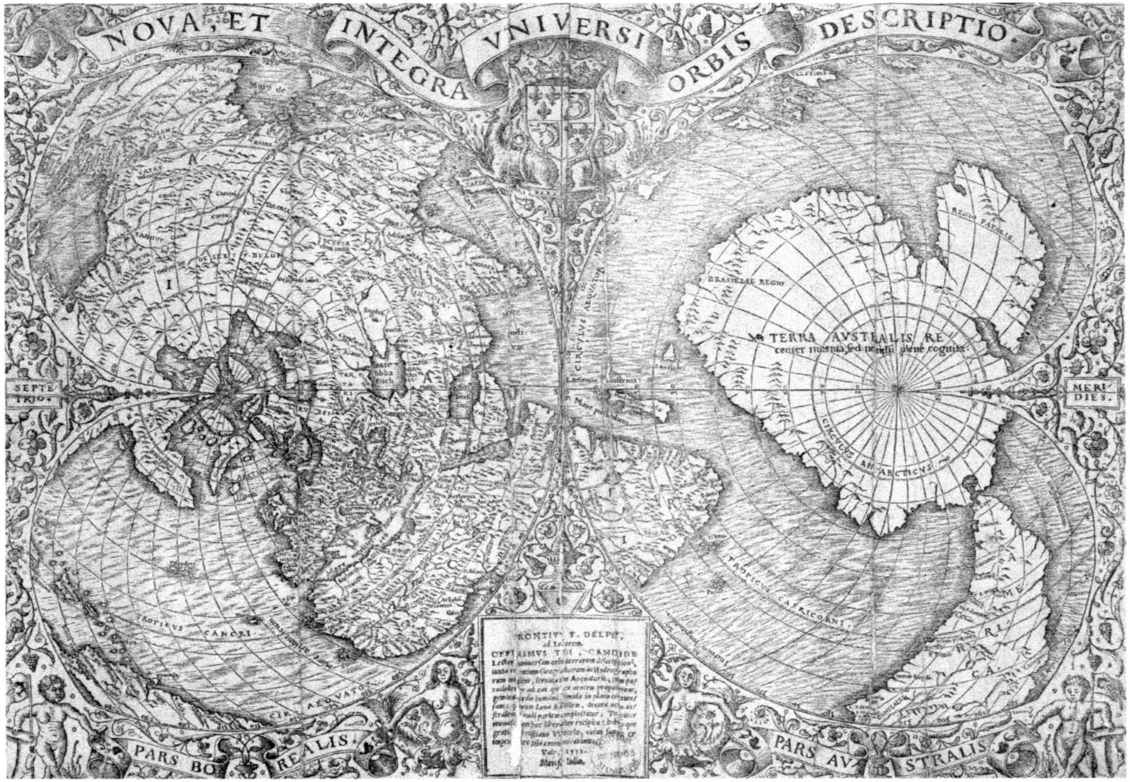

Figure 2.6 Oronce Fine, 'Nova, et integra universi orbis descriptio' (1531). London, British Library, Maps 920 (39). By permission of The British Library

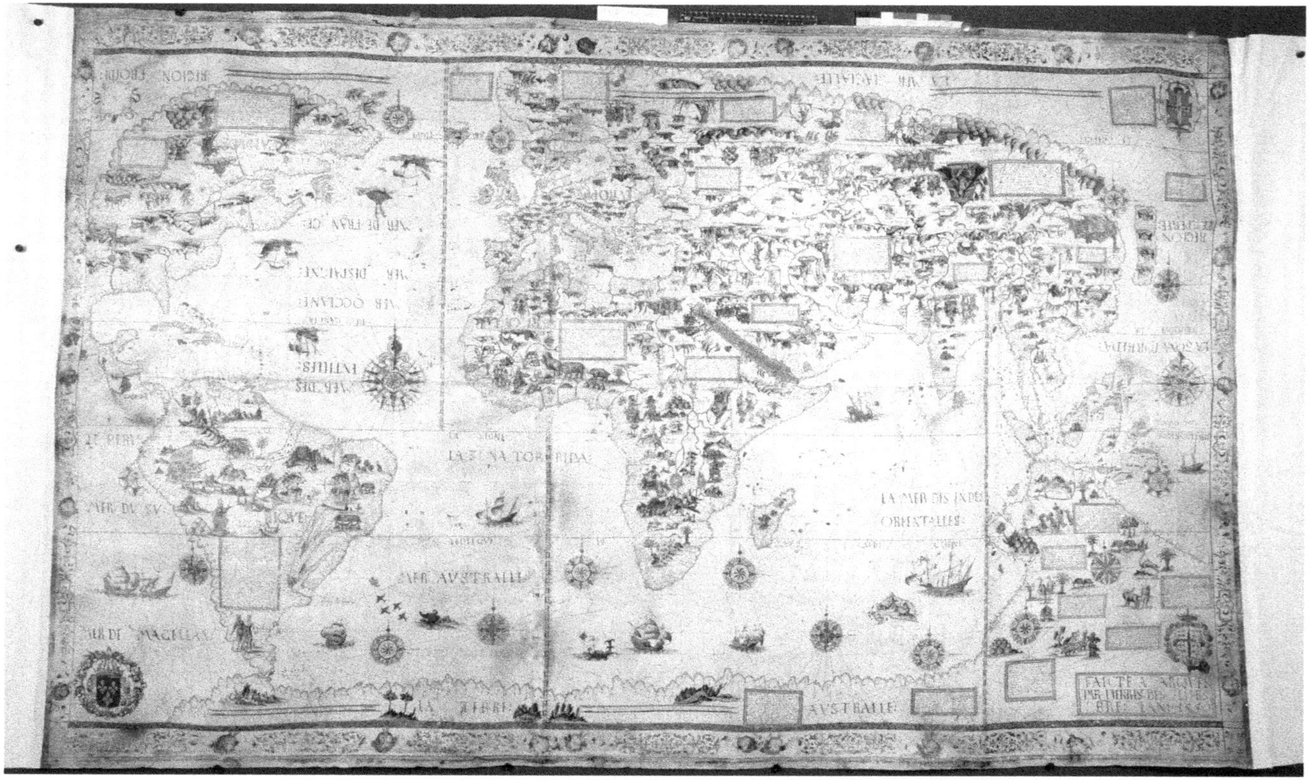

Figure 2.7 World map of Pierre Desceliers (1550). London, British Library, MS Additional 24065. By permission of The British Library

by the Dieppe school of mapmakers around the middle of the sixteenth century the southern land (sometimes called Java or Jave-la-Grande) enjoyed not only an extensive coastline, but also exotic interiors, models for which were drawn from South America, Asia, and from the mapmaker's imagination (Figure 2.7).[57] A 'Regio Patalis', meanwhile, visible on some representations of the continent, went all the way back to Pliny's *Natural History*, where Patale is an island in the Indus river.[58] *Terra Australis* was thus a manifestation of the age of exploration, with its rapid, and ongoing, charting of the Americas and southeast Asia. But it was also a product of the collection and transmission of knowledge across many centuries: of the encyclopaedia from Pliny through Isidore of Seville, to its geographic manifestation in Abraham Ortelius' *Theatrum Orbis Terrarum*.

The creative potential of *Terra Australis* was as vast as its coastline, and its evocative space prompted some of the most visually striking maps in the history of European cartography. Jodocus Hondius' 'Christian Knight Map of the World' used the southern continent much in the manner of a stage to present an allegory of the Christian Knight or Spiritus – whose features may resemble the Protestant Henri IV of France – in combat with vice, and accompanied by verses from Ephesians 6:18 (Figure 2.8).[59] Sin, Diabolus and Death menace the Knight,

---

chapter 27 of Lodovicus Patricius of Rome will easily believe that there are most vast regions here' (Vastissimas hic esse regiones facile credet qui 11 & 12 caput lib: 3 M: Pauli Veneti legerit collato simul 27 capite libri 6 Lud: Rom: Patricij); Mercator was referring to di Varthema's *Navigationes*, in Simon Grynaeus, *Novus Orbis Regionum ac Insularum Veteribus Incognitarum* (Paris and Basle, 1532), book 6, chapter 29.

57 Examples include The Vallard Atlas of 1547 (with Sumatran images on the southern land), Pierre Desceliers' world map of 1550 with text boxes and illustrations representing aspects of southeast Asia, and the Norman pilot Guillaume Le Testu's *Cosmographie Universelle selon les Navigateurs, Tant anciens Que modernes* of 1556, which replicated South American features in its striking interiors of the southern land. Le Testu stated that the southern land was depicted 'only by imagination': Frank Lestringant, *Mapping the Renaissance World: The Geographical Imagination in the Age of Discovery*, trans. David Fausett (Berkeley, CA, 1994), p. 133. On the Dieppe 'school' of cartography see Sarah Toulouse, 'Marine Cartography and Navigation in Renaissance France', in Woodward (ed.), *The History of Cartography*, vol. 3.2, pp. 1550–68; Helen Wallis (ed.), *The Maps and Text of the Boke of Idrography presented by Jean Rotz to Henry VIII* (Oxford, 1981); A. Anthiaume, *Cartes marines* (2 vols, Paris, 1916), vol. 1.

58 Pliny the Elder, *Naturalis historiae libri xxxvii*, ed. C. Mayhoff (6 vols, Leipzig, 1892–1909), 6.23. 'Regio Pathalis' found its way onto the map of the southern land by way of Roger Bacon's *Opus Maius*, vol. 1, pp. 308–10, and Pierre d'Ailly's *Ymago Mundi*, ed. Edmond Buron (3 vols, Paris, 1930), vol. 1, pp. 260–62.

59 'Typus Totius Orbis Terrarum, in quo et Christiani militis certamen super terram (in pietatis studiosi gratiam) graphicè designatur. à Iud. Hondio caelatore'. For the interpretation of the tableau as religio-political propaganda see Peter Barber, 'The Christian Knight, the Most Christian King and the Rulers of Darkness', *The Map Collector*, 52 (1990): 8–13.

Figure 2.8 Jodocus Hondius, 'Christian Knight World Map' (1598). London, British Library, Maps 158.K.1 (56). By permission of The British Library

Figure 2.9 'Crapulia' from Joseph Hall, *Mundus Alter et Idem* (1605). London, British Library, 684.d.5. By permission of The British Library

who tramples Flesh and resists the temptations of World. If the religio-political reading of this image is correct, Mundus represents the Whore of Babylon and therefore, for the Protestant Hondius, the Papacy. Just a few years later *Terra Australis* was deployed in a similarly striking though generically distinct manner in the novel of Joseph Hall, *Mundus Alter et Idem*. Here, rather than allegory, Hall uses the southern land for satire. His narrator, Mercurius Britannicus, discovers a *Terra Australis* divided into the human vices (Figure 2.9). A region of 'Moronia' abuts provinces of 'Gluttony' and another of 'Intoxication'. Nearby lands include those of the Viragos (Viraginia), and thieves. Hall revels in the construction of toponyms appropriate to his satire: the city of Mortadella, the foul smelling village of Fourmagium, and the metropolis of Artocreopolis.[60] Yet crucial to the satire is the presence of the known, mapped world – of southern Africa and South America, and of Europe itself. For Hall's satire is also staunchly anti-Catholic: 'Il Buffonio Otimo Massimo', a thinly veiled reference to the Pope, is the leader of Moronia. In a famous phrase, Abraham Ortelius styled geography as the 'eye of history', but here it appears rather as the eye of infamy.

As I have presented it so far, the southern land – whether termed *Terra Australis*, 'Jave-la-Grande', or 'Magellanica' – was the product of a mixture of conjecture and scraps of information, the descendant of a speculative geography always conditioned by rigorous principles of symmetry and mathematical precision. But what if the extensive coastline of *Terra Australis* was something rather different: what if, as has been suggested on several occasions, Jave-la-Grande is in fact a distorted representation of the northeastern coastline of Australia, charted by Portuguese or perhaps Spanish or French explorers in the early sixteenth century?[61] Or what if, as in a recent and startling re-elaboration of this theory, a Chinese fleet had mapped Australia just under a century earlier, then passed on this and other information to European savants, monarchs and navigators?[62]

---

60 Joseph Hall, *Mundus Alter et Idem Siue Terra Australis ante hac semper incognita longis itineribus peregrini Academici nuperrime lustrata* (Frankfurt, [1605]); *Another World and Yet the Same: Bishop Joseph Hall's* Mundus Alter et Idem, ed. and trans. John Millar Wands (New Haven, CT, 1981).

61 Helen Wallis, 'Java-la-Grande: The First Sight of Australia?', in Wallis (ed.), *The Maps and Text of the Boke of Idrography*, pp. 58–67; Roger Hervé, *Chance Discovery of Australia and New Zealand by Portuguese and Spanish Navigators between 1521 and 1528*, trans. John Dunmore (Palmerston North, 1983); Elizabeth Bonner, 'Did the French Discover Australia? The First French Scientific Voyage of Discovery, 1503–1505', in David W. Lovell (ed.), *Revolution, Politics, and Society: Elements in the Making of Modern France* (Canberra, 1994), pp. 40–48.

62 Gavin Menzies, *1421: The Year China Discovered the World* (London, 2002), pp. 186–214; Menzies, *1434: The Year a Magnificent Chinese Fleet Sailed to Italy and Ignited the Renaissance* (London, 2008).

It is not my intention to dwell on these interpretations of the southern land, which in my view have been rebutted successfully elsewhere.[63] But I would make two observations. The first is that people who look at early maps through the prism of twentieth- and twenty-first-century cartography – those who superimpose on early modern cartography the image of the earth we know today – will be inclined to discount the presence of hypothesis, to dismiss the possibility that theory rather than direct observation might contribute to the representation of the world. Yet it is important to resist anachronism of this kind. *Terra Australis* should not be equated with modern Australia: it was always something much more, and much less. Its proponents would have been disappointed, and their theories of continental equipoise upset, to have seen 'Magellanica' reduced to the size of Australasia as we know it. And at the same time the Schöners, Fines, Mercators and Orteliuses knew they were dealing in conjecture, and not fact. That is why they termed the continent '*nondum cognita*'.

My second observation is that those who, under the guise of contesting Eurocentrism, urge the fifteenth-century Chinese mapping of Australia are the true heirs of early modern travel narratives such as *The Mighty Kingdom of Krinke Kesmes*, in which Europeans stranded in *Terra Australis* discovered that the land had already been colonized by Asiatic civilizations. To such theories we can respond 'not so'; but we should also respond 'so what'. Why is it important to establish which European nation first reached the shores of Australia or New Zealand? What if Chinese did 'get there first' and establish a tenuous presence, of which even its most ardent supporters can find only the merest trace? Why do we indulge in this discourse of priority about lands that were *already inhabited*? The indigenous inhabitants of modern-day Australia did not need to have their land discovered, and for them such debates must seem trivial at best. Rather than who saw what first, it might be more profitable for us to think about the impact of discovery. If the Chinese did discover Australia before Europeans, it was a discovery of lasting insignificance.

'Alien Eyes from Another Century'

What of the idea of the antipodes after the Middle Ages? Embedded within so much of the classical and medieval European inheritance, the antipodes were not

63 William A.R. Richardson, *Was Australia Charted before 1606?: The* Jave la Grande *Inscriptions* (Canberra, 2006) brings together Richardson's arguments against the hypothesis that Jave la Grande is an early representation of Australian coastlines. For refutation of Menzies see in particular the reviews of Felipe Fernández-Armesto, *Literary Review* (Dec 2002/Jan 2003): 21–2 and Robert Finlay, *Journal of World History*, 15 (2004): 229–42; Richardson, *Was Australia Charted before 1606?*, pp. 97–100.

simply subsumed into the idea of *Terra Australis*. The term entered vernacular languages, and it continued to enjoy literary resonance. Within the works of Shakespeare there are five references to the antipodes; two of these may be taken to illustrate the enduring currency of the concept, as well as the contrasting uses to which it could be put. In Act 2 of *Much Ado about Nothing* Benedick, recently put down by his sparring partner and future wife Beatrice, and still protesting his loathing of her, invokes the idea of the antipodes as refuge. Shakespeare in fact gives us a series of geographical allusions: Benedick complains that if Beatrice's 'breath were as terrible as her terminations, there were no living near her, she would infect to the north star'. Informed by Don Pedro that she approaches, Benedick becomes desperate:

> Will your Grace command me any service to the world's end? I will go on the slightest arrand now to the Antipodes that you can devise to send me on; I will fetch you a toothpicker now from the furthest inch of Asia, bring you the length of Prester John's foot, fetch you a hair off the great Cham's beard, do you any embassage to the Pigmies, rather than hold three words' conference with this harpy. You have no employment for me?
> *Don Pedro*. None, but to desire your good company.[64]

The references here would have been as meaningful two centuries earlier as they were when the play was completed in 1598. Prester John, the mythical Christian ruler in Asia or according to some claims central Africa, Pygmies and the 'Cham' (the Khan, ruler of the Mongol empire) were as familiar to Mandeville's or Chaucer's audience as they were to Shakespeare's. In Benedick's formulation the antipodes function within the figure of hyperbole, suavely undercut by Don Pedro's modest reply: it is good company that the play will celebrate, rather than the trappings of political ambition represented by the trophies and embassies Benedick comically offers. So the antipodes feature momentarily as Benedick's escape from the woman he shortly thereafter describes as 'Lady Tongue' – an escape from their verbal sword-play, but also from language, wit and, we might suspect, sexuality.

Of Shakespeare's other antipodal references, three use the trope of the opposition of day and night in antipodal places.[65] The most subtle and least

[64] *Much Ado About Nothing*, II.i.263–73, in *The Riverside Shakespeare*, ed. G. Blakemore Evans et al., 2nd edn (Boston, MA, 1997). All subsequent references to this edition.

[65] *The Merchant of Venice*, V.i.127; *A Midsummer Night's Dream*, III.ii.55 (Hermia states 'I'll believe as soon/ This whole earth may be bor'd, and that the moon/ May through the centre creep, and so displease/ Her brother's noontide with th' Antipodes' than that Lysander would leave her'); *Richard II*, III.ii.49 (Richard states that Bolingbroke 'all this while hath revell'd in the

easy to interpret of these comes in Act V of *The Merchant of Venice*. Portia has returned to her home at Belmont from triumph over Shylock in Venice; it is still night, about two hours away from dawn, but we are told at the beginning of the act that the moon is bright. So bright that Portia observes

> This night methinks is but the daylight sick,
> It looks a little paler. 'Tis a day,
> Such as the day is when the sun is hid.
> *[Enter Bassanio et al.]*
> *Bassanio*. We should hold day with the Antipodes,
> If you would walk in absence of the sun.
> *Portia*. Let me give light, but let me not be light,
> For a light wife doth make a heavy husband ...[66]

Whereas the antipodes for Benedick signalled escape from the known world and the all-too well known charms and contagions of Beatrice, Bassanio's reference is to the reversal of perspective. Portia emits her own light: the light, based on her encounter with Shylock, of wisdom, justice and mercy. She lights up the night to the extent that it becomes day, enabling people to 'hold day with antipodeans'. Night for day of course is a trope that extends back at least as far as Seneca's jibe about Romans who sleep in the day being 'antipodeans in our own city'. Here, in a scene preoccupied with light, and hierarchies of light ('So doth the greater glory dim the less'),[67] Portia quickly moves the word-play on, punning on light as the antithesis of heavy, rather than day. But if Portia is night's day, she surely also figures the light of Christ, in distinction to the half-light of Judaism represented by Shylock.

The concept of the antipodes in these ways retained functions that *Terra Australis* could not. Benedick could have demanded an embassy to the unknown southern land, just as Bassanio could have referred to day in Java Maior. But what seems to appeal to Shakespeare are the resonances of the antipodes, the classical and medieval reference held within the term: its conjuring of people on

---

night,/ Whilst we were wand'ring with the antipodes'). This line, absent from the First Folio, is often understood as an assimilation of Ireland to the antipodes. Seen in context, however, Richard is figuring himself as the sun, which, circling the globe, has 'wandered with the antipodes' during the northern hemisphere night; Richard/the sun is now 'rising in our throne, the east'. The fourth, *3 Henry VI*, I.iv.135, simply takes the antipodes as a simile to illustrate opposition: *York*. [to Queen Margaret] 'Thou art as opposite to every good/ As the antipodes are unto us,/ Or as the south to the septentrion.'

66 *The Merchant of Venice*, V.i.124–30.

67 *The Merchant of Venice*, V.i.93.

the other side of the earth, their feet pressed to ours – people at once extreme, yet opposite and identifiable, Europeans turned upside-down. *Terra Australis* may have seemed both too protean and too recent for Shakespeare's purposes: locatable on the map, known and not yet known, more New World than Old.

Shakespeare was not alone in his deployment of the antipodes. In *El Nuevo Mundo descubierto por Cristóbal Colón* his great Spanish contemporary, Lope de Vega, has Christopher Columbus insisting on the possibility of 'Hombres a nuestras plantas contrapuestos' and promising Ferdinand and Isabella that he will evangelize the Indies in their names.[68] Within Stuart drama, however, allusions to the antipodes are associated particularly with the lying traveller. In the anonymous *Timon*, possibly performed as early as 1600, Pseudocheus gulls the credulous Gelasimus into believing he possesses a ring given by the King of the Antipodes.[69] Similarly, Aston Cockayne's *The Obstinate Lady* sees the gallant Lorece impressing the young widow Vandona with a ridiculous account of his travels, including the information that:

> They at the Antipodes hear with their noses, smell with their ears, see by feeling, but taste with all their senses, – for they are the most insatiable gluttons under the cope, – and they feel not anything, for they cannot be hurt.
> *Vandona*. This is wonderful![70]

In James Shirley's *The Humorous Courtier* (1631), Volterre is described by the misogynist Orseollo (eager to know if any woman in the entire world can be described as virtuous) as a man who has lived out something resembling Benedick's embassy:

> Thou hast beene a travailer, and converst
> With the Antipodes, almost put a girdle
> About the world, taken dimensions
> Of every nature, tasted all aires, and canst
> Distinguish em to an atome ...[71]

The culmination of the use of the antipodes in Stuart drama as a sign of falsehood, or at least exaggeration, is doubtless Richard Brome's *The Antipodes* (acted in

---

68 Lope de Vega, *El Nuevo Mundo descubierto por Cristóbal Colón*, ed. J. Lemartinel and Charles Minguet (Lille, 1980), I.i.385–421, I.i.906–47.

69 *Timon*, ed. J.C. Bulman and J.M. Nosworthy (Oxford, 1980), I.iv.258–399.

70 Aston Cockayne, *The Obstinate Lady*, ed. Catherine M. Shaw (New York, 1986), II.i.75–80.

71 *James Shirley's The Humorous Courtier*, ed. Marvin Morillo (New York, 1979), I.i.85–9.

1638), a play in which Peregrine, a melancholy young man with a fixation on *Mandeville's Travels*, is tricked into believing that he has voyaged to 'the antipodes of England'. Most of the play takes place in the mock antipodes with rather heavy-handed comedy generated by Peregrine's parents, the jealous Joyless and the vapid Diana, along with Hughball, the scheming doctor who orchestrates the business. Here again the reversal of social, and particularly sexual, roles is emphasized, as in Hughball's explanation of the antipodes to Diana:

> But there the women overrule the men.
> If some men fail here in their power, some women
> Slip their holds there. As parents here and masters
> Command, there they obey child and servant.
> ... But there the maids do woo
> The bachelors, and 'tis most probable,
> The wives lie uppermost.
> *Diana*. That is a trim,
> Upside down, antipodean trick indeed![72]

It is notable that, precisely as exploration of antipodal regions was actually taking place, the English stage renditions of the idea deployed familiar tropes in which the concept continued to function as an interior space, one located more accurately in the imagination than on the map. The antipodes and antipodeans are the conjurations of liars and hoaxsters, rather than, as in de Vega's play, a reality perceived by a visionary explorer who intends to go beyond the world known to the ancients. Attainable only through a lie, the semiotic charge of the antipodes as the epitome of opposition, of reversal and of self-reflection remained potent – more potent, as far as early seventeenth-century dramatists were concerned, than the emerging evidence of their actual nature.

As the preceding discussion has indicated, the idea of the antipodes generated a conceptual and representational space most unusual within the history of European thought. The antipodes were necessary for the dominant modes of describing the world in its entirety, but for obvious reasons such a space could be apprehended only by means of reasoning: by hypothesis, but also by imagination. This curious position exterior to observation gave the concept its potency, but also its danger: to draw the wrong conclusions from the antipodes ran the risk, as Augustine and others noted, of contradicting scripture. Perhaps as a consequence, representation of antipodal spaces in the Middle Ages

---

72 Richard Brome, *The Antipodes*, in *Three Renaissance Travel Plays*, ed. Anthony Parr (Manchester, 1995), pp. I.iii.; 123–6; 139–41.

seems notably restrained. A period that was perfectly willing to depict the more extraordinary manifestations of the natural world in parts of the known world refrained, by and large, from such activity in the antipodes. It took the age of exploration and the redrawing of the world map to sweep away such restraint, but even then classical and medieval paradigms were maintained. Indeed, the emergence of New Worlds, and new maps thereof, merely served to intensify the other side of the earth's rich potential for satire, and the questions it posed about the legitimacy of European political ambition.

In the twenty-first century the idea of the antipodes continues to spark creative expression around the idea of dislocation. No longer simply representative of a north–south dichotomy in which the term is used to invoke an emergent, non-European identity,[73] the concept of antipodes may be regaining some of its flexibility. In 'The Antipodes and the Century', a short story published in 2001 by the Mexican author Ignacio Padilla Suárez, a Scottish academic is assaulted by bandits in the Gobi desert; delirious, lost, but convinced that he has returned to Edinburgh, he is taken by a nomadic tribe to be a prophet. The professor's raving – yet precise – descriptions of his home town are recorded and, eventually, given physical form: the city of Edinburgh is rebuilt in the desert, and its inhabitants develop a perverse hatred of the English, despite dressing like them and 'speaking their language with a Caledonian cast'. After the professor's death the city is levelled by a sandstorm, but with the passing of time its memory acquires a legendary and mystical quality, such that it is sought out by travellers.

> One after the other, men and animals would collapse exhausted on the dunes, gasping out a kind of prayer as they gave up the ghost – although no one could make out what language that wandering band had chosen to speak. All of a sudden, their eyes, dry now for so long from the wastes of stone and salt, filled once again with tears, in which Scotland's capital gleamed for a moment, a palace preserved in amber. It was as if someone had intruded into their retinas an elephantine fortress of streets, bridges, and windows which, like alien eyes from another century, would watch their own eyes close in final peace in their tombs of sand.[74]

---

[73] In the ringing assertion of the authors of 'The Antipodean Manifesto' (1959): 'To the ancients the antipodes was a kind of nether world; to the peoples of the Middle Ages its forms of life were monstrous; and for us, European by heritage (but not by birth) much of this strangeness lingers. It is natural therefore that we should see and experience nature differently in some degree from the artists of the northern hemisphere' (Bernard Smith, *The Antipodean Manifesto: Essays in Art and History* (Oxford, 1976), p. 166). One notes the erroneous nature of the compressed history in this statement, particularly with regard to the Middle Ages.

[74] Ignacio Padilla Suárez, *Las Antípodas y el siglo*, trans. Alastair Reid (London, 2004), p. 3.

The city, dislocated, replicated at its antipodes, is also chronologically antipodal: watching, a teary mirage, 'alien eyes from another century'. These words seem at the same time an apt description of the way we look at medieval and early modern images of antipodal land, and how they seem to us. Our own world image contains no *terra incognita*, or hypothetical zones of inhabitation. *Terra Australis* is consigned to history as a non-place, once a land of 'not yet', now land that never was. But it is the case that in studying the history of non-places such as the classical, medieval and early modern space of the antipodes, we glimpse something of the complexity of European understanding of the world beyond its borders. And in an age of globalization across stubborn local and national identities, we may also start to understand more of our own, equally complex, antipodal dislocations.

Chapter 3
The Roman South

Bill Leadbetter

If the Romans have preserved a reputation into the contemporary world, it is probably as empirical technocrats, who conquered their world by building a professional army, the roads for it to march upon, the ships for it to sail in, and the economic infrastructure to house it, feed it and pay it. Roman achievement in engineering, governance, military tactics, town planning, agriculture and the like was in large part the consequence of a happy (rather than perfect) marriage between theory and practice. Much of the theory was itself the product of Hellenistic speculative science. Roman thinkers and writers in turn drew upon and advanced that intellectual heritage, making it their own. The mathematical and geometrical precision of Roman engineering was not accidental. It was the product of an ongoing dialogue between tradition and innovation in a way that is familiar to the practices of contemporary scholarship. The Emperor Trajan's court architect, for example – one Apollodorus, famous for the construction of a major military bridge across the Danube and the Markets of Trajan, the world's first great shopping mall – once mocked a young man for his preference for sketching what he called 'gourds'. The young man was the future emperor Hadrian; the 'gourds' were domes, first used by Hadrian in the reconstruction of the Pantheon; and Apollodorus allegedly regretted his sneer with his life.[1] Apollodorus' gibe is the kind of comment that a man recognized as a master of his craft might make to a young enthusiast, brimming with new ideas. It represents the common conflict between old and new in Roman intellectual life, a conflict pursued just as intensely in the contemporary academy.

Roman geography, however, was not subject to these same intellectual practices. Theory and practice did not meet in a happy and creative union in Roman geography. Rather, they seem to have existed completely independently of one another. Insofar as they mused upon it, the Romans lived within a geographical paradox. The world in which they lived, travelled and traded was an evident quadrilateral that surrounded the Mediterranean, the waterway that

1 Dio Cassius 69.4.

the Romans called *mare nostrum* – 'our sea'.[2] This rectangle was fringed by the cold Baltic to the north; to the west, the vast Atlantic; to the south, the heat of the African desert; to the east, the lands of the Persians, Indians and Seres. When Virgil, in Book 6 of the *Aeneid*, hymns the universal rule of the Augustan state, he writes in lines 787–810 of the limits of that world:

> Here is Caesar and all the great line of Iulus;
> Destined for the heaven of heavens.
> And here is one, this man, of whom you have often heard promised:
> Augustus Caesar, born of a god,
> founder of a golden age once again in Latium,
> where Saturn once ruled.
> He will spread his rule over the Garamantes and India,
> Even to a place outside the stars and the tracks of year and sun,
> Where heaven-bearing Atlas
> turns the heaven of burning stars upon his shoulder.
> Even now, at expectation of his arrival
> The kingdoms of the Caspian and Maeotis
> Quake in trepidation,
> And the sevenfold mouths of Nile shudder in alarm.[3]

This passage is an important exposition of Augustan imperial ideology. It is also a clear reflection of Virgilian cosmology. Virgil's world is a world with borders: the Garamantian and Indian; the Mede and Scythian dwell at its edges. Yet beyond even these distant realms lies another world. Some have seen this as a reference to the unreachable antipodes.[4] The text probably does not bear that view, although it does seem to imply that, if Augustus had space ships and warp drive, there would be no limit to the extent of his rule. Virgil's world-view is poetic and Homeric; he consciously and consistently asserts a flattish earth. In his *Georgics*, a series of pastoral poems written to celebrate the victory and

2 Peregrine Hordern and Nicholas Purcell, *The Corrupting Sea: A Study of Mediterranean History* (Oxford, 2000), pp. 11–12.

3 'Hic Caesar et omnis Iuli progenies magnum caeli ventura sub axem. Hic vir, hic est, tibi quem promitti saepius audis, Augustus Caesar, Divi genus, aurea condet saecula qui rursus Latio regnata per arva Saturno quondam, super et Garamantas et Indos proferet imperium: iacet extra sidera tellus, extra anni solisque vias, ubi caelifer Atlas axem umero torquet stellis ardentibus aptum. Huius in adventum iam nunc et Caspia regna responsis horrent divom et Maeotia tellus, et septemgemini turbant trepida ostia Nili.'

4 See, for example, Alfred Hiatt, *Terra Incognita: Mapping the Antipodes before 1600* (Chicago, IL, 2009), p. 25.

apparent world-rule of Augustus, he cites and adapts the *Hermes* of Eratosthenes in a description of the heavens (1.233–9):

> Five zones hold the firmament: one of these is always red with blazing sun and ever parched by fire; about it, to the left and right deep blue extremities are extended, [all] solid ice and black storms: between these and the middle, two have been granted by divine gift to enervate mortals, and a dividing road between the two wherein the slanting zodiac might turn. As the world rises up sharply towards Scythia and the Rhipaean heights, so it sinks downwards in the Libyan south. One pole is always high above us, while the other deep below our feet sees dark Styx and the spirits of the dead. Here Lord Draco slides his sinuous coils, river-like, between and around the two Bears, the Bears fearful to be drenched by the waters of Ocean ...[5]

There is a key difference between Eratosthenes' poem and Virgil's. Virgil is describing the heavens; Eratosthenes' description is of the earth:

> Five encircling zones were girt around it: two of them darker than greyish-blue enamel, another one sandy and red, as if from fire ... Two others were there, standing opposite one another, between the heat and the showers of ice; both were temperate regions, growing with grain, the fruit of Eleusinian Demeter; in them, men dwelt, antipodal to one another ...[6]

Virgil's quotation of Eratosthenes reveals his own deep knowledge of the way in which the intellectual world of Hellenistic philosophy and science viewed the world, as well as his own rejection of it. The received wisdom of his time asserted that the world was a kind of globe. The principal figures here were Greek scholars: the pre-socratic Parmenides, the great Aristotle, and the Hellenistic thinkers

---

5 'Quinque tenent caelum zonae; quarum una corusco semper sole rubens et torrida semper ab igni; quam circum extremae dextra laevaque trahuntur caeruleae, glacie concretae atque imbribus atris; has inter mediamque duae mortalibus aegris munere concessae divum, et via secta per ambas, obliquus qua se signorum verteret ordo. Mundus, ut ad Scythiam Rhipaeasque arduus arces consurgit, premitur Libyae devexus in austros. Hic vertex nobis semper sublimis; at illum sub pedibus Styx atra videt Manesque profundi. Maximus hic flexu sinuoso elabitur Anguis circum perque duas in morem fluminis Arctos, Arctos Oceani metuentis aequore tingui.' This passage is replete with references to both extant and lost texts. See Richard S. Thomas, 'Virgil's Georgics and the Art of Reference', *Harvard Studies in Classical Philology*, 90 (1986): 171–98, 195–7; also James S. Romm, *The Edges of the Earth in Ancient Thought* (Princeton, NJ, 1992), pp. 128–9; Hiatt, *Terra Incognita*, p. 23.

6 Eratosthenes, *Hermes*, fr.16, ll 3–5, 16–19, translation James S. Romm, *The Edges of the Earth*.

Eratosthenes and Posidonius. These intellectuals constructed their knowledge from a varied combination of the theoretical and empirical and all believed, on the basis of both observed evidence and reason, that the world upon which they stood was an elliptical sphere. Virgil's contemporary, Strabo, in his surviving work, the *Geographika* sets forth a long description of the world as he understood it. In the second book of the *Geographika*, he outlines both how he understands the world to be and the methods that have caused him to reach those conclusions. His key sources here are Posidonius, Eratosthenes and the astronomer Hipparchus (whom he mostly dismisses). It is clear that Strabo's understanding of the world is founded upon two assumptions: that the world itself is a kind of flattened sphere in which the 'inhabited world' (*oikoumene*) is a longish rectangle bounded on all sides by the ocean; and that the *oikoumene* is located in a northern temperate zone, between a frigid Arctic Zone and a scorching Torrid Zone.[7] Strabo sets out this assumption more fully at 2.3.1–3, where he rebuts Polybius' argument for six global zones, and again, Strabo sets out his rationale for five zones, including a single broad equatorial Torrid Zone, although he glosses this with the addition of two new zones of his own (2.2.3), which he describes as habitable torrid zones.

This view of the world as, at least, spheroidal (if not actually spherical) is reflected in the kinds of terminology that we find in Latin prose, and even on coinage, to refer to the earth as a whole. The expression *orbis terrarum* which we find used by Cicero, Caesar and Livy, three of our more extensive earlier Latin prose writers, is generally translated as 'the whole world', but means, more literally 'the circuit of lands'.[8] It is worth noting in this context that, of these three, only Livy was a contemporary of Virgil, while Cicero and Caesar belong to an earlier generation. Another of this generation was the Roman mint official, Pomponius Musa, who in 66 BC, issued a series of denarii depicting the Muses, thereby punning on his own surname. One of these was Urania, the Muse of Astronomy, who is depicted contemplating a globe, plainly that of the earth.[9]

7 Strabo, *Geographika*, 1.1.20 (where he describes the world as 'spheroidal'); 2.1.1–32.

8 See, for example, Caesar, *de bello gallico*, 7.29.6; *de bello civili*, 3.72.4, 3.43.2; Livy, *Historiae*, 1.16.6, 5.3.47, 5.51.8, 7.25.8; Cicero, *pro Archio*, 7.232, 4.21; *pro Fonteio*, 1.87, 5.99, 4.57; *pro Sulla*, 5.67, 1.15; *pro Cluentio* 17.53; *de imperio C. Pompei*, 36.103, 45, 131; *pro Flacco*, 26.63, 41.103; *pro Rabirio perduellionis*, 12.32; *contra Rullum*, 1.1.2, 1.3.9, 1.4.11, 1.5.15, 2.10.26, 2.13.33, 2.14.37, 2.17.45, 2.24.64, 2.28.76, 2.36.98; *pro Deiotaro*, 5.15; *pro Milone*, 27.72; *Phillipicae*, 2.7, 3.14.35, 4.16.14, 7.7, 8.3.10, 10.5, 11.13.34, 13.15; *in Verrem*, 2.36.88, 4.10.23, 4.38.82; *pro Balbo*, 6.15, 28.64; *pro Sestio*, 30.66, 31.67, 61.129; *de domo*, 28.75, 42.110; *de haruspiciis responsis*, 13.29; *ad Atticum*, 1.19.7, 4.1.6, 14.5.2; *ad Brutum*, 1.16.2; *ad familiares*, 2.16.4, 3.8.4, 4.1.2, 4.7.4, 4.12.3, 5.7.3, 10.8.3; *de oratore*, 3.131.

9 Trustees of the British Museum, *Coins of the Roman Republic in the British Museum* (London, 1970), Rome #3628; Michael H. Crawford, *Roman Republican Coinage* (2 vols, Cambridge, 1974), I, p. 438, #8.

Musa's coin is not an isolated example of the globe used in this context, merely an early one. Then young Augustus developed an elaborate visual language of power featuring the star-sign Capricorn. On coins, and in other visual media, Capricorn is depicted as a horned goat with a globe between its forelegs.[10] The Augustan depiction of the goddess Victoria normally has her as a winged woman, holding out a wreath while standing on a globe. This may be a representation of the statue of the goddess that Augustus brought from Tarentum and placed in the Senate House, where it remained until controversially removed by the emperor Gratian in the later fourth century. Other examples of this figure with the globe remain extant from the early imperial period. Elsewhere in Augustan iconography, the globe is used in representations of the triumphant ruler, who stands with his foot atop a subject earth.[11] So pervasive was this imagery that it became natural for his successors to take it over, even develop it further. In later decades and centuries, it became normative for depictions of imperial fairness (AEQVITAS AVGVSTI) and foresight (PROVIDENTIA AVGVSTI) to include a globe, presumably as symbolic of the universal nature of the imperial virtue.[12] It is also used, from time to time, in imperial statuary. Pieces of a porphyry statue of the emperor Galerius were found, at the site of his intended retirement villa at Gamzigrad, which included a hand clutching an orb (Figure 3.1). Some emperors of the late third century were even known, somewhat grandiosely, by the title *restitutor orbis*, 'restorer of the entire world'.[13]

As we have already noted, it is one thing to believe, and relatively widely, that the world is, at least, spheroidal; it is quite another to ascribe qualities to that shape. Roman geography did so with the application of zone theory. Strabo's description of the world is neither isolated nor uncommon. Cicero, who, like Strabo, owed a great intellectual debt to the philosopher Posidonius, shared

---

10 See here Tamsyn Barton, 'Augustus and Capricorn: Astrological Polyvalency and Imperial Rhetoric', *Journal of Roman Studies*, 85 (1995): 33–51, 49–50.

11 On the globe generally, see Stefan Weinstock, *Divus Julius* (Oxford, 1971), pp. 42–53; Paul Zanker, *The Power of Images in the Age of Augustus*, trans. Alan Shapiro (Ann Arbor, MI, 1988), p. 83 f.; Robert Alan Gurval, *Actium and Augustus* (Ann Arbor, MI, 1995), p. 61 f.; on the Statue of Victory, see H.A. Pohlsander, 'Victory: The Story of a Statue', *Historia*, 18 (1969): 588–97.

12 Seth W. Stevenson, *A Dictionary of Roman Coins* (London, 1884; repr. 1964), pp. 420–21 (sv. *Globus*).

13 See Bill Leadbetter, 'Imperial Ideology and the Christians in the Late Third Century', in M. Dillon and L. Garland (eds), *Religion in the Ancient World* (Amsterdam, 1995), pp. 535–45; Clifford Ando, *Imperial Ideology and Provincial Loyalty in the Roman Empire* (Berkeley, CA, 2000), pp. 333–5.

Figure 3.1 Porphyry fragment of a statue of Galerius from Gamzigrad. National Archaeological Museum, Zaječar. Reproduced with permission

that image of the world in which the great globe can be divided into zones.[14] A famous and influential passage in his 'Dream of Scipio' (*Somnium Scipionis*) sets out his geographical understanding of the world. The 'Dream of Scipio' is not a stand-alone text, but an extract from Cicero's constitutional dialogue 'On the Commonwealth' (*de re publica*). This text is a discourse on politics and civics,

[14] On Cicero and Posidonius, see Cicero, *de natura deorum*, 1.6. Plutarch, *Life of Cicero*, 4.4. Strabo (16.2.10) regarded Posidonius as 'the most learned philosopher' (*tōn … philosophōn polumathestatos*).

and was written as a dialogue with a number of famous Roman stoic aristocrats as the interlocutors. The most notable of these is Scipio Aemilianus who relates the tale in Book Six that we know as 'The Dream of Scipio'. The point of his narrative is to provide a kind of context to the social strivings discussed in earlier books, since what happens in his dream is that Scipio is taken away from the world by his adoptive grandfather Scipio Africanus to gaze upon it from afar, as if to note just how tiny and insignificant the world that is Rome is. Africanus says to his famous descendant:

> Then Africanus said 'Once more I see you looking at the abode of men. If it seems as small to you as it actually is, gaze ever then upon the heavens; set aside that which is human. For can you expect great celebrity among such people, or any glory worth pursuing? For you see that the inhabited places on the earth are few and small in scope, and between these "spots", so to speak, that are inhabited, there lie such vast wildernesses, that those who live upon the earth are so isolated that it is not possible to communicate between them. Some of these stand at right-angles to you; other across while others lie directly opposite; amongst such you can certainly anticipate no glory.
>
> You also see some belts encircling and surrounding the earth, and that the two with the greatest distance between them lie, frozen with frost, beneath the poles of the firmament, while the one in the middle, the largest, is burned by the heat of the sun. Two of them are habitable: the southern one, where people, when they walk, put their feet opposite to yours and have nothing to do with your kind; the other, in the north, where the people obey you – but see how small it is! For the entire world, within which you live, is narrow in the north, wider at the sides, a little island surrounded by that sea that is sometimes called "the Atlantic", sometimes "the Great Sea" or sometimes "the Ocean", yet you see how, despite this grand name, how little it all is ...'[15]

[15] 'Tum Africanus: "Sentio," inquit, "te sedem etiam nunc hominum ac domum contemplari; quae si tibi parva, ut est, ita videtur, haec caelestia semper spectato, illa humana contemnito! Tu enim quam celebritatem sermonis hominum aut quam expetendam consequi gloriam potes? Vides habitari in terra raris et angustis in locis et in ipsis quasi maculis, ubi habitatur, vastas solitudines interiectas eosque, qui incolunt terram, non modo interruptos ita esse, ut nihil inter ipsos ab aliis ad alios manare possit, sed partim obliquos, partim transversos, partim etiam adversos stare vobis; a quibus exspectare gloriam certe nullam potestis.

Cernis autem eandem terram quasi quibusdam redimitam et circumdatam cingulis, e quibus duos maxime inter se diversos et caeli verticibus ipsis ex utraque parte subnixos obriguisse pruina vides, medium autem illum et maximum solis ardore torreri. Duo sunt habitabiles, quorum australis ille, in quo, qui insistunt, adversa vobis urgent vestigia, nihil ad vestrum genus; hic autem alter subiectus aquiloni, quem incolitis, cerne quam tenui vos parte

This text gives a very clear idea of what might be regarded as the conceptual or theoretical geography of the ancient Romans. This is the world that they understood and knew, with a frozen world of ice and snow to the north, and, to the south, unknown lands beyond the burning and lethal heat barrier of the torrid zone. Strabo, and other Roman writers on geography (or what they called chorography, that is a description of known lands), such as Pomponius Mela, concentrated principally upon describing the world that they could reach. The part beyond the torrid zone was beyond them, and so they did not engage in idle speculation as to its nature or society. Mela did prefer to use the term "antichthon" over the word "antipodes", which I suspect emphasizes his refusal to speculate about what he thought could never be perceived.[16] He preferred not to posit any kind of symmetry of hemispheres, perhaps wishing to leave the question open. Nevertheless, such writers provided something of a resolution of the apparent paradox of living in a rectangular and yet simultaneously spherical world. The theory of zones allowed the Romans to live in a northern temperate rectangle, balanced by a southern, but unreachable, southern temperate rectangle.[17] For some Romans, this notion of symmetry went even further. According to the architectural writer Vitruvius, who was a contemporary of both Caesar and Augustus, within the temperate zones itself, which alternated from a colder north to a hotter south, Italy was most perfectly located in the midst of this zone, most perfectly combining the character advantages of both extremes:

> For the peoples of Italy are absolutely the most perfectly constituted for courage, both in bodily form and the vigour of their spirits. So the same way, just as the star of Jove [i.e. Jupiter] is temperate, its course running between Mars, which is extremely hot, and Saturn, which is extremely cold, so Italy, for the same reason, lying between the north and the south is a mixture of each, and her glory is proper and cannot be vanquished. So counsels shatter the valour of the barbarian, and might of arms confounds the craft of the southerner. It was, therefore, the divine mind that set the city of the Roman people in a place both prominent and temperate, so that it might come to command the whole world.[18]

---

contingat! Omnis enim terra, quae colitur a vobis, angustata verticibus, lateribus latior, parva quaedam insula est circumfusa illo mari, quod 'Atlanticum', quod 'magnum', quem 'Oceanum' appellatis in terris; qui tamen tanto nomine quam sit parvus, vides ..."' (Cicero, *de re publica*, 6.6.1–3).

16 F.E. Romer, *Pomponius Mela's Description of the World* (Ann Arbor, MI, 1998), p. 34, n. 4.

17 Romm, *Edges of the Earth*, pp. 41–4.

18 'Namque temperatissimae ad utramque partem et corporum membris animorumque vigoribus pro fortitudine sunt in Italia gentes. Quemadmodum enim Iovis stella inter Martis

To Vitruvius, symmetry in all things had real power. Symmetry was the principle by which things held in place. As an architect and a designer he depended upon geometric balance in order to create large structures. Roman architecture, of which Vitruvius remains the main acknowledged authority, developed and depended upon the arch, not merely as a decorative feature, but also as a means of spanning space whether as bridges or barrel vaults for roofing, and also as an internal buttress for large brick-built structures. In Vitruvius' view, the architectural principle of symmetry, met most obviously in the arch, applied universally, including in geography.[19]

To others, the symmetry that Vitruvius used to posit innate Roman superiority was used to generate fictive worlds that existed almost parallel to the empirical world of experience. These lands at the fringes of the inhabited world (*oikoumene*) housed cultures and beings alien to actual experience and so were the subject both of speculation and of fantasy. In the south, there lived the Blemmyae. The Blemmyae sustained a kind of double identity. To some, they were understood as a fabulous community of headless people whose eyes were in the midst of their chests. Others defined them, less mythically and more prosaically, as a group of desert nomads who lived around Meroe.[20] Beyond them were the Ethiopians, who were known since Homeric times to dwell at the southern-most fringe of the *oikoumene*. In both the *Iliad* (1.424–5) and the *Odyssey* (1.22–3), gods are depicted as being either distracted or simply unavailable for counsel because of their absence in the far south to feast with the Ethiopians. To Herodotus, the Ethiopians dwelt at the farthest reaches of the earth. Confronted by Cambyses, the Persian conqueror of Egypt, they mocked and humiliated him. They ridiculed the rich gifts that his ambassadors brought, provoking the Persian king into a rash and poorly provisioned march southwards to their land during which so many in his army perished from hunger that he was obliged to turn back. Herodotus also reports the wonders of Ethiopia, most notably 'the table of the sun' which was, each day, furnished with foods of every

ferventissimam et Saturni frigidissimam media currens temperatur, eadem ratione Italia inter septentrionalem meridianamque ab utraque parte mixtionibus temperatas et invictas habet laudes. Itaque consiliis refringit barbarorum virtutes, forti manu meridianorum cogitationes. Ita divina mens civitatem populi Romani egregiam temperatamque regionem conlocavit, uti orbis terrarum imperii potiretur' (Vitruvius, *de Architectura*, 6.1.11).

19 Vitruvius, *de Architectura*, 1.2.4.

20 On the Blemmyes in general, see Robert T. Updegraff, 'The Blemmyes I: The Rise of the Blemmyes and the Roman Withdrawal from Nubia under Diocletian (with additional remarks by L. Török)', in Hildegarde Temporini (ed.), *Aufstieg und Niedergang der Römischen Welt*, vol. 2.10.1 (Berlin and New York, 1988), pp. 44–106; on the mythical Blemmyes, see Pomponius Mela, 1.48; Herodotus, 4.191.3; Pliny, *Natural History* [hereafter *NH*], 7.23.

kind upon which all comers might feast.[21] To the Hellenistic novelist Heliodorus, Ethiopia was a Swiftian setting for a significant part of his romance novel the *Aethiopica*, in the distant land of the African south. His Ethiopians are depicted as much the same as everyone else, neither especially wise, nor by any measure unhuman. They are simply a different community upon whom Heliodorus can project otherwise unknown customs (such as the human sacrifice of the first enemy captives taken in war) and the possession of various magical implements, most notably a golden griddle that will instantly burn up any who stands upon it who has ever had sexual intercourse.[22] What is interesting about these fictional communities is that, while they are in the south, they are not too far to the south. They are not cut off from the *oikoumene* by the impenetrable equatorial heat barrier, but are accessible from it, if only by long and arduous travel.

These fictional worlds stand alongside the world of experience. While Roman fabulists might write of the strange wonders to the south, and geographical theoreticians try to compute the structure and dimensions of the great globe itself, others did what we might expect any pragmatic Roman to do: they went to have a look. They went as traders and as scouts, sometimes a little of both. Towards the end of the first century BC, the Romans gained control of Egypt and therefore the lower Nile valley. The Nile had long been a highway of sorts between the Mediterranean world and the deeper interior of Africa. Augustan ideology required that highway be exploited and, by the turn of the millennium, two major military expeditions had been dispatched into the deep south. The first was that of Virgil's friend, Cornelius Gallus; the second, that of his successor as Prefect of Egypt, C. Petronius. Gallus, apparently, led his troops as far south as Meroë, which to the Romans was the very heart of Ethiopia. His inscription at Philae, somewhat bombastically, claimed that not only never had there been a Roman army so far south before, but neither had any of the Pharaohs achieved such a feat. Petronius did not get so far, marching to, then sacking and burning Napata, some two hundred kilometres north of Meroë.[23] Some years later, in the time of Nero, a small expedition headed by two centurions of the Praetorian Guard was sent up the Nile. The purpose of this reconnaissance was twofold: to explore the lands to the south, and to report upon their suitability for conquest. They returned with a

---

21 Herodotus 3.22–4; see Romm, *Edges of the Earth*, pp. 55–60.

22 Heliodorus, *Aethiopica*, 10.

23 *CIL*, 3.14147. His temerity cost him the favour of Caesar, and his life. See E.A. Judge, 'Veni. Vidi. Vici, and the Inscription of Cornelius Gallus', in E.A. Judge and James R. Harrison (eds), *The First Christians in the Roman World: Augustan and New Testament Essays* (Tübingen, 2008), pp. 72–84. On the expedition of Petronius, see Dio, 54.4.5 ff.; Strabo, 17.819 ff.; Pliny, *NH*, 6.181 ff.; Shelagh Jameson 'The Chronology of the Campaigns of Aelius Gallus and Caius Petronius', *Journal of Roman Studies*, 58 (1968): 71–84.

bleak report of deserts and marshes, which ultimately proved impassable. Aspects of that report are recorded by Seneca, and circumstantial detail implies that they made it as far south at least as the marshes of the Sudd.[24]

Roman military interest in the south was not limited to the Nile Valley. They were also aware of the communities of the deep desert, both those of the Egyptian oases, and also the cultures further to the west, most particularly that of the Garamantes.[25] The Romans knew of them as a community dwelling on the fringes of the distant south, and provided an image in the *Aeneid* of the assertion of Roman power at the distant fringes of the world (*Aeneid* 6.692). The historical reality was a truly remarkable society of the deep desert. It was able to grow and flourish through the exploitation of fossil water channelled into underground canals.[26] In 19 BC, Cornelius Balbus led an expedition against them which succeeded sufficiently for him to be granted a triumph – the last celebrated in classical times by a person who was not a member of the imperial family.[27] This expedition makes it clear that there was a reasonably regular trade between the Libyan coast and the Garamantes, and from time to time the Garamantes made a nuisance of themselves. Conflicts with the Garamantes are also recorded in 15 BC, AD 22 and AD 69 when Garamantian troops were present in a force that raided Lepcis Magna, which led in turn to a punitive expedition in the following year.[28] While the Garamantes were never incorporated into the empire as such, they were content for some centuries to be understood as loyal Roman clients.

They were also a useful staging post for some commercial exploration. At some point, probably late in the first century AD, two different travellers, Septimius Flaccus and Maternus, apparently embarked on two different journeys south, either through or with the Garamantians. Our principal source for this is a reference in Claudius Ptolemy's highly influential *Geographika*.[29] Ptolemy's

---

24 Seneca, *Natural Inquiries*, 6.8.3–5; also Pliny, *NH*, 6.181, 184–6; M. Cary and E.H. Warmington, *The Ancient Explorers*, revised edn (Harmondsworth, 1963), p. 212 f.; see also Jehan Desanges, 'Les relations de l'Empire romain avec l'Afrique nilotique et érythréenne, d'Auguste à Probus', in Hildegarde Temporini (ed.), *Aufstieg und Niedergang der Römischen Welt*, vol. 2.10.1 (Berlin and New York, 1988), pp. 3–43, 15–17.

25 Indeed, this community was known to Herodotus (4.183), whose description of them, while semi-fabulous, was widely circulated in the classical world. See also Daniel F. McCall, 'Herodotus on the Garamantes: A Problem in Protohistory', *History in Africa*, 26 (1999): 197–217.

26 Louis Werner, 'Libya's Forgotten Desert Kingdom', *Saudi Aramco World*, May/June (2004): 8–13.

27 Pliny, *NH*, 5.36–7.

28 Tacitus, *Annals*, 3.74; 4.23–6; see R.C.C. Law, 'The Garamantes and Trans-Saharan Enterprise in Classical Times', *The Journal of African History*, 8.2 (1967): 181–200.

29 Ptolemy, *Geographika*, 1.8.

attempt to describe the world is heavily based upon the earlier work of Marinus of Tyre. Both display a fundamentally empirical approach, deploying evidence from explorers, travellers and others who have described their journeys. While he is sometimes critical of the earlier work of Marinus, those criticisms are limited to Marinus' conclusions, rather than the means used to derive them.[30] Ptolemy's discussion of the travels of both Flaccus and Maternus is essentially a quotation from and critique of the work of Marinus in his calculation of the southernmost limits of the *oikoumene*.[31] While very little is known of either of these men and their travels, it is likely that both made their journeys towards the end of the first century AD, between the completion of Pliny the Elder's work and that of Marinus, that is, between AD 75 and 100.[32] Moreover, it is probable that the story of Septimius Flaccus' march is to be identified with a known military campaign in AD 86 of Suellius Flaccus against the Nasamones, a northern desert neighbour to the Garamantians.[33] According to Ptolemy, Flaccus reached the Garamantians and then went beyond their lands, reaching the land of the 'Aethiopians' after a march of a further three months. While this is unlikely for an army, it is not for a scouting party under Flaccus' authority. On the other hand, Maternus is depicted as a single traveller, who travelled initially to Garama, and then a further four months to the south with the Garamantian king. According to Ptolemy, he reached a land called Agisymba where the Athiopians dwelt and where the rhinoceros gathered together.[34]

Their journeys caused Marinus a little puzzlement. The travellers had themselves measured their journeys and had come to a final calculation of the location of Agisymba, that Marinus initially accepted, 24,680 stades (4,442.4 km) south of the equator. Some critical reflection followed, during which Marinus apparently realized that his globe was too small to permit this calculation to be correct. Ptolemy, when he had composed his *Almagest*, had accepted Eratosthenes' calculation of the size of the globe, which was in fact a little too large. By the time he had come to write his *Geographika*, he had revised this view and considered the globe somewhat smaller, a view that Marinus seems to have shared. When confronted with the reports of the distances travelled by Dioskoros and Maternus, he either had to increase the size of his globe (which he did not do) or arbitrarily reduce the estimated distance to Agysimba and

---

30 J. Lennart Berggren and Alexander Jones, *Ptolemy's Geography: An Annotated Translation of the Theoretical Chapters* (Princeton, NJ and Oxford, 2000), pp. 23–30.

31 Ibid., pp. 145–6.

32 On Marinus' *floruit*, see ibid., pp. 23–4.

33 Dio, 67.4.6–7; Brian W. Jones, *The Emperor Domitian* (London and New York, 1992), pp. 139–41.

34 Ptolemy, *Geographika*, 1.8; Berggren and Jones, *Ptolemy's Geography*, pp. 145–7.

Cape Prason. This led him to re-estimate the distance of Agysimba as 12,000 stades (or 2,162 km) south of his calculation of the equator, which was slightly south of Cape Guardafui.

Moreover, this was also consistent with what Roman sailors were finding as they travelled along the eastern coast of the African continent. The Red Sea had long been a trading highway of sorts.[35] It had also been discovered, at some indeterminate point during or slightly after the Hellenistic period, that it was possible to catch the monsoon from the southern end of the Red Sea and make a quick passage to the Indian ports.[36] Some voyages did not go as planned. On one occasion, in the reign of Claudius, a freedman of the Roman businessman, Annius Plocamus, was blown off course and made landfall in Sri Lanka. This resulted in an embassy from a Sri Lankan king to Claudius and the establishment of more direct Roman trade there.[37] South-east was not the only direction in which navigational problems occurred. At some point before the end of the first century AD, a sea captain named Diogenes, who sailed the India route, met contrary winds as he returned home and, instead of entering the Red Sea, was driven south for a total of 25 days. Ptolemy describes the place that he reached as nigh to the lakes from which the Nile flows and slightly north of 'Cape Rhapton'.[38] This was already a well-travelled coast and a description of it, as far as the mart of Rhapta, appears in the anonymous *Periplus of the Red Sea*. This text describes a journey quite a long way south with a list of landing and market places between Aromata (Cape Guardafui) and Rhapta. It also describes a large offshore island which is called Menouthias, and has been identified with either Pemba or Zanzibar.[39] What Diogenes, and another sailor named Dioskoros, provided, was evidence of distance. For his part, Dioskoros sailed even further south, as far as a cape he calls Cape Prason. Dioskoros himself estimated the distance between Rhapta and Cape Prason as 5,000 stades (about 900km). On this basis, Marinus had estimated a long distance along this coast, and then reduced it for the same reason as he reduced his inland measure of the

35 Desanges, 'Les relations de l'Empire romain'; Steven E. Sidebotham, 'Ports of the Red Sea and the Arabia-India Trade', in Vimala Begley and Richard Daniel de Puma (eds), *Rome and India: The Ancient Sea Trade* (Madison, WI, 1991), pp. 12–38; Gary K. Young, *Rome's Eastern Trade: International Commerce and Imperial Policy 31 BC–AD 305* (London and New York, 2001), pp. 28–89.

36 *Periplus of the Erythraean Sea* [hereafter *PME*], 57; the Romans named the wind after him (Pliny, *NH*, 6.26).

37 Pliny, *NH*, 6.24. One other consequence of this was the bending of the Roman mental map of India, which generally associated it with 'the east', further to the south.

38 Ptolemy, *Geographika*, 1.9.

39 *PME*, 7–18.

south: that it resulted in a location that was so far to the south that it ought to be approaching the colder regions of the southern hemisphere.[40] Ptolemy (1.8) not only accepts Marinus' emendation but also provides a rationale for it.[41] In particular he notes that the initial distances would have placed Agysimba and Prason so far to the south that the climate ought to have been as cool as that enjoyed by the Scythians and Sarmatians who lived north of Lake Maeotis (i.e. the southern Baltic). Ptolemy followed Marinus' reasoning that the climate of both Prason and Agysimba was continually hot and therefore the estimated distance had to be shortened appropriately. It was through these calculations that Ptolemy constructed the map that remained influential for a millennium. Much of his version of Africa disappears into an unknown south, merging with a great southern land mass, that in turn links with China and turns the Indian Ocean into a vast inland sea. This new map quietly disposed of an ancient notion of a great encircling ocean. Moreover, and of some significance to this study, it used the empirical evidence of travellers to extend a description of the world to a point well south of the equator – both the actual equator and the line (mis) calculated by Ptolemy and Marinus.

The line of that equator, as defined by Ptolemy and Marinus, ran along a parallel roughly nine degrees north of the actual equator.[42] South of their equator, there stretched a long trading coast and a deep hinterland that included the sources of the Nile. While the work of geographers like Marinus and Ptolemy extended the Roman intellectual understanding of their physical world, it also clearly controverted a central tenet of zone theory: that the antipodes were unattainable, cut off from the *oikoumene* by a wide belt of unbearable heat. Now, instead, there was a known and an attainable world beyond the equator, and accessible both through the heart of Africa and along its eastern coast. Zone theory nevertheless held it as axiomatic that, as the focus of the encircling torrid zone, the equator had to mark the hottest and most uninhabitable portion of the globe. The philosophers upon whom Cicero and Virgil had based their very powerful images had asserted that the far south was unreachable because it was unendurable. Ptolemy and Marinus of Tyre described a world in which it wasn't. What is more, the nameless freedman of Annius Plocamus, Septimius Flaccus, Diogenes, Dioskoros, and all of the merchants who read the Periplus and plied their wares as far south as the coast of modern Tanzania showed that it wasn't.

It would be salutary to say that this insight fundamentally altered the Roman world view, and that the realization that the south was attainable after all resulted

40 Ptolemy, *Geographika*, 1.9.

41 See also Susan P. Mattern, *Rome and the Enemy: Imperial Strategy in the Principate* (Berkeley, CA, 1999), pp. 31–2.

42 Berggren and Jones, *Ptolemy's Geography*, p. 138, plate 7a.

in its more sustained and directed exploration and exploitation. But, for the most part, Roman intellectuals either missed this work or preferred the versions of Virgil and Cicero. By the fourth century, the anonymous author of the *Expositio Totius Mundi et Gentium* could describe an Africa with neither Garamantes nor an eastern coast, merely a vast inner emptiness sparsely populated by the miserable Mazicae and Aethiopians.[43] Although writing in a Christian Empire, the author of the *Expositio* was a pagan, heir both to the traditions of Virgil and Cicero, especially the more esoteric of their works, and also the mystical aspect of classical paganism that had enjoyed especial emphasis in the fourth century.[44] It was this that shaped his writing far more than the empirical school of Alexandrian geographers exemplified by Ptolemy and Marinus. Likewise, Macrobius, an author of far greater erudition (and slightly later date) passes over the works of these geographers in silence (although not ignorance).[45] His estimate of the size of the globe, for example, derives directly from Eratosthenes, as if there had been no revision by Marinus.[46] For Macrobius, the most southerly point of navigation, and indeed habitation in the *oikoumene*, lies in what he calls the 'cinnamon country', that is, India.[47]

Arguably, Macrobius' position derives from a clear polemical desire to assert the essential value of classical texts in an increasingly Christian environment. It also asserted the same supremacy of theory over experience that is still discernible in contemporary ideological discourse. What is clearly absent is that sense of the primacy of experience over theory with which this chapter began. From the late third century onwards, it would be fair to say that exponents of classical learning and philosophy felt under serious threat from an emerging, and divergent, Christian world view. This was a world view that exalted a completely different set of texts from those by which ancient Graeco-Roman intellectuals had defined themselves and in which they had, for centuries, sought authoritative statements about the world. The fall back into the scholarly bunker was not without deftness or a certain creativity. What was missing was a sense of debate and contestability. Hellenistic and Roman scholars had created knowledge through both discourse and exploration. Inherent in such a process is the possibility that things might be other than they are believed to be. Late Roman intellectuals,

43 *Expositio Totius Mundi et Gentium*, 62. For the date, see Jean Rougé, *Expositio Totius Mundi et Gentium: introduction, texte critique, traduction, notes et commentaire*, Sources Chrétiennes, 124 (Paris, 1966), pp. 9–26.

44 Rougé, *Expositio*, pp. 48–55.

45 William Harris Stahl, *Commentary on the Dream of Scipio by Macrobius: Translated with an Introduction and Notes* (New York, 1952), p. 164, n. 6.

46 Macrobius, *Commentary on the Dream of Scipio*, 2.6.2.

47 Ibid., 2.7.1–4.

faced with a tectonic threat to their entire world view, retreated into simple, text-based (and often false) knowledge. The vitality that had once fuelled the challenging of certainties was now channelled into their defence. The irony here is that their rivals offered little that was different, save the texts themselves. There is not much evidence that ancient Christians thought about the physical world in which they lived. When they did, they based their propositions on scriptural authority. In the sixth century the Christian divine Kosmas Indikopleustes, who had actually sailed to India and back, formulated an argument (despite his own empirical experience) that there was no distant south, that the antipodean concept was frankly silly, and that the world was, in fact, flat and really did have four corners. While his ideas did not catch on as decisively as some have thought, they nevertheless reversed the position of centuries of Roman and Hellenistic thought. While Kosmas might have expected some little intellectual storm over his arguments, they seem to have been received so quietly and unremarkably that the distant south of Ptolemy and all his predecessors might never have been.

Chapter 4

Meanings of the South: From the *Mappaemundi* to Shakespeare's *Othello*

Christopher Wortham

As a contribution to a volume about perceptions, this chapter has for its basis a number of concepts that are ontologically prior to perception: it asks, how did European people think about the south before they reached there? It also suggests in a speculative way some thoughts surrounding a second question: in what ways did preconceptions affect or even condition what those European people saw when they arrived? Perception and conception overlap in that perception primarily concerns sensory experience but in a secondary sense perception also denotes mental understandings gained from thought processes. So there is little point in distinguishing too precisely between concept and perception.

Given that *Terra Australis* simply means 'southern land', what did medieval and early modern Europeans understand by the concept of the geographical 'south'? We must also ask ourselves: what traces lingered of classical Greek and Roman understandings, and in what ways were these understandings developed, modified or discarded through the interaction of thought and experience? It is my purpose here to suggest some responses to these questions with reference to two seemingly disparate sets of artefacts: maps and plays. In referring maps to plays and plays to maps, I am following what has become a practice in the group of approaches gathered together under the label of New Historicism, namely to consider how cross-reference between different domains of cultural activity may result in a kind of illumination that a mono-disciplinary approach would not be able to achieve. Maps and plays are categories of artefact that seem diverse in purpose, provenance and cultural significance, yet in many ways they are intimately related. It was for good reason that Shakespeare and his theatrical colleagues gave their new theatre on the Bankside the name *The Globe*.

The adjectival term of *australis* derives from the Latin name of a southerly wind that was named in Rome after Auster, one of the wind gods. Auster was identified with one of the Greek gods of the winds, the Anemoi, called Notos or Notus. Both Notos and Auster later become identified with the Sirocco. This wind begins as a hot, dry wind over north Africa, but as it is drawn across the

Mediterranean by easterly-moving low pressure systems it becomes cool and damp over Europe as the bringer of drenching autumnal rains.

The changing nature of this southerly wind may well have contributed towards a belief widely held in Europe until the late fifteenth century that the equatorial region was too hot for humans to endure. By extension it may well have been argued that, if the wind cooled to a great extent as it crossed from north Africa to Europe, the equatorial region must be proportionally even hotter than the Saharan. One of the early references to the south wind as being extremely hot and equatorial in provenance is Aristotle's *Meteorologia*. Aristotle says:

> So the south wind clearly blows from the torrid region. Now the sun is so near to that region that it has no water, or snow which might melt ... But because that place is far more extensive and open the south wind is greater and stronger and warmer than the north and penetrates father to the north than the north wind does to the south.[1]

Despite pervasive beliefs in the extreme heat of the torrid equatorial region, Western travellers had ventured into it, though probably not as far as the equator itself, in pursuit of the spice trade many centuries before European ships reached that supposedly torrid zone. Accounts of such journeys date back into Roman times. And much later, in the thirteenth century, Marco Polo may have travelled as far southward as Java. Inhibitions about reaching the equator itself remained intact until the second half of the fifteenth century, when the fall of Constantinople to the Ottomans in 1453 led European powers to look for an alternative to the land route to the east by attempting to find a route around Africa. Bartolomeu Dias sailed around the southern coast of Africa in 1478 and ten years later Vasco da Gama followed and reached India by sea. These accomplishments meant that concepts of southern landforms had to be reappraised.

Terra Australis had been configured in several variants within the Roman world. Predominantly, the literati of Rome accepted that the earth was spherical, or at least a spheroid. The idea of five zones consisting of cold regions at each pole, intermediate temperate regions in both hemispheres and a fiery central equatorial region was a commonplace in classical thought and an extension of Greek philosophical and scientific theory about the rotundity of the earth. In practical terms, however, the Romans constructed the world, as far as it was known to them, in an approximately rectangular form. The long northern margin extended laterally from the British isles to Asia Minor; the shorter longitudinal

---

[1] Aristotle, *Meteorologica*, 2.5. Online at: www.greektexts.com/library/Aristotle/Meteorology/eng/print/486.html [accessed: 20 November 2010].

side extended down to the east shore of Africa; the long southern margin followed a latitude of approximately 15 degrees north across to the western extremity of Africa; and the western longitudinal line reached up to include the British isles.[2]

Mappaemundi

Two distinctly different Roman models co-existed until the late medieval period: they were the so-called 'T-O' map and the zonal map. We shall consider the T-O genre first. The precise origins of this form of map are uncertain and may well antedate Christianity, though the genre becomes strongly Christianized. This genre adumbrates the Roman understanding of the known world in quasi-rectangular form and is extended to cover more known regions and some imagined ones too, acquiring a circular shape for the whole. This circular shape of the T-O incorporates much of the three known continents of Europe, Asia and Africa and although its limits extend far beyond those of the Roman world its assumptions and centre of interest owe much to Roman thought. In the most characteristic form of the T-O map, the continents are separated by a 'T' inscribed within the 'O' of the circumference. The 'O' represents the river or sea called *Ocean* that was thought to encircle the entire world in ancient Greek tradition. The vertical down-stroke of the letter 'T' represents the Mediterranean Sea; the portion of cross-stroke on the European side represented either the River Don or, in some versions, the Black Sea; and the portion of the cross-stroke on the African side represented the River Nile. At the junction between the down-stroke and the cross-stroke of the 'T', or very close to it, lies the holy city of Jerusalem. This genre of map is heavily stylized and more symbolic than representational, though it has a basic representational element in the relative positioning of the three known continents of Europe, Asia and Africa. Within the T-O genre there were many variants and sub-genres.[3] A characteristic instance of the basic form is represented in an early printed version of the encyclopaedic *Etymologies* of Isidore of Seville (Figure 4.1).

In this most simple form of the T-O map, Europe, Africa and Asia are seen shown from the viewpoint of someone in western Europe, someone seeing the map primarily as an aid to Christian devotion. At the centre of the circle, or very close to it, is the holy city of Jerusalem. Just as most medieval churches in western Europe were oriented – literally – to Jerusalem, so one reads the map

---

2 See Bill Leadbetter's chapter 'The Roman South' in this volume for a more detailed account of Roman geography and cartography.

3 For various instances within the T-O genre see P.D.A. Harvey, *Mappa Mundi: The Hereford World Map* (Toronto, 1996).

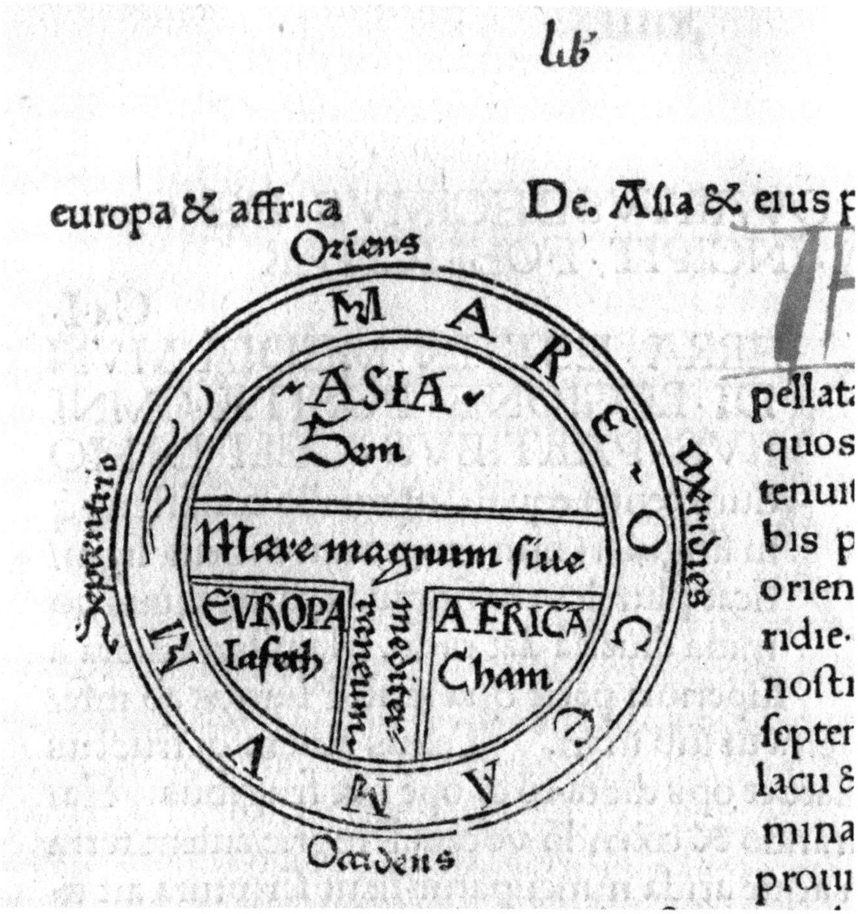

Figure 4.1 T-O Map from a printed edition of the *Etymologies* of Isidore of Seville (1472). Reproduced with permission of the Harry Ransom Center, University of Texas

from west to east in order to find Jerusalem at the centre. In such a map one is likely to find *Oriens*, the name of an early Greek sun god, appearing at the top to indicate the eastern extremity of the world. In more developed forms of the T-O map, Jerusalem is strongly marked at the centre and an even more holy place than Jerusalem is found at the eastern perimeter, namely the lost paradise of Eden. And beyond Eden at the top of the map is the imagined habitation of God. Only in the late fifteenth century would the navigational primacy of north begin to take precedence over east as topmost point of reference in cartographic representations of the known world. At the foot of Isidore's T-O map is *Occidens*, the western periphery where the sun and the stars are seen to set. The name itself comes from the Latin verb *occidere*, to fall or set.

The other two cardinal points are marked on Isidore's map as *Septentrio* and *Meridies*, respectively representing north and south. *Septentrio* was the Roman equivalent of the Greek god of the icy north wind, *Boreas*. The name of *Septentrio* comes from the seven stars of the constellation of Ursa Major, which surrounds the north pole. The cardinal point of the south is at first sight somewhat surprising. It is not marked by the south pole, as one might expect, but by the equator. *Meridies* primarily refers to noon or midday, but early classical usages suggest an extended cartographic sense signifying the furthest known point south, which in Roman terms would have been within the equatorial region. In medieval and early modern times, usages of the term *meridies* become more diversified and diffuse: *meridies* sometimes signifies the point at which the sun is at its zenith, i.e. the equator, but occasionally it signifies the south polar region. However within the *mappamundi* tradition, the equatorial meaning prevails.

Whatever the origins of the T-O map, which may go back to pre-Christian times, as a cartographic form it comes to be strongly marked within a religious paradigm during the Christian era. While its origins are unclear in date or place, it certainly came to be identified with imposing a Christian meaning on geographical terrain. The T became a representation of the Greek tau cross, without an upper vertical extension, which in turn was identified with the cross of Christ in early iconography. Isidore's map is one of those in which the centre is not in Jerusalem, as it came to be in a number of T-O maps, but in the Mediterranean rather than on the mainland. A possible explanation is that the marine centre is thought of as being the island of Cyprus rather than Jerusalem: for 300 years Cyprus would be the home of the exiled Christian kings of Jerusalem after Jerusalem itself had fallen to the army of Saladin at the end of the third crusade in 1192. The Isidore map seems to affirm this understanding.

During the later medieval period, T-O maps became increasingly representational, while maintaining the basic tripartite layout of the known continents. The later maps within the T-O genre often allow for the central point of the circle to be Jerusalem itself. Examples of this variant are to be found in the Hereford *Mappamundi*, the Ebsdorf Map and the thirteenth-century Psalter map in the British Library. These three maps exhibit further commonality in another important feature: it is that within the area of the far south, on the far side of the River Nile, there lurk many monstrous and deformed humanoid figures. In all of these maps the great length of the Nile is represented as being curled along the southern perimeter.

So, if we follow the line of the Hereford and other *mappaemundi* around the circumference from west through south to east, we encounter a progression

Figure 4.2 Psalter World Map, mid-thirteenth century. British Library, MS Additional 28681, f. 9r. By permission of The British Library

from the known world of normality (west) through an unknown zone of extreme abnormality (south) to that imagined perfection (east) which is the lost earthly paradise of Eden. It is a progression from marvels of the natural world, like the pillars of Hercules in the Straits of Gibraltar, to monsters and thence to the idyllic. Some of the monsters might simply be termed wonders and as such morally neutral in significance, but since the *mappaemundi* are so charged with spiritual values it seems more likely that a pejorative significance is implied. In short, farthest south is farthest from God. Physical Africa is imbued with spiritual meanings imposed by Europe. In developed T-O maps there is often a mixture of marvels and monsters in the distant regions of both the South and the North, but the monstrous are more densely concentrated along the southern extremity within Africa. This concentration is strongly marked in the maps I have mentioned: the Hereford, Ebsdorf and psalter maps. The psalter map emphasizes the monstrous inhabitants of the south particularly clearly (Figure 4.2).

There is little basis for concluding from the disc-shaped maps of the T-O genre that there was any general belief that the earth was flat. Aristotelian and a number of Greek mathematicians had long since disposed of any idea of a flat earth: the T-O map and its derivatives would have been widely accepted as a two-dimensional representation of a three-dimensional reality, though limited in offering a diagrammatic account only of the known world. A particularly significant extension of Aristotelian thought was to be found in the global paradigm proposed by Crates of Mallos, some 200 years after Aristotle. In Crates' paradigm, the area that would later become the three continents of the *mappamundi* tradition in fact constituted only one quarter of the whole world and constituted the known half of the northern hemisphere that the Greeks had called the *oikoumene* (or *oecumene*).[4] The region of the *antipodes* (as the place of diametrically opposed feet) was not in the quadrant immediately below the known world represented in the T-O map but to be found in the south on the other unknown side of the world. The southern region beneath the known northern one represented in T-O maps was known as the *antoikoi* (or *antoeci*) (Figure 4.3). It is important to be aware of the adjoining southern regions of the

---

[4] Alfred Hiatt, *Terra Incognita: Mapping the Antipodes before 1600* (Chicago, IL, 2008), pp. 17–20. The alternative spellings in brackets in this paragraph are those used by E.L. Stevenson in his diagram, from *Terrestrial and Celestial Globes* (New Haven, 1921), vol. 1, Fig. 5. See Figure 4.3 in this chapter. For another recent work on the history of concepts and representations of the antipodes, see Douglas Mark Goldie, *The Idea of the Antipodes: Place, People and Voices* (New York, 2010). Goldie's book was not yet available when this chapter was written. Goldie examines classical, medieval and early modern instances, bringing to them a postcolonial critical perspective. It is thoroughly recommended as parallel reading with Alfred Hiatt's book.

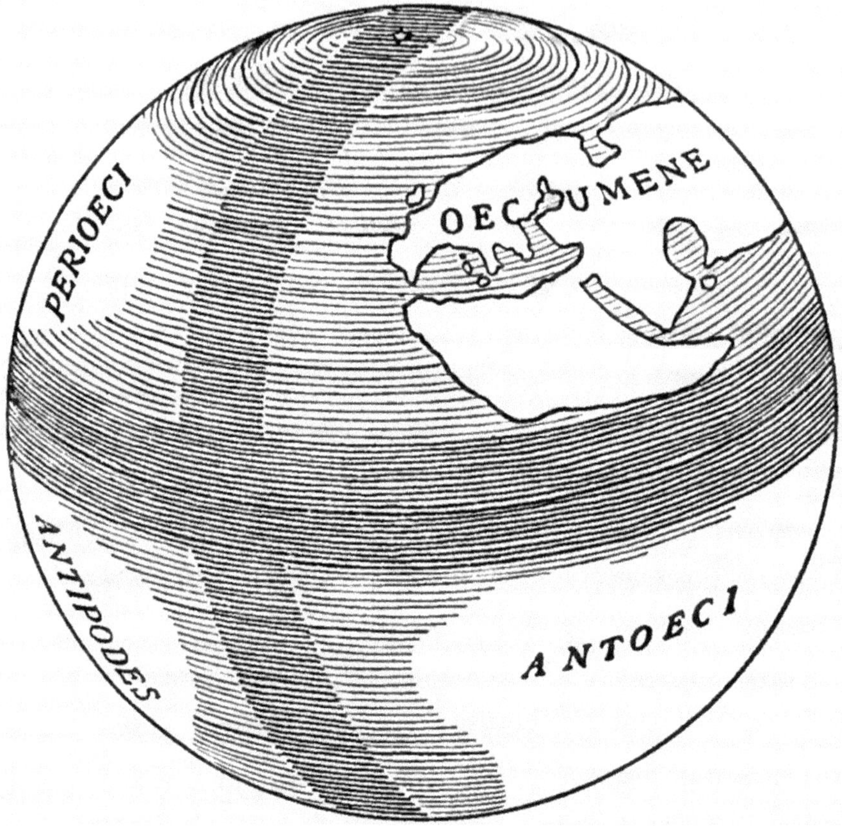

Figure 4.3 Diagram showing the division of the earth into four parts by Crates of Mallos. Original line drawing from E.L. Stevenson, *Terrestrial and Celestial Globes* (New Haven, CT: Yale University Press, 1921), vol. 1, fig. 5. Reproduced with permission

antiokoi and the *antipodes*. Joined together as two quadrants of 180° each, they would give rise to the massive and unbroken southern land projected later by such early modern cartographers as Ortelius. We shall return to Ortelius later.

As we have noticed, *mappaemundi* were primarily devotional. The Hereford map is a case in point. As part of a devotional triptych, contained within the kind of frame that often served for an altarpiece, the map invites being read as though it were an altarpiece.[5] As with many such devotional triptychs, the central panel

[5] Whether the map actually served as an altarpiece or was designed to be read as though it were an altarpiece is a matter of continuing debate. See Martin Bailey's essay 'The Discovery of the Lost *Mappamundi* Panel: Hereford's Map in a Medieval Altarpiece?', in

contains the devotional core. The two complementary wings – long lost and only recently recovered – could either be opened to show supplementary matter or closed to protect the central panel. The central panel of the Hereford map consists of the circular map. Above the map there rises a pointed arch in which Christ is seen presiding over the Last Judgement. The map, seen in its full original context, shows a symbolic distribution of space. We may think of the entrance to the map as beginning at the foot of the map, where there is a narrow entrance to the Mediterranean. Directly in front of that entrance, on the surface of the altar itself, the elements of bread and wine will stand, awaiting consecration in the Mass. If worshippers, as imaginary pilgrims, meditate upon the map they will see a direct path upwards to the devotional centre of the map, the city of Jerusalem. As their eyes are drawn further upward with the skyward thrust of the cathedral's Gothic arches, from the middle of the map they will pass to the lost garden of Eden at the topmost (easterly) point of the map. Directly above that, across the line of the circle enclosing the material world, there is the spiritual kingdom of heaven. In short, the way through the middle of the map is the way to godliness and ultimately to God.

In this privileged path from the western perimeter of the familiar world through the central point to the apex we find ourselves in the middle way through the earth's middle sea, literally the Mediterranean. The straight way is to Jerusalem: all deviations from this path lead to unholiness of one kind or another. And the straight way beyond points towards the divine. In this journey we have both philosophy and theology as our twin companions. The philosophical way of moderation tells us to avoid both the cold of the north and the heat of the south and the theological way of direction towards God tells us that the north and the south will lead us away from our salvation.

Plays

There is one medieval play that shows astonishing affinity with the spiritual geography of the *mappamundi* journey towards God. This is the morality play, *The Castle of Perseverance*. This play, about the salvation of the individual human soul in the face of multiple temptations from the sins congregated under the influences of the world, the flesh and the devil, draws upon the significance of the cardinal points of the compass as the domain of these groupings. *The Castle* is unique in that the manuscript includes a stage design for its performance. That design creates an acting space modelled upon the *mappamundi* tradition (Figure 4.4).

P.D.A. Harvey (ed.), *The Hereford World Map: Medieval World Maps and Their Context* (London, 2006), pp. 79–93.

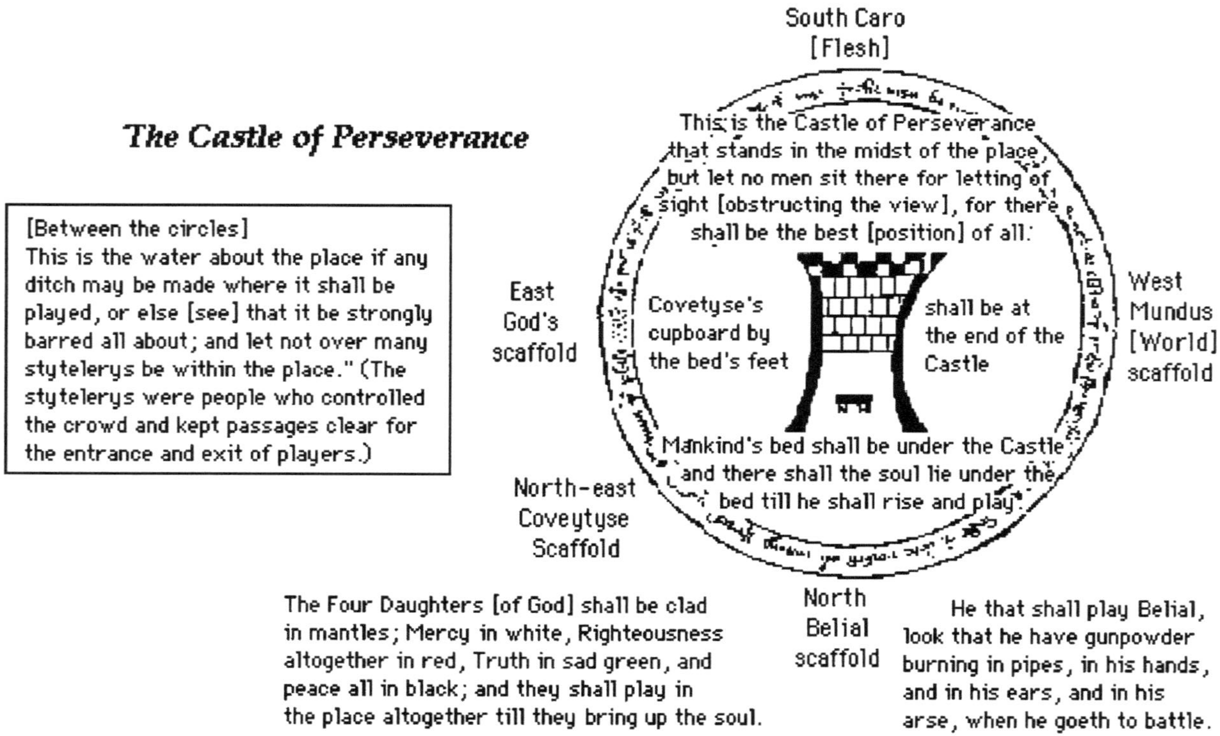

Figure 4.4 Diagram of Performance Space for *The Castle of Perseverance*. Michael Best, 'The Staging of a Morality Play (2)', *Internet Shakespeare Editions*, University of Victoria. Online at: http://internetshakespeare.uvic.ca/Library/SLT/stage/early stages/castle2.html [site updated November 2010]. Reproduced with permission

At each of the four cardinal points of the compass there is a scaffold. Respectively they represent the attributes we would expect from a reading of the *mappaemundi*: on the eastern scaffold is God; on the western is Mundus, the embodiment of worldliness; on the northern is the Devil; and on the southern scaffold we find Caro, the flesh. Our habitation in the west is the starting point for the play's action. Gathered there are those sins that represent the temptations of material life, which is inimical to the life of the soul. The protagonist, who represents humanity both individually and at large, moves on from the scaffold of the world to that of the flesh in the hot south. Here he encounters personifications of gluttony, lechery, sloth and other forms of depravity. Of these, lechery is the most insistent. Coming down from the southern scaffold and showing off her wares to Mankind, the female representative of lechery says:

> Ya, when thy flesh is fair fed,
> Then shall I, lovely Lechery,
> Bounce up and down with thee in bed.
> For this serve food and drinks so dry –
> In love thy life shall be led.
> Be a lecher 'til thou die:
> Thy needs shall be the better sped
> If thou give thee to fleshly folly
> 'Til death's river thou shalt leap!
> Lechery, since the world began,
> Hath advanced many a man;
> Therefore, Mankind, my dear lemman
> Into my cunt thou shalt creep![6]

The south, as place of bodily transgression in *The Castle*, is conceptually the same domain as that of the monstrous races of the *mappaemundi*. As this chapter has already pointed out, monstrous forms are not exclusive to the south – in the *mappaemundi* they occur also in the north among the Scythians and towards the east in India – but the greatest concentration is within the south, where the greatest extremes of abnormality are found too. Although it is possible in the specific case of the Hereford map that the monstrous forms represented there were simply introduced into the map as curiosities recycled from mythologies and pseudo-histories of the ancient world, the influence of Christianity is pervasive and has coloured the reception of the mythic. Whether any of the

[6] Modernization by Sandra F. Johnston. For her full text see http://homes.chass.utoronto.ca/~ajohnsto/cascomp.html [accessed: 20 November 2010].

monstrous races of Africa may be candidates for salvation is a vexed question and the issue is treated differently in different sources; however, these strange beings are often identified with exclusion from normality and concomitantly exclusion from salvation.[7]

Into Shakespeare's time the connection between physical deformity and moral enormity was pervasive enough to ground a theatrical convention wherein the state of the soul was mirrored in the shape of the body. For example, much is made in *Richard III* of the hunchback protagonist as a villain whose twisted body matches his twisted nature. There are numerous instances of such symbolic associations in other Shakespeare's plays. Another example: in *Othello* a recital of transgressive enormity at the very beginning of the play is made to seem particularly ominous as Iago and Roderigo hurl insults and innuendoes into the night air. Later on, the already tainted Othello, explaining to the Duke and the assembled senators how Desdemona has come to love him, gives an account of his journey through the south with its marvels and monsters. By ironic inference Othello, in rebutting the accusations against him, serves only to vindicate them, for the journey he recounts – and into which he has already drawn Desdemona – is an excursion into the realm of taboo. Brabantio has just accused Othello of stealing his daughter by using witchcraft, saying that she has erred 'against all rules of nature' in having come 'to fall in love with what she feared to look on', as though he were something monstrous himself, when Othello responds:

> Her father loved me, oft invited me,
> Still questioned me the story of my life
> From year to year – the battles sieges, fortunes
> That I have passed.
> I ran it through, even from my boyish days
> To the very moment that he bade me tell it;
> Wherein I spake of most disastrous chances,
> Of moving accidents by flood and field,
> Of hair-breadth scapes i'th'imminent deadly breach,
> Of being takn by the insolent foe
> And sold to slavery; of my redemption thence,
> And with it all my travels' history:
> Wherein of antres vast and deserts idle,
> Rough quarries, rocks, and hills whose heads touch heaven,
> It was my hint to speak – such was the process:

7 Naomi Reed Kline, *Maps of Medieval Thought: The Hereford Paradigm* (Woodbridge, 2001), gives an account of various medieval arguments in a section entitled 'The World of Strange and Monstrous Races', pp. 141–64.

> And of the cannibals that each other eat,
> The Anthropophagi, and men whose heads
> Do grow beneath their shoulders. This to hear
> Would Desdemona seriously incline;
> ...
> She loved me for the dangers I had passed,
> And I loved her that she did pity them.
> This is the only witchcraft that I have used.[8]

What Othello says both vindicates him and vilifies him. His rhetorical elegance is great, and as the Duke says in response to the speech: 'I think this tale would win my daughter too.' And yes, he has taken Desdemona with him on an imagined journey to the far south: the 'antres' (caves), 'hills whose heads touch heaven', the 'Anthropophagi' and the 'men whose heads do grow beneath their shoulders' are all landmarks on a journey through the *mappamundi* to the source of the Nile. So together Othello and Desdemona have entered the realm of the fleshly lusts that will soon destroy them both. Their subsequent shared journey to Cyprus, the Jerusalem in exile, along the spiritual path from Venice towards Jerusalem, will not save them. In part they are already doomed by their physical desire in a situation that demands the military man have no distraction from his appointed tasks; and in part their inward susceptibility will be exploited and abetted by another form of inward evil: the voice of Iago from within the society, who is far more dangerous to them than the Turks outside it. In fact, Iago will escort Desdemona to the 'citadel', mentioned several times in the early moments after arrival on Cyprus. In any warrior society the citadel is both a military stronghold and a religious shrine, so it is doubly dangerous that Iago lurks here and that within this stronghold Desdemona is in the process of inverting due precedence by becoming 'her captain's captain'.[9] Iago is dangerously evil, but Othello and Desdemona have become vulnerable to his machination by taking their journey, literally by Othello and imaginatively by Desdemona, to the forbidden place that is the south.

In *Othello* Shakespeare plays relentlessly upon cultural anxieties. In the opening scene Iago and Roderigo disturb the quiet night, hurling insults from the street up to the window of Desdemona's father, Brabantio. Proclaiming the elopement of Othello with Desdemona, they cry out in language laced with taunts about race, religion and the age difference between the couple. Not only is Othello a seducer, they proclaim, but he is a Moor – the word is used

[8] *Othello*, ed. Norman Sanders ('The New Cambridge Shakespeare'; Cambridge, 1984), I.i.127–68.

[9] *Othello*, II.i.74.

repeatedly without mentioning Othello by name – or rather *the* Moor: he is a singular oddity in Venetian society. Iago shouts into the night air:

> Even now, now, very now, an old black ram
> Is tupping your white ewe. Arise, arise;
> Awake the snorting citizens with the bell
> Or else the devil will make a grandsire of you.[10]

This warning tells Brabantio and the English audience in Shakespeare's theatre that Europe sleeps while Africa awakes to libidinous perfidy. Othello is 'the thick-lips' and is thus identified at once by his race and his supposedly erotically excessive disposition as 'a lascivious Moor'. Brabantio is told that 'you'll have your daughter covered with a Barbary horse': she will be having sexual intercourse in an act of bestiality with one who is not only a north African barbarian but an animal. Othello does indeed descend to irrational sexual excess that manifests itself in extreme jealousy when he murders Desdemona. We know that his credulity has been practised upon by Iago, the white villain of the piece, but what the play tells us, nevertheless, is that age-old European perceptions of the south, as embodied in the nether reaches of the continent of Africa, are still readily identifiable. The audience of Shakespeare's time is still engaged with preconceptions that had become entrenched within a medieval world vision encoded in the cartographic and cultural genre of the *mappamundi*. As I shall suggest in concluding this chapter, such preconceptions have survived into our own time.

Cartography and Cultural Inheritance

The T-O map and the zone map were concurrent in the European consciousness and each had its purpose. The T-O map had strongly encyclopaedic associations and in its most developed form served as a memory bank of history, mythology and religion. The zone map was more speculative than denotative, for it constructed imagined regions as yet unknown, suggesting a more speculative view of the world. Cicero's *Dream of Scipio*, which was the basis for the genre of zone map realized by Macrobius in the fifth century AD, shows the elder Scipio showing his young descendent the smallness of the earth as seen from a distance in space. T-O maps and zone maps broadly agree over the placement of land masses for north, west and east, but differ radically in their understanding of the south. In the T-O map the south is identified with equatorial heat, but

10 *Othello*, I.i.89–92.

in the zone map with polar cold. This radical difference in the understanding of the south was to have consequences for later conceptions, perceptions and representations of *Terra Australis*.

The early medieval zone map illustrating Macrobius's commentary on Cicero's *Dream of Scipio* shows a land mass to the south of the equator. In line with much classical Greek thought, it posits a southern set of three climatic regions – cold, temperate and hot – mirroring those of the *oikoumene* that fall within the T-O concept of a world stretching from the north pole to the equator.[11] Extending the T-O tradition of a circling ocean around the *oikoumene*, both northern and southern landmasses are depicted in the Macrobian map as being surrounded by ocean (Figure 2.3 in Chapter 2).

Of particular interest in the Macrobian map is the polar zone identified as *Frigida Australis*. At first sight it looks like a separated southern continent, but what appears to be a narrow sea separating it from the southern temperate zone of *Temperata Antikorum* is only a broad line representing the boundary latitude of the polar region. What is most remarkable about the Macrobian map is the difference in size of the southern land mass, much larger than that of the northern hemisphere. Africa is seen as a strip along the Mediterranean seaboard, terminating in a vast equatorial ocean. But then, on the southern side of the equatorial ocean, there is an undivided continent of massive size, extending from the tropics to the south pole. It is from this paradigm that later cartography drew when zone maps began to supersede the T-O form at the time of the great explorations in the last quarter of the fifteenth century. Ortelius's world map of 1570, made almost a hundred years after the first European rounding of the southern tip of Africa, incorporates the findings of those southern journeys of exploration but still posits a great southern continent in the unexplored Indian-Pacific region that extends from the pole almost to the equator. So the Macrobian assumption remains intact, though modified to allow for the established fact of global circumnavigation around southern Africa and South America.[12]

Back to the developments of the century preceding Ortelius: in 1477, at the very time when long-distance nautical exploration was about to begin in the Atlantic region, a world map along the lines taken by the Alexandrian geographer and astronomer Claudius Ptolemaeus, generally known in English as Ptolemy, appeared.[13] This was in the first printed text of a great atlas published in order to illustrate the text of Ptolemy's *Geographia*. Only the text of Ptolemy's original manuscript, compiled long before in the second century AD, had survived, so

11 See Hiatt, *Terra Incognita*, pp. 4–6 for an account of the provenance of zonal maps.

12 See the Ortelius map entitled 'Typus Orbis Terrarum', Figure 7.2 in this volume.

13 See the Ptolemy world map, Figure 2.5 in this volume.

the map in the 1477 *Geographia* is essentially a fifteenth-century reconstruction. In this map there is no vast equatorial ocean dividing north Africa from the southern landmass. Instead, this landmass is seen as huge and unbroken, extending right through the equatorial zone at it southern limits, reaching across to join China in the far east. The Indian Ocean becomes surrounded by land, very like the Mediterranean though much larger.

In some versions of the Ptolemy world map the winds are shown around the perimeter, with Auster at due south. There is some variation in early editions. In some copies, including the one pictured in Figure 2.5 in Chapter 2 the legend *Terra incognita secundum Tholomeum* is written across the land along the eastward-extended southern boundary of the Indian Ocean. This caption occurs within the longitude and latitude of the continent we know today as Australia. This nexus does not suggest that any of the cartographers who reconstructed the Plotemaic world map believed that there was a southern continent precisely in this location; nor does it suggest that the legend of *terra incognita* in proximity to the wind god Auster would necessarily result in the eventual naming of that continent as Australia. But what it does suggest is that the potential name of Australia for a landmass in the Indian-Pacific region was already within the ambit of European consciousness.

Only a decade after the first publication of the *Geographia*, Bartolomeu Dias rounded the Cape and a decade after that Vasco da Gama completed the journey to India. So the new *Geographia* was quickly rendered obsolete, especially in relation to the idea of the Indian Ocean as an inland enclosed sea. However, the cultural memory of a southern continent survived from the medieval zone map into world maps that proliferated in the wake of these voyages of discovery. The exploits of Dias and da Gama were soon to be followed by even bolder voyages of global circumnavigation, beginning with Magellan's in 1519–21. The great world map of Ortelius entitled *Typus Orbis Terrarum* (1570) (see Figure 6.2. in Chapter 6), shows an enormous southern landmass that stretches right around the globe, covering the entire Antarctic polar region and extending upwards into the tropics at its eastern and western extremities. Ortelius names it thus: *Terra Australis Nondum Cognita* (southern land not yet known). There is unresolved debate as to whether the tip of this landmass that reaches almost up to Java on the Ortelius map may have been the land we now know as Australia.[14]

What we may deduce with certainty from the Ortelian map and other world maps of the time – such as those of Plancius, van den Keere and Hondius – is that the concept of the Macrobian zone map remained strong. As I have

[14] Leslie Marchant argues controversially for early French exploration of the coast of Western Australian before the arrival of the Dutch in his book *France Australe: The French Search for the Southland* (Perth, c. 1998).

already suggested, the vast lateral expanse of the southern landmass seems to owe something to an amalgamation between the two southern quadrants of the *antoikoi* (immediately south of the known north) and the *antipodes* (under the unknown northern quadrant). The vertical extent may also be explained by an amalgamation, though of a different kind: a hypothetical southern land extending from the pole almost to the equator in the Indian-Pacific sector may owe less to substantive reports of the large landmass of the as-yet undiscovered Australia than to an accommodation between the two conflicting medieval perceptions of the south, terminating in the pole, as in the Macrobian zone tradition or at the equator, in the T-O construction.

While Ortelius labels his southern land *Terra Australis Nondum Cognita*, Van den Keere and Plancius label theirs alternately as *Magallanica* (after the first circumnavigator) and *Terra Australis*. Other maps of the southern land by Catalan cartographer Juan Martines and Desceliers (see Figure 2.7 in Chapter 2) of the Dieppe School also identified a southern land.[15] However, some cartographers of the post-Ptolemaic period do not indicate the presence of a great southern land.[16]

In the new cartography of landmasses that became dominant during the first century of the age of exploration, as a result of the voyages of Dias and da Gama, Africa became distinct from any physical concept of *Terra Australis*. It had not been so hitherto: as we have noticed, in the 1477 *Geographia* of Ptolemy Africa was not yet detached from *Terra Australis*. It may be argued that although Africa became detached from *Terra Australis* in the new age of discovery, a long-entrenched set of culturally encoded understandings about the south in general survived. Africa still was the south and the south was still Africa in that both embodied the dangerous otherness of the equatorial region. Africa had been separated from *Terra Australis* physically but not imaginatively.

What took much longer to determine, both physically and conceptually, was the distinctness and separateness of the landmasses we now know as Antarctica and Australia. It was not until early in the nineteenth century – with the voyages of Flinders and Baudin, who circumnavigated Australia in opposite directions to meet at the place named Encounter Bay – that the southernmost boundaries of the continent we now know as Australia became established, differentiating it conclusively from the larger *Terra Australis* as it had been imaginatively constructed by Ortelius and others. Until the end of the eighteenth century the name *Terra Australis* continued to be ascribed both to Australia and Antarctica

15 The history of representations of the southern land has been covered in detail by Hiatt in *Terra Incognita*. See especially the section 'Ethnographic Dislocation: The Dieppe School', pp. 236–44.

16 The world map of Battista Agnese (c. 1540) is an early example. By the mid-seventeenth century many cartographers no longer showed a great southern land.

in an undifferentiated way. Pedro de Quirós, a Portuguese explorer, landed at Vanuatu in 1606 and declared it to be part of what he called *Terra Austrialia del Espiritu Santo*. In the mid-seventeenth century the name of New Holland was informally given by Dutch voyagers to the western shores of Australia when they made purposeful or accidental landfall on the coast on their way to their East Indies, but as yet they had no certainty of its southern limits or whether it might be attached to a greater southern land.

The transposition of the name from an indeterminate *Terra Australis* to designate the specific confines of Australia came late. In 1793, George Shaw and Sir James Smith published their *Zoology and Botany of New Holland*, describing the territory as 'the vast island, or rather continent, of Australia, Australasia or New Holland'.[17] Matthew Flinders, when preparing his manuscripts and charts for the publication of his 1814 collection of papers under the title of *Voyage to Terra Australis*, wrote: 'Had I permitted myself any innovation on the original term, it would have been to convert it to Australia; as being more agreeable to the ear, and an assimilation to the names of the other great portions of the earth.'[18] In the event, it was not until 1824, as a result of persistent advocacy by Governor Macquarie, that the Admiralty finally agreed that the continent should be known officially as Australia.

Culturally influenced responses to the land that is now Australia were mixed. Though the western side of Australia was colonized last by Europeans, the western seaboard became known long before European sailors reached the east coast. Early reports were not encouraging. To the Dutch, who were the chief group to experience the west coast, the land looked arid and unpromising compared with the tropical lushness of the East Indies. The island just off where modern Perth stands today they named *Rottnest*, which meant 'Rat's nest' and they were referring, evidently with some revulsion, to the diminutive marsupials that populated the island's seemingly barren shores. Dutch experience of the tricky coastline was generally accidental and sometimes catastrophic: only ships bound for the Indies that were blown off course ventured there. The wreck of the Dutch East India Company flagship, the *Batavia*, was a striking instance. The tale of planned dereliction, mutiny and murder that ensued from that wreck in the year 1629 quickly became public knowledge. Few came out of it well and the worst casualty may be thought of as the reputation of the land on which the massacre had taken place. The story of the *Batavia*, together with the legacy

---

17 John Alexander Ferguson, *Bibliography of Australia: 1784–1830* (Canberra, 1975), vol. 1, p. 77.

18 Matthew Flinders, *A Voyage to Terra Australis; Undertaken for the Purpose of Completing the Discovery of that Vast Country, and Prosecuted in the Years 1801, 1802 and 1803, in His Majesty's Ship, the* Investigator (London, 1814).

it left for later centuries of extrapolation and myth-making, has been traced in this volume by Leigh Penman in his chapter entitled 'The Wicked and the Fair'. As Penman demonstrates, both historical and fiction writing surrounding the region have taken up the fate of the *Batavia* as a narrative of dystopia. Penman begins his discussion of the *Batavia* disaster with reference to an article by the West Australian novelist, Randolph Stow, entitled 'The South Land of Antichrist'.[19] Stow reviews past perceptions of Australia as ambivalent, with the land being variously interpreted as either prison or as paradise. Penman affirms Stow, citing de Quirós's idealized appellation for his landfall as *Terra Austrialia del Espiritu Santo* as being countered by other visions of a dystopic 'southland of Antichrist'.[20] Of the tradition that has emerged in representing Australia through the lens of the Batavia calamity, Penman concludes:

> Throughout the history of its interpretation, the wreck of the *Batavia* has functioned as a mirror which has reflected back at its audience a surprisingly diverse and durable set of myths and historical realities, which, in a very real sense, are intimately concerned with mapping out the contours and topography of a *Terra Australis*. This particular *Terra Australis*, however, is not that of the physical world, which explorers and cartographers of the past pursued so intently, but one of the imagination.[21]

We do not know whether any specific prompting of previous European experience governed the opinion formed by William Dampier when he first encountered northwestern Australia and its indigenous people some 70 years after the Batavia disaster, but he was clearly unimpressed. Of these people he famously wrote in his diaries published as *New Voyage Round the World* (1697):

> The inhabitants of this Country are the miserablest People in the World. The Hodmadods of Monomatapa, though a nasty People, yet for Wealth are Gentlemen to these; who have no Houses and skin Garments, Sheep, Poultry, and Fruits of the Earth, Ostrich Eggs, etc as the Hodmadods have: And setting aside their Humane Shape, they differ but little from Brutes.[22]

---

19 See Penman's chapter in this volume.

20 The land actually seen by de Quirós was the Melanesian archipelago now known as Vanuatu, but at that time the geographic boundaries of southern lands were still unidentified. For a more detailed discussion see the chapter by Camino, 'Ceremonial Encounters: Spanish Perceptions of the South Pacific, 1567–1794', in this volume.

21 Penman's chapter in this volume.

22 W. Dampier, *Dampier's Voyages*, ed. J. Masefield (2 vols, London, 1906), vol. 1, p. 453. For further interpretation of this view, see the chapter by Michael McCarthy in this volume.

Dampier's account, which looks back to the pessimistic views of Thomas Hobbes rather than forward to the more optimistic ones of Voltaire and Rousseau, is particularly notable for its comparison with the inhabitants of central Africa, whom he accounts a 'nasty People'. It indicates a lingering association between the peoples of Africa and the dark-skinned indigenes of other southern climes.

In this study I have not been concerned with facts about societies in the southern hemispheres but with perceptions based on concepts instilled long generations ago in Europe. Ian Crawford, a prominent Western Australian anthropologist, has succinctly summarized the issue:

> When the first European visitors and settlers arrived in Australia they were so heavily conditioned by their own cultural concepts that they had great difficulty in recognizing any cultural environment among Aborigines. Europeans perceived the ideal man in their own image both physically and culturally, and because Aborigines were so different, they seriously debated whether or not they were human.[23]

The doubt about the humanity of indigenous people is one aspect of the cultural conditioning brought to Australia from Europe. That conditioning may well have its roots in early perceptions of the south that found expression in the *mappamundi* tradition and in plays such as *The Castle of Perseverance* and *Othello*.

Should there be doubt about the enduring nature of the thought patterns inscribed in early maps, it would be apposite to conclude with a final observation drawn from the T-O map from Isidore's *Etymologies* in Figure 4.1. If one looks at the three continents, under the name of each is the name of one of the sons of Noah: Sem for Asia, Japheth for Europe and Cham (or Ham) for Africa. In the Bible the youngest son of Noah is severely punished for prurient indecency in looking upon the naked body of his drunken father and in bringing his brothers to share the sight. He is punished by being cursed when Noah awakes to learn what has happened. 'A servant of servants shall he be unto his brethren.'[24]

Colonial intrusions into the lives of indigenous peoples of southern Africa and Australia were predominantly Protestant in nature. An extreme instance: the Afrikaner members of the Dutch Reformed group of churches in white South African society took the strictures of the Old Testament very seriously. Not only was the curse of Noah upon the sons of Ham in the book of Genesis interpreted literally, but it was also rigidly applied. The Prohibition of Mixed Marriages Act of 1949 prevented marriage between white people and those of other races

---

[23] Ian M. Crawford, 'Aboriginal Cultures in Western Australia', in C.T. Stannage (ed.), A *New History of Western Australia* (Perth, 1981), pp. 3–34, 3.

[24] Genesis 9:25.

based on this and other biblical texts, and in the following year any inter-racial sexual contact was prohibited in the Immorality Act. Afrikaner Calvinism in South Africa gave the ideological basis for the doctrine of apartheid and that doctrine persisted until it was finally dismantled in 1994. While Australia never subscribed to such extreme and inhumane views and laws, Australian society was contaminated by old European thought to the extent that Aborigines were not regarded as citizens until 1969 and until as recently as 1992 the land itself was declared to have been *terra nullius*, that is, owned by no-one, until the arrival of the Europeans. Whether in relation to Africa or Australia, old ideas about the nature of the south have died hard.

Chapter 5

Terra Australis, *Jave la Grande* and Australia: Identity Problems and Fiction

W.A.R. (Bill) Richardson

Preliminary Note: The Effects of Transmission Errors in Marco Polo's *Travels* and Nomenclature Confusion

One of the major identification problems on late fifteenth- and early sixteenth-century maps covering eastern and southeast Asia is due to cartographers' reliance on Marco Polo's *Travels*. He followed Arabic twelfth-century usage, which referred to Java as *Java Major* and to Sumatra as *Java Minor*. His erroneous statement that the former was the largest island in the world led many to acquire an exaggerated idea of its size. Some of the Dieppe cartographers applied *Jave la Grande*, a translation of the former, to Java, some to the enigmatic landmass south of it; some applied *Petite Jave* (*Java Minor*) to Java, some to Sumbawa. Alonso de Santa Cruz applied *Java Minor* to Borneo. Owing to misprints in the Paris and Basel editions of Marco Polo's *Travels* (1532), Gastaldi, Mercator, Ortelius and others invented the outlines of a landmass to accommodate such countries as *Beach* and *Maletur* south of Indonesia to represent them. Islands, especially *Java Minor*, and others whose names, in Marco Polo's spelling, were not recognized, were similarly imaginatively misplaced there.[1] Most modern writers apply *Jave la Grande* to the controversial, apparent continent extending southwards from Java and Sumbawa on the mid-sixteenth century Dieppe maps.[2]

---

1 Simon Grynaeus and Johann Huttich, *Novus Orbis Regionum ac Insularum Veteribus Incognitarum* (Paris and Basel, 1532); see W.A.R. Richardson, 'Mercator's Southern Continent: Its Origins, Influence and Gradual Demise', *Terrae Incognitae*, 25 (1993): 67–98, especially pp. 74–8.

2 The most significant Dieppe cartographers are Johne Rotz, Pierre Desceliers, Nicolas Desliens and Guillaume Le Testu, as well as two anonymous ones, responsible for the Harleian or Dauphin map, and the 'Vallard' atlas. For details, see W.A.R. Richardson, *Was Australia Charted before 1606? The* Jave la Grande *Inscriptions* (Canberra, 2006), p. 96 or Helen Wallis (ed.), *The Maps and Texts of the Boke of Idrography Presented by Jean Rotz to Henry VIII* (Oxford, 1981), pp. 85–8.

I

Jave la Grande: *Coastal Outlines and Inscriptions*

The sixteenth-century 'Portuguese discovery of Australia' hare, which Alexander Dalrymple set up in 1788, has had a long run. Advocates of the theory have included R.H. Major (1860s), A.F. Calvert (1893), George Collingridge (1895), Kenneth McIntyre and Ian McKiggan (1977), Roger Hervé (1955 and 1982), Helen Wallis (1981–95), Lawrence FitzGerald (1984) and, most recently, Peter Trickett (2007). Gavin Menzies (2002) claimed that the Chinese mapped Australia in the 1420s, and that the Portuguese copied their charts. All the above identified *Jave la Grande* as Australia, but there is remarkably little consensus in their interpretations of the same supposed evidence on the Dieppe maps. Only two paid any real attention to inscriptions, Peter Trickett and myself. Only two completely different explanations have been forthcoming, Robert King's (see below) and mine, in *Was Australia Charted before 1606? The* Jave la Grande *Inscriptions*.[3] The latter is outlined below, in a very abridged version, for direct comparison with the theories of Trickett and King.

The Dieppe Maps and Jave la Grande

The outline of *Jave la Grande*, similar on all the Dieppe maps, is attached to Sumbawa, and on some to Java itself as well. Figure 5.1 shows it as it appears on the Harleian or Dauphin map (c. 1546), superimposed on the modern map, assuming that what is identified as Java on both maps should coincide, and that necessarily both latitudes and longitudes are the same on both; in fact they are not.[4] The large, very prominent, triangular east coast feature tipped by *cap de fremose* is shown on the actual maps in about 45°S latitude, which would be just south of Tasmania's Southeast Cape (43° 49'S).

If the islands identified as Java and Sumbawa are Java and Sumbawa, then the two southward-trending coasts must either be fictitious, or real coastlines which have been misplaced. Rather than treat them as enigmas requiring investigation, the proponents of the Portuguese discovery theory jumped to the conclusion that *Jave la Grande* must be Australia, and then sought to prove it. However, numerous factors militate against their preconceived conviction, such as its position, and above all the *cap de fremose* promontory. Hypothetical explanations have been concocted in attempts to account for these anomalies. Its

3 See Richardson, *Was Australia Charted.*

4 British Library, Additional MS 5413. Reproduced in Richardson, *Was Australia Charted*, plate 3.

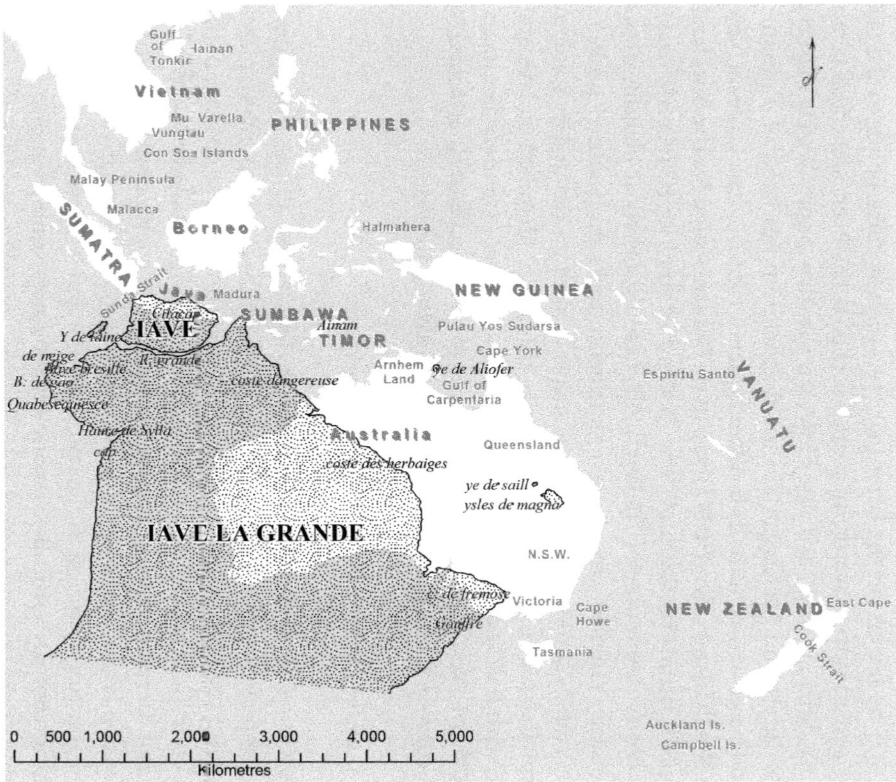

Figure 5.1 Superimposition of the Harleian or Dauphin map's *Jave la Grande* on a modern map. Composed by Robert Keane. By permission of the National Library of Australia

supposed misplacement has been said to be due to: the difficulty of measuring longitude until the late eighteenth century; its intentional removal westwards to make more of it appear to be in the Portuguese sphere of influence under the Treaty of Tordesillas (1494); and confusion of the outlines of Java and Sumbawa with those of Arnhem Land and Cape York Peninsula. None of these suggested explanations is convincing. Though the northern part of the east coast of *Jave la Grande* does perhaps look a bit like the northern part of Australia's east coast, there is no feature on the latter that even vaguely corresponds to the *cap de fremose* promontory towards its southern end. It has been identified in no less than six different ways; as Cape Howe on the New South Wales/Victoria border (McIntyre and McKiggan), as the northeastern headland of Tasmania (FitzGerald), as its most southerly headland (Hervé 1955), as East Cape on New Zealand's North Island (Hervé 1982), as part of Tasmania joined to the Auckland and Campbell Islands far south of New Zealand by solid ice (Gavin Menzies), and as Wilson's Promontory (Trickett). All these hypothetical identifications have necessarily entailed tampering unjustifiably with the feature's position and orientation as shown on the Dieppe maps. Coastlines claimed by some to be Arnhem Land and Cape York Peninsula are clearly identified as Java and Sumbawa, with Bali and Lombok between them. The *Jave la Grande* 'landmass' has no southern coast, but on several Dieppe maps it is speculatively attached to the north coast of Tierra del Fuego, both eastwards and westwards.

Inscriptions on the Dieppe Maps' Enigmatic 'Continent'

Inscriptions on *Jave la Grande* were virtually ignored by all the Portuguese discovery proponents, except Peter Trickett, probably because toponymy, place-name study, is a very demanding discipline, requiring historical knowledge of the languages concerned, an ability to read early handwriting, detailed knowledge of fifteenth- and sixteenth-century maps and charts, and of relevant travel literature, not to mention lateral thinking. Moreover, owing to the fact that place-names were very liable to corruption in the course of oral and written transmission, it is essential to compare critically as many versions as possible of a given name; single examples are extremely unreliable. Specific, unambiguous identifications are needed to establish an enigmatic coast's identity. Identifications such as *coste dangereuse* (dangerous coast) on the east coast of *Jave la Grande* as the Great Barrier Reef are not acceptable, unless that coast has already been proved to be Australia's east coast; there are hundreds of dangerous coasts in the world.

We have no idea how legible the handwriting of the archetype cartographer(s) was. The original inscriptions must have been in Portuguese, or were Portuguese phonetic attempts to record local place-names. The French were undoubtedly copying very early Portuguese sketch charts of coasts that were previously unknown to them. There was no consistency in spelling in any language in the sixteenth century. Words *sounding* like words in a cartographer's own language, or in writing *looking like* them, were liable to have those words written down; this sometimes led foreign cartographers to 'translate' them. Words which did not make sense to copyists were sometimes imaginatively 'edited' by them until they did.

Jave la Grande*'s West Coast*

The key west coast inscription appears in several different forms on different maps. On the Harleian map (Figure 5.1) it reads *Quabesequiesce*. As it stands, this means nothing in Portuguese or French, and it suggested no place name I could think of, or find. Its length suggested that it could be one of the not infrequent cases where two or more short words had been miscopied and run together. The first six letters, the word *quabes*, was eventually found, in the 1590s English translation of Jan Huyghen van Linschoten's *Itinerario*. In the original Dutch it had read *quabeb*, a now little-known spice called *cubeb* in English. It then grew in one place only, Sunda, the western third of Java. If *quabes* was the miscopied name of a spice, what about the rest of the inscription? Comparing it with similar ones on the famous '*Cantino*' map of 1502, it had clearly consisted of two Portuguese words *aqui esta* (is here), which the French had miscopied and run together. Confirmation of this interpretation was obtained from an inscription on a chart of 1522 by the Spanish cartographer, Nuño García de Toreno. An island in Sunda Strait, between Java and Sumatra, is identified by the 'word' *biliqualiabjquj*. The middle of this 'word' was fairly evidently a miscopying of *quabeb* (cubeb), followed by *aqui* (here). The first four letters meant nothing in French or Portuguese. They are almost certainly the Malay word, now spelled *beli*, meaning 'buy'. Thus 'buy cubeb here'. The whole 'word' thus identified Sunda which, on a number of early maps, was mistakenly depicted as a separate island.

Other inscriptions confirm the identification of the west coast as part of Java. The word *gao*, attached to a bay, is an old rendering of the Portuguese word now spelled *jau* (Javanese). The word recorded as *negre* or *noirre* (black), or *neige* (snow) on different maps, is attached to an island adjacent to the *gao* feature; these French variants were evidently 'edited' versions of the Malay word *negri* (town, village); the Portuguese original had probably been *negri gao* (Javanese town).The word *bresille* attached to a bay is a French translation of the Portuguese word *brasil*, a southeast Asian dyewood related to the one that gave its name to Brazil. The inscription *Haure de Sylla* (harbour of Sylla), together with the adjacent nameless *cap* (cape), is a misunderstood French rendering of

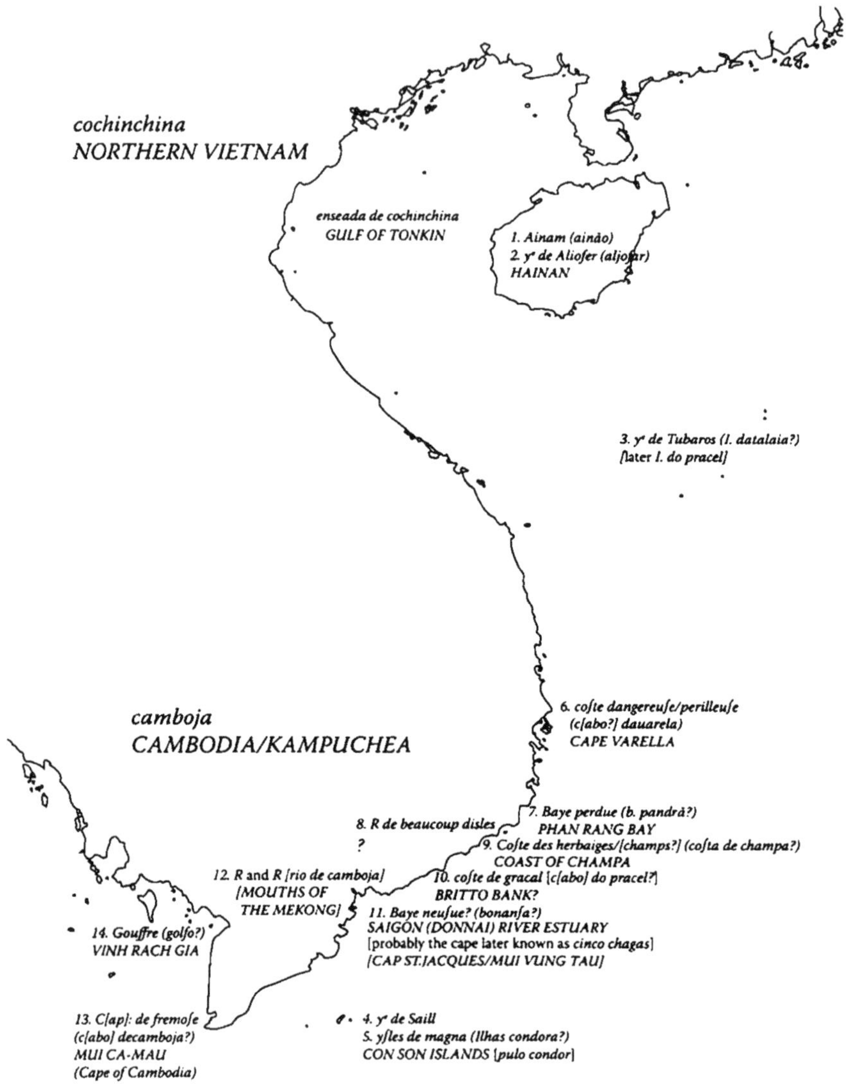

Figure 5.2 The coast of Vietnam and adjacent islands in the South China Sea, with features roughly corresponding to those on *Jave la Grande* and the islands east of it. Composed by Jens Smith and Robert Keane. By permission of the National Library of Australia

the name of the only significant port on Java's south coast, spelled *Tjilatjap* by the Dutch, and in Indonesian now spelled *Cilacap*.

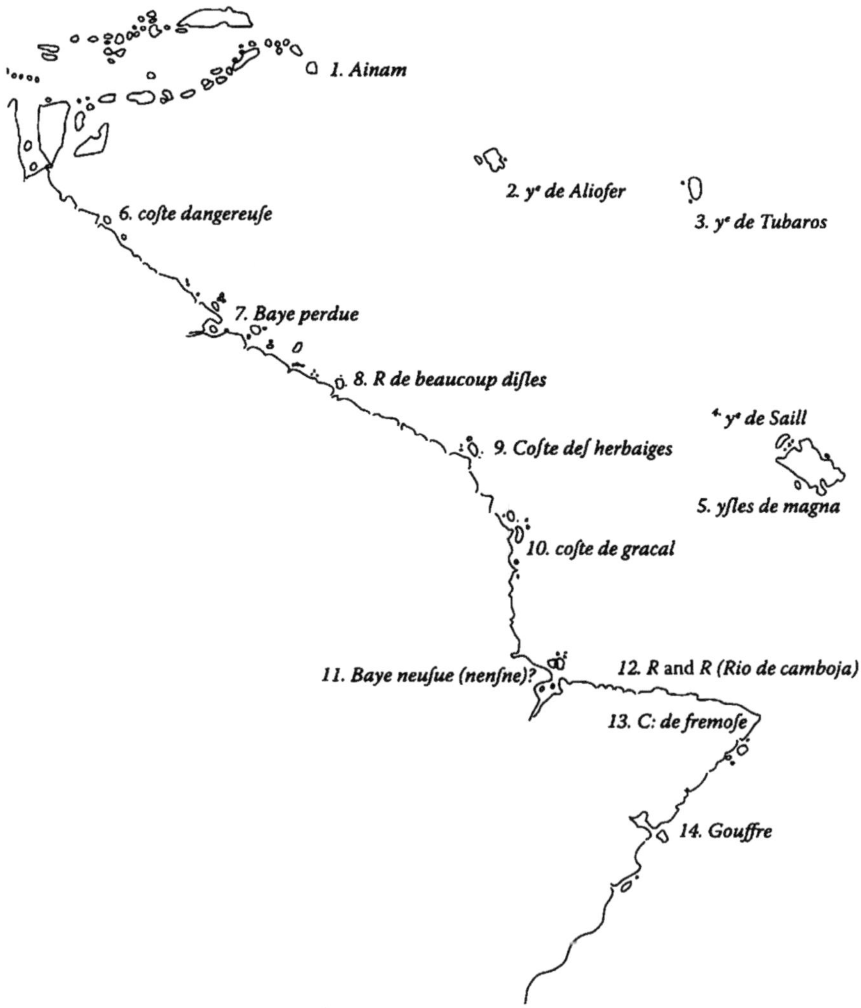

Figure 5.3 East coast of *Jave la Grande*. Composed by Jens Smith and Robert Keane. By permission of the National Library of Australia

Islands off Jave la Grande*'s East Coast*

Since at least four inscriptions on *Jave la Grande*'s west coast identify it as part of western Java itself, even though misaligned north/south rather than west/east, and evidently on a vastly larger scale than the main map, it seemed even less likely that the east coast was Australia's east coast. Only one southeast Asian coastline, together with the islands off it, seemed reasonably similar, the southern part of

Vietnam, especially the triangular Mekong delta. Figure 5.2 shows the coast of Vietnam, and, numbered 1–2 Hainan Island, 3 the Paracels, and 4–5 the Con Son Islands. Potentially confirmatory inscriptions are shown on Figure 5.3, the east coast of *Jave la Grande*. The islands numbered 1–5 correspond to those on the previous map, as the inscriptions confirm.

The most northerly island off the *Jave la Grande* coast (No. 1), from its position looking as though it is an Indonesian one, is named *Ainam*, a Portuguese version of Hainan Island, off the south coast of China, and north of Vietnam. Its pearl fishery was so famous that the Portuguese word *aljofar* (now *aljófar* 'pearl') was sometimes attached to it on Portuguese maps, and never to any other island. Some spelling of *aljófar* appears on several Dieppe maps, attached to an island southeast of *Ainam*, as here (No. 2). This suggests that the original cartographer knew both names, but didn't realize that they both indicated the same island, so depicted Hainan twice. Southeast of *aljófar* on several Dieppe maps is *y*ᵉ *de tubaros* (No. 3), conceivably a version of Portuguese *tubarões* (sharks). But on other sixteenth-century maps other names appear: *atabalero* (kettledrummer), *atalaia/atalaya* (shoal), and *atabaca* (?) on Jean Rotz's map of southeast Asia, probably a misrendering of *atalaia*. It seems probable that this word was on the original sketch chart, and was later replaced by a synonym, *pracel*, or *parcel*; those islands and shoals are still known as the *Paracels*.

The islands of *Saill* and *Magna* (Nos. 4 and 5), correspond roughly in position to the islands known to the Portuguese by their Malay name *pulo condor* (gourd island), now called Con Son, just off the Mekong delta. The Portuguese knew this name, in an Arabized form, *fulucandora*, as early as 1502 on the 'Cantino' map. How can it have turned into *Saill* and *Magna* by the mid 1540s? Compared with later, non-Dieppe maps, the group appears to have been inserted upside down. Could the manuscript inscription itself possibly have been read upside down by the French? Since *i*, *y*, and *j* were virtually interchangeable in spelling at the time, the original could have appeared to the French to read *sayll*, with a *y*, even though they transcribed it as an *i*. A manuscript *sayll* upside down could easily be read as *Ilhas*, Portuguese for 'islands'. Figure 5.4 provides an admittedly contrived explanation of what could have happened. The *Ilhas / sayll* inversion would appear obviously possible. Readers not convinced about the likelihood of the *candora / magna* inversion, may be less sceptical after looking at Figure 5.5. It shows the Arabized version *fulucandora*, on Martin Waldseemüller's famous 1507 world map, both the right way up and upside down; surely the inverted form of this printed version of *candora* bears a remarkable resemblance to *magna*?[5] A manuscript version might well have been even more convincing.

5 Reproduced in Joseph Fischer and F.R. von Wieser, *The Oldest Map with the Name America of the Year 1507 and the Carta Marina of the Year 1516 by Martin Waldseemüller* (Innsbruck, 1903).

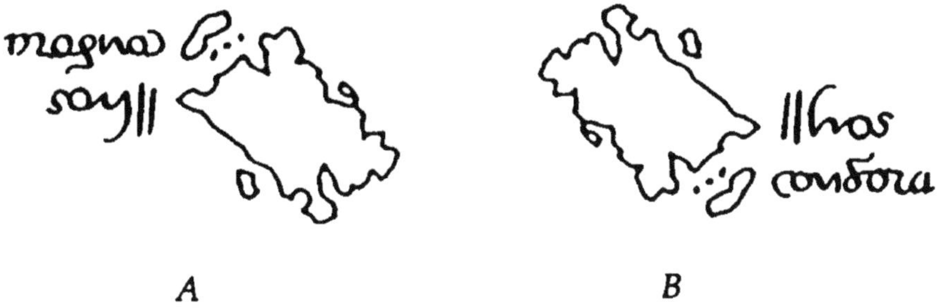

Figure 5.4 The outline of the *saill* and *magna* group, with hypothetical versions of the inscriptions. Traced by Jens Smith, with script by Helen Richardson. By permission of the National Library of Australia

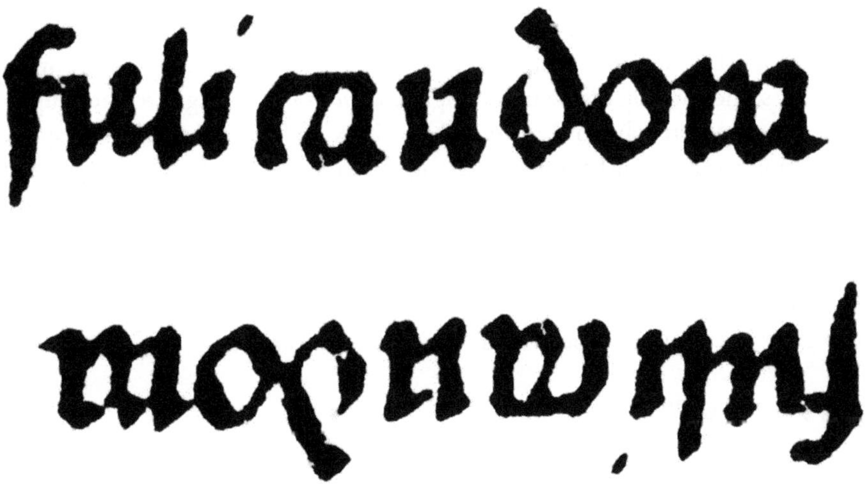

Figure 5.5 The inscription *fulicandora* on Martin Waldseemüller's 1507 world map the right way up and inverted. By permission of Flinders University Library

Jave la Grande*'s East Coast*

The most northerly inscription on the mainland coast is *cofte dangereufe* (No. 6). No Portuguese word meaning 'dangerous' appears on Portuguese charts depicting Vietnam, but one name could well have led the French to 'edit' it to that. The most easterly cape on the coast of Vietnam was known by a Malay word, transcribed as *uarela* (*varela*) meaning pagoda or idol, because of the shape of a rock formation on it. On Portuguese charts it was spelled *c.* for *cabo* (cape) *da uarela* or *dauarela*, with *da* (of the), prefixed to the following noun, a common sixteenth-century practice; the abbreviation *c.* could also, of course, stand for *costa* (coast). The manuscript letters *u* and *n* were frequently confused, as were *l*, *f*, and the old long letter *ſ*. The apparently meaningless *dauarela* could easily have been taken by a Frenchman to be a miscopying of *dangereuſe* and so 'corrected'. As it stands, *Coſte des herbaiges* (coast of pastures) (No. 7), is too unspecific for identification. But in a roughly equivalent position on Portuguese charts covering Vietnam, there appears *coſta de champa* (coast of Champa), then a kingdom in central and southern Vietnam. *Champa* is only one letter different from the French word *champs* (fields), a near synonym for *herbaiges*.

(No. 8) *Baye perdue* (lost bay) is a strange name to give to a bay on an apparently hitherto unknown coast. The old capital of Champa, originally called *Panduranga*, is now called Phan Rang, but the adjacent bay and cape both still go by the name of *Padaran*; the most common Portuguese versions of the name were the similar-sounding *pandeirão* and *pandrã*; the French seem to have 'made sense' of some spelling of the unintelligible *pandrã* by editing it to the roughly similar-looking French word *perdue*. (No. 12) *R* and *R*, for 'rivers', correspond in position to the mouths of the Mekong. There is evidently something wrong with (No. 13) *C: de fremoſe*, literally 'cape *of* beautiful'. It appears to be a misread, edited version of *cabo de camboja* (Cape of Cambodia), for it was then in Cambodia. An early sixteenth-century manuscript chapter heading, *Reyno decamboja* (Kingdom of Cambodia), in the *Suma oriental* of Tomé Pires shows how easily *dec* could have been misread as *fr* and the *j* as a long *s* (*ſ*).[6]

Why These Two Charts Were Misplaced

The north coast of *Jave la Grande* is exactly what its shape and inscriptions say it is, the north coasts of Java and Sumbawa. The French apparently failed to identify the two southward-trending coasts, either from their outlines, or their inscriptions. They may possibly have recognized the original 'west' coast as being part of Java, if the inscription *iaua maior*, which they translated as *Jave la Grande*, had been on it; clearly they did not recognize the names on the original 'east' coast, or the islands off it, or they would not have 'edited' them. Both primitive sketch charts must have lacked indications of orientation, scale and latitude. However, believing them genuine, the French speculatively attached them where they did for three reasons, two already mentioned above:

6 Richardson, *Was Australia Charted*, fig. 34.

1. There was incredible confusion over the identities of *Java Major* and *Java Minor*.
2. There were widely-held, potentially related beliefs in the existence of a large southern continent, and in the enormous size of *Jaua Maior* (*Jave la Grande*).
3. On sixteenth-century maps, the south coasts of Java and Sumbawa were sometimes indicated hypothetically; on Portuguese charts they were usually left blank, as there was no record of their having sailed along them. This invited the attachment of unidentified coastlines, for the north coast of Tierra del Fuego, discovered by Magellan in 1520, was widely believed to confirm the existence of *Terra Australis*. Several Dieppe cartographers evidently guessed that these two charts could be part of such a *terre australle*, and so depicted it hypothetically; more circumspect ones, the author of the Harleian map and Jean Rotz, left these two southward-trending coasts unfinished.[7]

A critical examination of *all* the Dieppe maps shows that no part of *Jave la Grande*, and not a single inscription can legitimately be shown to have anything whatsoever to do with Australia. There is no need to tamper unjustifiably with its coastlines. The apparent landmass of *Jave la Grande* is not a landmass at all, but consists of two unrelated, differently-scaled, speculatively-sited coastlines actually attached to Sumbawa, and on some maps to Java itself as well.

II

Peter Trickett: Radical Cosmetic Surgery

Peter Trickett, however, in his *Beyond Capricorn*, published in 2007, some four months after my book, claims to have proved that the Portuguese *did* reach and chart both Australia and New Zealand.[8] Genuine research for any thesis requires that *all* relevant evidence be examined and weighed in the balance, not only that which can, by some means, legitimate or other otherwise, be presented as supporting it. The omission of relevant evidence is not legitimate.

Trickett's 'Realigned' East Coast of 'Australia'

Trickett virtually ignores all the other Dieppe maps, relying on just three in one atlas, the 1547 'Vallard' one, in which all the maps are orientated

---

[7] Reproduced in Richardson, *Was Australia Charted*, plates 3 and 5.

[8] Peter Trickett, *Beyond Capricorn: How Portuguese Adventurers Secretly Discovered and Mapped Australia and New Zealand 250 Years before Captain Cook* (Adelaide, 2007).

south.[9] To make them readily recognizable, he legitimately inverts all three so that north is at the top. But he then solves the crucial, triangular *cap de fremose* promontory problem on the 'Vallard' east coast map of *terra jaua* (*Jave la Grande*) to his own satisfaction by hypothesizing that the French incorrectly joined two Portuguese charts at the northern base of the triangle. On that completely unjustified pretext, which he treats as though proven, he rotates the promontory some 90° clockwise, thus 'converting' *cap de fremoſe* into 'Wilson's Promontory'. It is significant that on his 'realigned' map Trickett omits the latitude scale.[10] It is also significant that on his supposedly parallel coastal sketches he omits any indication that would reveal the very considerable latitude differences between the east coasts of *Jave la Grande* and Australia.[11] As a result, he doesn't realize that the point from which he rotates the *Cap de fremoſe* promontory, about 42°S, is actually south of the latitude of the *real* Wilson's promontory (39°), so *his* 'Wilson's Promontory' is much too far south (Figure 5.6).

Trickett's 'North Island of New Zealand'

To the southeast of the coastline that Trickett has 'realigned', together with the islands to the east of it, he identifies the *Illa do magna* as New Zealand's North Island, but as the part north of 'Auckland' is missing, he adds it hypothetically.[12] However, he does not explain how the navigator responsible can have missed it, nor how he can have sailed through 'Cook Strait' without seeing and recording the northern shore of 'South Island'. Then, having manufactured his defective 'North Island', he 'confirms' its identity, by 'translating' *magna* as though it were *mogno* (mahogany), and presenting it as New Zealand's *totara* tree.[13]

Trickett's 'Realignment' of Jave la Grande*'s West Coast*

Trickett treats the 'Vallard' west coast of *Jave la Grande* in much the same way as he does the east coast. He uses the same hypothetical pretext of the French having incorrectly joined Portuguese charts. He effects two major 'realignments'. A comparison of the 'Vallard' west coast with Trickett's 'realigned' version of it on his supposedly parallel sketch map reveals the extent of this unjustifiable

9 Trickett, *Beyond Capricorn*, plates 3, 4 and 5. Huntington Library, San Marino, California, MS. HM 29.

10 Trickett, *Beyond Capricorn*, compare plate 3 with plate 4, and with Figure 5.6 in this chapter.

11 Ibid., pp. 101, 122, 173, 182.

12 Ibid., especially pp. 230–32.

13 Ibid., p. 233.

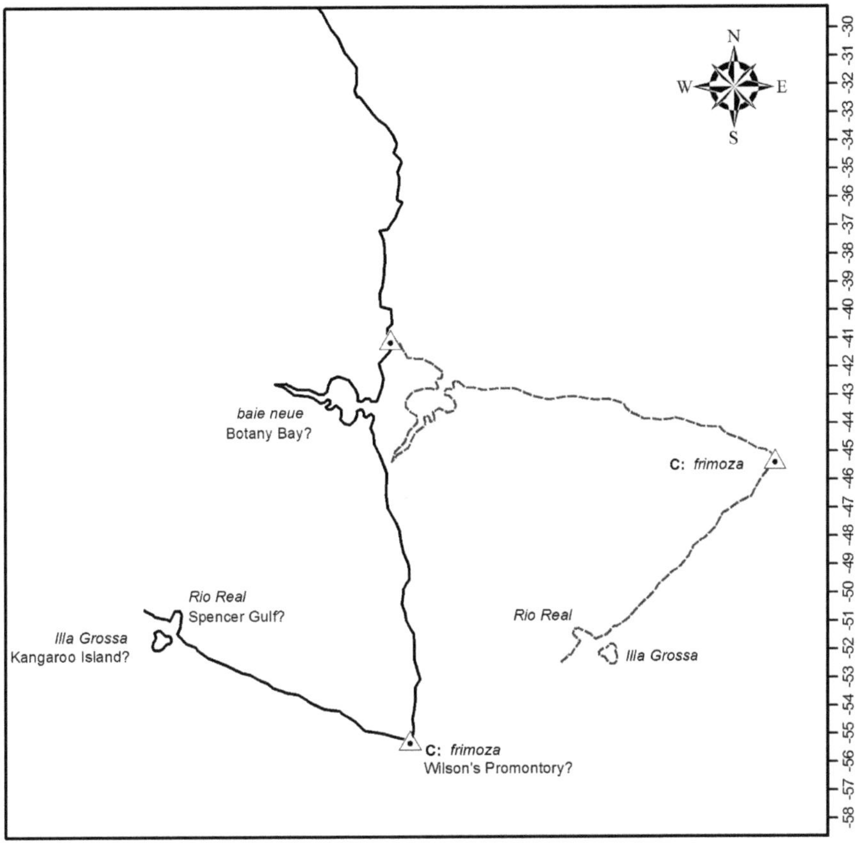

Figure 5.6 The outline of *Jave la Grande*'s east coast on the 'Vallard' atlas map, compared with Trickett's speculative 90° rotation of the southern part of it. Composed by Robert Keane. Reproduced with permission

tampering with primary evidence.[14] By straightening out that section of coast, he extends it to about 44°S, to an 'unmistakable ... broad headland' that he illustrates prominently on his sketch map, and says represents Capes Naturaliste and Leeuwin.[15] Cape Leeuwin is 34°22'S. A careful comparison of this area on the 'Vallard' chart with Trickett's version of it further reveals that this 'unmistakable ... broad headland' does not actually exist. The whole of the area where the southwest coast ends is covered by a large, indigo-coloured landscape.[16]

14 Ibid., plate 5 and p. 314.

15 Ibid., p. 327.

16 Ibid., compare sketch map on p. 314 with plate 5.

Trickett's North Coast of 'Australia'

Having inexcusably tampered with the outlines of the east and west coasts of *Jave la Grande* to make them more closely resemble the 'corresponding' coasts of Australia, how does he deal with its north coast? He has recourse to the rather naïve solution proposed by some earlier 'Portuguese discovery' advocates; its absence on the Dieppe maps is due to 'the cartographer having wrongly assumed that the north coast of Australia was the same as the north coast of Java and its nearby islands'.[17] Gavin Menzies states that what are clearly identified as Java and Sumbawa, with Bali and Lombok between them, actually *are* Arnhem Land and Cape York; and he 'translates' selected Javanese place-names as though they were 'mediaeval' Portuguese.[18]

Trickett's Treatment of the East Coast Inscriptions

It was pointed out above that it is *the* basic principle of place-name research that as many versions as possible of a given place-name should be examined and compared. Without doing so, the elucidation of inscriptions on early maps is little more than guesswork, very liable to be influenced by preconceptions, as in the case of the *magna / mogno* case above. Trickett *only* looks at the 'Vallard' names; but the 'Vallard' atlas is unique among the Dieppe maps in showing over 50 inscriptions on the west coast, and over 60 on the east coast. The other similarly-scaled Dieppe maps only record about a dozen, in varying spellings, on each coast. Trickett presumably assumed that the 'Vallard' cartographer had access to much more information than the others did. However, cartographers making sumptuous maps for wealthy patrons were sometimes tempted to try to impress them by adding inscriptions on little known, or hitherto unknown, coastlines which were not over endowed with them. This must be one such case. Trickett, wrongly assuming that the inscriptions are all in French or Portuguese, does his imaginative best to match a number of them up to features on the Australian coast. However, any Romance language expert would realize that the 'Vallard' cartographer was a very poor linguist, mixing up different languages, not even making some definite articles and adjectives agree with their nouns.[19] His east coast map does include versions of some significant east coast names found on other Dieppe maps (*costa dangeroza, baie perdeue, baie neue* and *c:frimoza* in three different versions), but among the other 60 or so additional inscriptions he includes seven saints' names; not one of

[17] Ibid., p. 348.

[18] Richardson, *Was Australia Charted*, pp. 99–100.

[19] W.A.R. Richardson, 'Toponymy and the History of Cartography', *Journal of the Royal Australian Historical Society*, 78, Parts 1 and 2 (June 1992): 125–9.

the other maps shows any, and the only saints' names on early Portuguese maps of eastern Asia have been shown to be misinterpretations of local native names. The 'Vallard' cartographer also includes *Illa do tubaros* (sic), *Illa do magna* (sic) and *Ila do aljofar* (sic) which are also on other Dieppe maps, but more correctly spelled. The word *aljófar* (pearl), being, as we have seen, *the* key clue to the east coast's identity, must surely be included in every Portuguese dictionary. Trickett, however, doesn't even recognize it as being a Portuguese word; suggesting that it is Polynesian-derived, he attaches it to what he claims is Tonga.[20] His disregard of latitude alone invalidates his speculative identifications.

Trickett's Treatment of the West Coast Inscriptions

Part of the western end of Java itself was quite wrongly represented by a straightish north/south coast on Portuguese charts throughout the sixteenth century, and the Dieppe cartographers copied it.[21] So it is not surprising that they also speculatively extended it further south, in view of the then widely current, false belief that *Jave la Grande* was joined via a *Terre Austral[l]e* to the north coast of Tierra del Fuego. Trickett does not mention that, on the 'Vallard' west coast chart, the *key* inscriptions, *Baye bresille, negre / noirre / neige, gao, Quabeſequieſce,* and *Haure de Sylla* and *cap*, which identify the west coast on the other Dieppe maps, are omitted.[22] Some 50 invented ones replace them, including two more saints, a *Rameriqua*, and that familiar peripatetic place called *patallis*.[23] Trickett does not include these two names on his west coast sketch maps, mistranscribes the former as *Rio Meriqna* in his text, and fails to mention the inconvenient latter one. He borrows a *Baye dangerrosa* (sic) inscription from Pierre Desceliers's maps and inserts it at the southern end of his west coast sketch map; it does not exist on the 'Vallard' west coast map.[24]

The 'Evidence' of Jean Alfonse

Peter Trickett, like Roger Hervé, made use of quotations from *Les voyages auantureux* (1559), somewhat dubiously attributed to Jean Alfonse, the

20 Trickett, *Beyond Capricorn*, p. 256.

21 Richardson, *Was Australia Charted*, pp. 68–74 and figs 23–5.

22 Trickett, *Beyond Capricorn*, on his sketch map, p. 298, actually adds *Baye Bresille* alongside *Rio Camor*.

23 Ibid., pp. 298 and 314; regarding the identity of *patallis* see Robert J. King, 'The Jagiellonian Globe: A Key to the Puzzle of Jave la Grande', *The Globe*, 62 (2009): 1–50, 6–7.

24 Trickett, *Beyond Capricorn*, p. 326, and compare sketch map on p. 314 with the original, plate 5.

Portuguese pilot and cartographer who defected to France, and is believed to have died either in 1544 or in 1549.[25] Despite being called a *routier*, a set of sailing directions, by Hervé, it was reprinted on at least ten occasions and is quite obviously a work of popular travel literature of the time, so is hardly to be relied upon. Trickett and Hervé also quote from Alfonse's *La cartographie, avec l'espère et régime du soleil et du Nord* (1544/45), which was extensively pirated from Juan Fernández de Enciso's *Suma de geographia* (1519), especially so far as eastern and southeast Asia are concerned. On three occasions, Alfonse states that *Jave la Grande* is joined both eastwards and westwards via a *terre austral[l]e* to the coast south of the Straits of Magellan (the north coast of Tierra del Fuego).[26] But what support does that provide to the contention that *Jave la Grande* is Australia? What Trickett and Hervé see as the key quotation from Alfonse is his statement regarding *la Grand Jave* (sic): '*Et quant est de ce que j'en ay veu, c'est terre ferme*' (lit. 'And from what I have seen of it, it is mainland').[27] Both Trickett and Hervé, moved by wishful thinking, and confusion over the meaning of *Jave la Grande*, have taken this *literally* as meaning that Alfonse had been to what *they* mean by *Jave la Grande* (Australia). For Trickett, this even suggests that Alfonse could have been on Mendonça's hypothetical voyage down the east coast of Australia.[28] However, when Portuguese used the term *jaua maior* (*Jave la Grande*) they applied it to Java itself; *jaua menor* to them was Sumbawa; one of Alfonse's sketch maps shows that this was his usage.[29] There is absolutely no evidence to confirm Alfonse's apparent claim that he had been to Java. Even if he had been there, why should that have persuaded him that Java was attached to a *terre austral[l]e*?

Trickett quotes other passages from *La cosmographie* that together clearly reveal where he got the *Jave la Grande / terre australle* link from.[30] One, in the original, reads: 'Le cap de la Tourmente est à quarante six degrez et demy de la hauteur du polle antartique' ('Cape Tourmente is 46½°S'). From the context, Georges Musset, Alfonse's editor, logically identifies this as the tip of the Tres Montes Peninsula on the far southwest coast of Chile, a typical case

---

25 The dispute between the French and the Portuguese over the nationality of Jean Alfonse (João Afonso) still continues.

26 Alfonse (1544/45), fols 147v, 150v, and 159r; Jean Alfonse, *La cartographie, avec l'espère et régime du soleil et du nord, par Jean Fonteneau dit Alfonse de Saintonge, capitaine pilote de François 1er*. Paris: Bibliothèque Nationale de France, MS français 676, ed. Georges Musset (Paris, 1904), pp. 388, 399 and 427.

27 Alfonse (1544/45), fol. 147 r; Musset, pp. 388–9.

28 Trickett, *Beyond Capricorn*, p. 282.

29 Alfonse (1544/45), fol. 147v; repro. in Musset, p. 391, in Trickett, *Beyond Capricorn*, p. 283, and in King, 'The Jagiellonian Globe', p. 28, fig. 11.

30 Trickett, *Beyond Capricorn*, pp. 282–4; Alfonse (1544/45), fol. 158v; Musset, pp. 425–7.

of miscopying or 'editing'; it is actually 46°50'S.[31] The second reads: 'Cape Tourmente and Cape Formose, which is to say Beautiful, which is on the terre australle on the west side of the Pacific sea, are west-north-west / east-south-east [of one another], and there are six hundred leagues between them' (author's translation); the distance cited cannot possibly be right.[32] The third reads: 'The Beautiful Cape [i.e. *cap de formose*] is thirty-nine and a half degrees south.' Trickett pronounces this to be conclusive proof that his identification of *cap de formose* as Wilson's Promontory (39°) is correct.[33] But the latitude of *cap de fremose* on the Dieppe maps is about 45°S, and Trickett's realigned Wilson's Promontory is many degrees further south (see Figure 5.6).

These passages must mean that Alfonse's opinion was based, not on what he had actually 'seen' in *Jave la Grande*, but on a 'Dieppe' map showing a *terre australe* with *Jave la Grande* and its *cap de fremose* triangle on it, for it is the only Cape *Formose* even vaguely in the area indicated, on any map. There are, or were, at least three other capes of that name; Columbus named one on the north coast of Cuba, one is on the coast of Nigeria, and one, mentioned by Magellan's chief pilot Francisco Albo (Alvo), was an alternative name for what is still called *Cabo Deseado* (Desired Cape) at the western end of the Straits of Magellan.[34]

Trickett's manipulation of both the east and west coasts on the specious excuse that the French must have incorrectly joined Portuguese charts, his coastline realignments, his disregard of latitude, his omission of the other Dieppe maps' significant place-name evidence, his misreading or miscopying of many inscriptions and his suspiciously inventive treatment of them, especially *aljófar*, do not inspire confidence in his thesis. Nevertheless, even though he clearly has not rewritten history, wishful thinking has led to his book being published in translation in Portugal.[35]

Conclusion Concerning the Portuguese Discovery Theory

Genuine evidence may yet emerge to prove that the Portuguese did reach Australia before the Dutch. But not one of the attempts made so far to force

31 Trickett, *Beyond Capricorn*, p. 282, miscopies *Tourmente* as *Tormente*, and so mistranslates it as *Cape of the Torment*.

32 The Spanish and Portuguese *legua* varied considerably over time, but was usually about 3 miles.

33 Trickett, *Beyond Capricorn*, p. 285.

34 Martín Fernández de Navarrete, *Colección de los viages y descubrimientos que hicieron por mar los españoles* (5 vols, Buenos Aires, 1946), vol. 4, p. 197 and Antonio Pigafetta, *The First Voyage Round the World by Magellan*, ed. Lord Stanley of Alderley (London, 1874), p. 219.

35 Trickett, *Para Além de Capricórnio* (Lisbon, 2008).

the apparent, but actually non-existent landmass of *Jave la Grande* into the straitjacket of Australia's coastline stands up to rigorous examination.

III

Robert J. King's Identification of Jave la Grande

In his generously illustrated article entitled 'The Jagiellonian Globe: A Key to the Puzzle of *Jave la Grande*', Robert King proposes a solution to the enigma completely divorced from those so far proposed in the last half century or so, namely Australia and Java/Vietnam.[36] He makes no attempt to disprove either of these solutions, nor the methods by which they were reached, but correctly states that 'definitive proof of it [an early sixteenth-century Portuguese voyage along the east coast of Australia] has remained elusive';[37] he makes no comment about the west and north coasts.

He states, 'The outlines of Jave la Grande and the Southern Continent on these mappemondes are drawn, not from actual discovery and exploration of these regions by Iberian, French or other voyagers, but from the cosmographical ideas of the time.'[38] By no means all early sixteenth-century maps depicted any southern continent, but so far as those which did are concerned, he is correct. Since they were all invented, they vary enormously, from adaptations of Ptolemy's version, such as those of Piri Reis (1513) and Lopo Homem (1519), to those on Johann Schöner's globes (1515, 1520 and 1533), and the related one on Oronce Fine's bicordiform world map (1531), those on some of the Dieppe maps, from 1546 onwards, the anonymous oval, Antarctic ring continent in the Vatican Library,[39] and the Marco Polo-inspired one on Mercator's globe gores of 1541.[40] Guillaume Le Testu, who admitted inventing several sections of his own fictitious southern continent in his *Cosmographie Vniverselle* of 1555

36 King, 'The Jagiellonian Globe', pp. 1–50.

37 Ibid., p. 1.

38 Ibid., p. 12. King echoes the theory of Gayle K. Brunelle.

39 Vatican City, BAV, MS Urb. Lat. 274 fols 73v–74r; see Alfred Hiatt, *Terra Incognita: Mapping the Antipodes before 1600* (Chicago, IL and London, 2008), pp. 210–13, and Chet Van Duzer, 'Cartographic Invention: The Southern Continent on Vatican MS Urb. Lat. 274, fols 73v–74r (c. 1530)', *Imago Mundi*, 59 (2007): 193–222. Despite being titled *Terra inchognita avstrale*, it has numerous place-name inscriptions, some given by Columbus to places in the Caribbean area.

40 Mercator's southern continent was later elaborated upon on his famous 1569 world map; see Rodney W. Shirley, *The Mapping of the World: Early Printed World Maps 1472–1700* (London, 1983), plates 68 and 102.

[1556],[41] abandoned it and *Jave la Grande* on his 1566 world map,[42] in favour of an adaptation of Mercator's above mentioned southern continent, which appeared long before the much better known revised version of it in 1569.

King reaches his conclusion regarding *Jave la Grande*'s identity by examining three interrelated factors. One is what he terms the 'bilocation' of certain place-names. The second is the temporarily simultaneous existence of two different estimates of the size of the world, that avowed by Ptolemy, and the one used by Columbus. The third is his emphasis on the supposed significance of one particular island appearing in the Indian Ocean exclusively on two globes, the Lenox one of 1503–1507, and the Jagiellonian one of c. 1510; unnamed on the former, it is surprisingly named '*AMERICA · NOVITER · REPERTA*' (America recently discovered) on the latter; it appears on no maps.[43] King claims that the Jagiellonian globe 'illustrates how geographers struggled to reconcile the discoveries of new lands with the Ptolemaic cosmography inherited from antiquity, and is a key to understanding how the Dieppe view of Java Major, *Jave la Grande*, grew out of the cosmographical concepts of the early sixteenth century'.[44]

Bilocation

The bilocation feature was much in evidence in late nineteenth- and early twentieth-century attempts to establish a 'discovery' of Australia prior to 1606. Richard H. Major tried to relate the *Baye bresille* on the northwest coast of *Jave la Grande* on the Harleian map to Oronce Fine's *BRASIELIE REGIO* inscription in a vaguely similar position, on the fictional southern continent on his 1531 map; this led Major to suggest that the vast gulf between Fine's *BRASIELIE REGIO* and *PATALIS REGIO* land masses could represent the Gulf of Carpentaria.[45] Armand Rainaud similarly picked on Guillaume Le Testu's version of the same bay, spelled *Baie braecillie*, as being '*sans doute*' derived from Oronce Fine's *BRASIELIE REGIO*.[46] Such an assumption is quite unjustified.

---

41 Paris: Château de Vincennes, Bibliothèque du Service Historique de l'Armée de Terre; D.L.Z. 14.

42 Paris: Bibliothèque Nationale de France; Rés. [Ge. AA. 62].

43 King, 'The Jagiellonian Globe', figs 4, 5, 6, 8 and 9.

44 Ibid., p. 1. For different reconciliation accounts, see W.A.R. Richardson, 'East and Southeast Asia: Cartographers' Attempts to Reconcile the Maps of Ptolemy and Martellus with Marco Polo's *Travels* and with Portuguese Charts', *Terrae Incognitae*, 32 (2000): 1–22 and Hiatt, *Terra Incognita*.

45 Richard H. Major, 'Further Facts in the History of the Early Discovery of Australia', *Archaeologia*, 44 (1873): 233–41, 237–8; see Figure 5.8 in this chapter.

46 Cited in King, 'The Jagiellonian Globe', note 89.

As suggested above, the most logical reason for a *Baye bresille* being recorded in *Jave la Grande* would appear to be because brazil dye-wood was found there, as indicated in several places in southeast Asia, on the 'Cantino' map (1502); it was the finding of a different variety of it in Brazil that gave rise, first to a *R: do brasil* there, and shortly thereafter caused the name to be applied to the whole country, and for a while to the whole of South America. James R. McClymont and Edward A. Petherick somehow managed to find some similarity between Brazil's coastline and *Jave la Grande*'s east coast; the former tried to confirm the suggested relationship by the appearance of an *m: fremoso* inscription on the Brazilian coast on the Dieppe maps, and *c: de fremose* on *Jave la Grande*'s east coast.[47] George Collingridge tried to equate the *Hame* (i.e. *Haure*) *de Sylla* region on *Jave la Grande*'s west coast with Behaim's *Egtis-Silla* on the southwestern part of the fictitious 'Dragon's Tail' peninsula on his globe of 1492, because of imagined resemblances between their coastal outlines and the place-names.[48] King himself suggests that two 'Brazilian' place-names on the 'Vallard' map's east coast, *IRio dernero* (sic) (*Rio de enero*?, Spanish for Rio de Janeiro) and *C:ap fria* (sic) (*Cabo Frio*?) mean that the cartographer concerned believed that 'his continental *Terra Java* [*Jave la Grande*] was essentially a theoretical creation, a duplicate of *Brazil* (that is, South America)'.[49] As pointed out above, the 'Vallard' cartographer was unique in that most of his *Terra Java* (*Jave la Grande*) place-names were invented, many ignorantly recorded, like the two cases above, and included saints' names not appearing on *Jave la Grande* on any other Dieppe map, and *patallis* and *Ramerigua*, not to mention 'African' names, *cap bon espoir* (Cape of Good Hope) and *cap de vert* (Cabo Verde).

It is impossible for such suggested, isolated place-name 'equivalences' to confirm reliably the suspected identities of enigmatic coastlines on early maps. One cannot but wonder why King pays such attention to those two names on the 'Vallard' east coast map, but ignores the multiple Javanese, Vietnamese and South China Sea names on the other Dieppe maps. Such hypothetical linkages were perhaps understandable at the end of the nineteenth century, before much relevant research material was as readily available as it is today, and before reasonably reliable standards of scholarship had developed in the fields of history of cartography and, especially, of toponymy. For example, several writers of the period cannot have been acquainted with such vitally significant works as Martin Waldseemüller's maps of 1507 and 1516, which were not discovered until 1901, or studied until the publication of Fischer and von Wieser's work in 1903.

---

47 Ibid., p. 2 and note 21.

48 George Collingridge, *The Discovery of Australia* (Sydney, 1895), p. 82.

49 King, 'The Jagiellonian Globe', p. 8 and fig. 22.

Representation of Eastern and Southeast Asia

At the end of the fifteenth century and the beginning of the sixteenth, many, including Columbus, believed the Far East and southeast Asia to be as depicted on the world maps of the German cartographer Enricus Martellus (c. 1489). On them, after Bartolomeu Dias's rounding of the Cape of Good Hope, he had opened up the vast, Ptolemaic Indian Ocean lake to the west, deleted his southern landmass, provided China with an east coast, and had turned his non-existent southward extension of China into a peninsula which is commonly referred to as the 'Dragon's Tail' peninsula.[50] As recently as the 1990s, some Latin American authors tried to present it as a pre-Columbian map of South America.[51] Despite having *Champa* just north of it, it is not Indo-China, as King calls it.[52] Martellus placed on its west coast some Ptolemaic place-names including the port of *Catigara*, and elsewhere a number of place-names derived from Marco Polo which really belonged in India. Off its east coast he placed (misplaced) invented representations of islands mentioned by Marco Polo, including his two Javas. Because of Columbus's false idea of the size of the world, he believed he had arrived at *Champa*. The result, on some maps, was to assume that the isthmus of Central America was part of Martellus's peninsula.[53] On others, Columbus's names for Caribbean coastal features, such as *Beragua* (*Veragua*), appeared on Martellus's *Champa*.[54]

Waldseemüller's 1507 world map reflects Martellus's ideas on southeast Asia;[55] his 1516 world map[56] reflects the 'Cantino' map's outlines.[57] His world map included in the Strassburg (1513) edition of Ptolemy (Figure 5.7) reflects an intermediate stage, where both the 'Cantino' and Martellus peninsulas are included.[58] King states that 'The maker of the [Jagiellonian] globe had put America in twice, in opposite hemispheres', and implies that this is due to the use of two

---

50 Behaim referred to it as *Hoch India* (*India Superior*); most other cartographers called it *India Meridional.*

51 See W.A.R. Richardson 'South America on Maps before Columbus? Martellus's "Dragon's Tail" Peninsula', *Imago Mundi*, 55 (2003): 25–37 and plate 8.

52 King, 'The Jagiellonian Globe', p. 12.

53 See the Bartolomé Colón (Columbus) / Alessandro Zorzi sketch maps, reproduced in King, 'The Jagiellonian Globe', figs 17 and 18.

54 See Francesco Roselli's map (c. 1508) in Shirley, *The Mapping of the World*, plate 32.

55 Shirley, *The Mapping of the World*, plate 31.

56 Ibid., plate 43.

57 Richardson, *Was Australia Charted*, plate 1.

58 Shirley, *The Mapping of the World*, plate 38 and entry 35.

different longitude scales.[59] The explanation is far more likely to be that, since South America itself is identified in its proper place by earlier names, *TERRA DE BRASIL*, *MONDVS NOVVS* and *TERRA SANCTAE CRVCIS*, yet another name for it, *AMERICA · NOVITER · REPERTA*, was placed in the Indian Ocean because the cartographer misunderstood a passage in Waldseemüller's *Cosmographiae Introductio* (1507). It consisted of a list of recently discovered lands: the farthest part of Africa, Zanzibar, Java Minor, Seula (Sri Lanka) and the fourth part of the world – America.[60] For the ill-informed, it could easily be inferred, from the order of the list, that America was in the Indian Ocean.

King insists further on the supposed influence of different longitude scales, saying that Waldseemüller's 1507 map uses both of them, thus representing 'the eastern coast of Asia twice'.[61] A comparison of the longitude scales on the main map and the small twin-hemisphere one shows that they are actually identical.[62]

Fictitious Southern Continents in the Early Sixteenth Century

Martellus removed Ptolemy's speculative southern landmass, but did not replace it by one of his own, nor did his imitators, but other cartographers did. For two decades the most influential outlines were those on Johann Schöner's globes of 1515 and 1520, and on Oronce Fine's bicordiform world map of 1531. On his 1515 globe, Schöner identified South America as *AMERI / CA*, and south of it, depicted a speculative ring continent named *BRASILIE REGIO*, owing to his misunderstanding of a confusing passage in a pamphlet concerning a voyage to the River Plate, *Newen Zeitung auss Presillg Landt* (New Tidings of the Land of Brazil).[63] South America on his 1520 globe was identified by three alternative names, *AMERICA vel* [or] / *BRASILIA sive* [or] / *PAPAGALLI / TERRA* [Land of Parrots], while a differently shaped continent south of it was named *BRASILIA INFERIOR*. Oronce Fine, on his 1531 world map (Figure 5.8), produced yet another southern landmass, larger than either of Schöner's first two, named *REGIO PATALIS* in the southern Pacific, and *BRASIELIE REGIO* (sic) in the Indian Ocean, while South America appeared just as *A / ME / RI / CA*; in lower case lettering it had an *R. brasil* on its east coast, and Ptolemy's *Cattigara* on its west coast. Fine's removal of Schöner's *BRASILIE REGIO* to

59 King, 'The Jagiellonian Globe', p. 3.

60 Ibid., p. 4 has the actual text.

61 Ibid., p. 7.

62 Shirley, *The Mapping of the World*, plate 31; for detail, see reproduction in Fischer and Wieser, *The Oldest Map*.

63 Mark Graubard and John Parker (eds), *New Tidings of the Land of Brazil* (Minneapolis, MN, 1957); see also Richardson, 'Mercator's Southern Continent', p. 82.

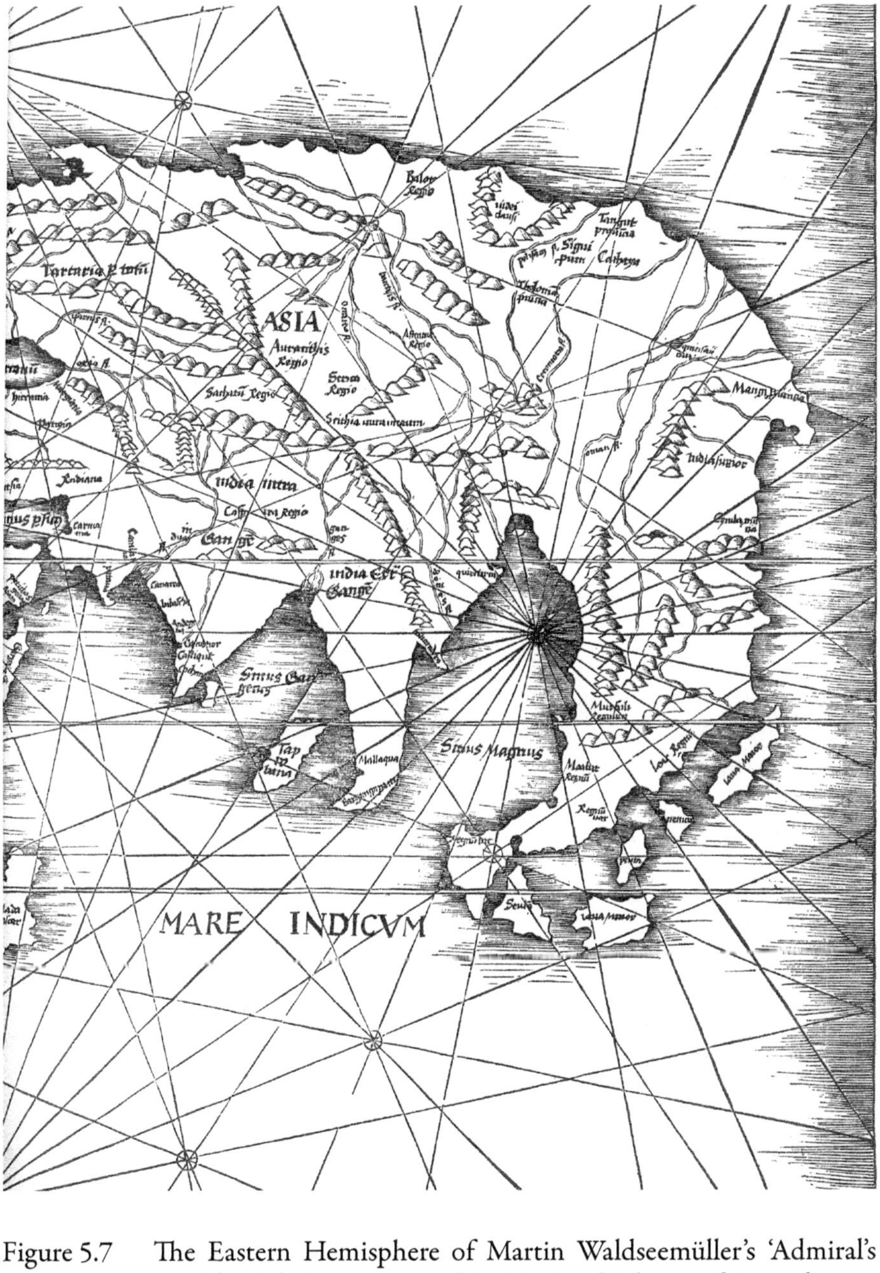

Figure 5.7 The Eastern Hemisphere of Martin Waldseemüller's 'Admiral's map' (1513). By courtesy of the National Library of Australia

the southern Indian Ocean was almost certainly due to the confusing variety of names given to Brazil itself, and misunderstandings arising from the letter sent from Lisbon by Giovanni Matteo (Il Cretico) to Venice in 1501; published in 1507, and often republished, it gave a somewhat garbled account of Pedro Álvares Cabral's voyage to India via Brazil. The confusion was elaborated on in Mercator's globe gores of 1541 and his world map of 1569.[64]

The 'Evidence' of Jean Alfonse

King, like many others, appears to be confused over *Java Major* and *Java Minor*. He says that Alfonse refers to Java as 'Jave Mynore' and the continent to the south as 'La Grand Jave' (sic), but one of Alfonse's sketch maps clearly shows *Java Minor* applied to Sumbawa, and *Java Major* to Java, and to its southward extension as well.[65] Like Hervé and Trickett, King quotes passages from Jean Alfonse's *La Cosmographie*, in which, as mentioned above, he states that *Jave la Grande* forms part of a *terre austral[l]e* (southern land) which extends in both directions to the coast on the southern side of the Straits of Magellan. The citations include those quoted above concerning the stated latitude of *Cap de fremose* (39½°S), and its stated bearing from *Cap Tourmente* (WNW), but he makes no comment on them, yet they must mean that Alfonse had seen a 'Dieppe' Map with *cap de fremose* on it.[66] It is not easy to see the relevance of Alfonse's writings for King's 'bilocation' thesis.

King's Theory and the Evidence of Portuguese and Spanish Cartographers

Having outlined King's theory, and the concepts he bases them on, it should be pointed out that, though mentioning the Portuguese 'Cantino' map of 1502, and the Caverio variant copy (1505), he virtually ignores Spanish and Portuguese cartographers. They based their maps and charts almost exclusively on actual voyages, paid scant attention to Ptolemy's portrayal of the world, or to adaptations of it made by Martellus, Behaim and their imitators, or the speculative depictions of places mentioned by Marco Polo in southeast Asia and

64 Richardson, 'Mercator's Southern Continent', especially pp. 76–82 and fig. 4.

65 King, 'The Jagiellonian Globe', p. 5, and fig. 11.

66 Ibid., p. 17, note 94. King appears not to accept *Cap Tourmente* as identifying the tip of the *Tres Montes Peninsula*, as Musset had logically suggested, for he translates it as 'Cap de Bon Esperance' (sic), a puzzling suggestion. Has he translated *Cap Tourmente* in imitation of the change from Bartolomeu Dias's *Cabo das Tormentas* (Cape of Storms) to João II's *Cabo da Boa Esperança*? Surely he can't mean *the* Cape of Good Hope, which is completely out of the geographical context?

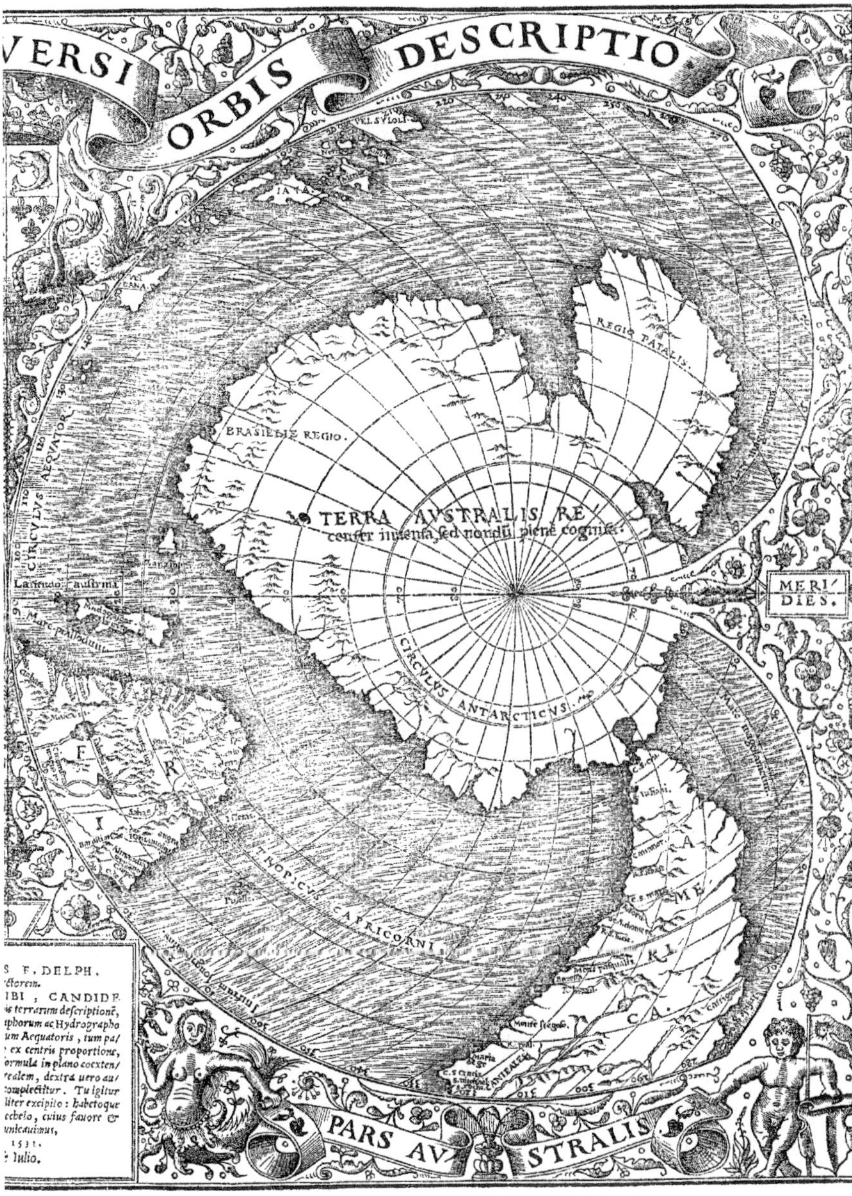

Figure 5.8 Oronce Fine, 'Nova, et integra universi orbis descriptio' (1531), detail. By permission of the National Library of Australia

the Indian Ocean. They utterly ignored the varying hypothetical landmasses invented by Schöner and Fine. Pigafetta, presumably reflecting Magellan's opinion, stated that if, on leaving the western end of the Straits of Magellan, one sailed due west, the first land one would find would be at the eastern entrance of those straits.[67] The inscriptions in the *Jave la Grande* area on the Dieppe maps show that its coastlines, and those of its offshore islands, are based on Portuguese material which the French 'edited', by miscopying, or otherwise tampering with inscriptions they could not understand, by inserting the fictitious, Marco Polo-inspired *Zanzibar*, by incorrectly inserting two differently-scaled Portuguese sketch maps and, on some maps, by joining them up by a hypothetical *terre australle* to the north coast of Tierra del Fuego. Even the French, Dieppe cartographers failed to include any element derived from their compatriot, Oronce Fine.

None of King's artificial construct theory seems compatible with the detailed evidence regarding *Jave la Grande*'s identity deduced from the Dieppe maps themselves. He presents *Jave la Grande*, part of which is indubitably Java and Sumbawa, as a northerly projection of a fictitious southern continent, presenting Schöner's 1533 *REGIO PATALIS*, adapted from Fine's one of 1531, as a 'forerunner' of *cap de fremose*.[68] Is it not more logical to present a hypothetical southern continent as a southward projection of clearly identified Java (*Jave la Grande*) and Sumbawa, via named, though speculatively misplaced, southward extensions?

Conclusion

Two factors will probably ensure that debates over *Jave la Grande* and *Terra Australis* will long continue. One is the 'best seller' popularity and commercial profitability of works such as Gavin Menzies' *1421: The Year China Discovered the World*, which present fiction as history, even though the theories on which they are based can be easily disproved by detailed critical examination. The other is the ever-present temptation for some to seek fame and fortune by claiming to have rewritten history.

Of the three explanations outlined above for the composition of the 'continent' of *Jave la Grande*, it is only the one arrived at primarily from place-name evidence that is free from any preconceptions regarding the identities of

---

67 Antonio Pigafetta, *Magellan's Voyage: A Narrative Account of the First Circumnavigation*, ed. and trans. R.A. Skelton (2 vols, New Haven, CT, 1969), vol. 1, p. 58; Pigafetta, *The First Voyage*, p. 66.

68 King, 'The Jagiellonian Globe', p. 11, and caption to fig. 25.

its coastlines.[69] The nomenclature, some of it necessarily elucidated, reveals that the 'enigmatic landmass' consists of Java and Sumbawa, speculatively extended southwards by the French. They attached to them two differently-scaled coastlines which they could not identify from the outlines and inscriptions, but were actually southwestern Java, and Vietnam and the islands east of it. The resulting *apparent* landmass is not, as both the Dieppe cartographers and subsequent commentators have assumed, really a landmass at all, but two unrelated coastlines that give that impression. Their misplacement south of Indonesia was due to the then common beliefs in the existence of *Terra Australis*, and the possibility that Marco Polo's misunderstood *JAVA MAJOR* ('Great Java' or *Jave la Grande*) might well be part of it.

[69] The other attempts to support the Australian identification are critically examined in Richardson, *Was Australia Charted*, pp. 42–55.

Chapter 6

Mapping *Terra Australis* in the French Seventeenth Century: The *Mémoires* of the Abbé Jean Paulmier

Margaret Sankey

In 1664, when the outlines of the fabled Great South Land or Lands[1] were beginning to disappear from European maps, and the contours of New Holland were beginning to be drawn in the southern hemisphere, the Abbé Jean Paulmier, Canon of the Cathedral of St Pierre in Lisieux in Normandy, published a submission to Pope Alexander VII proposing that the *Terres australes* be evangelized by the French. Paulmier's initiative was to have repercussions on French exploration of the southern seas over the following century, as well as on an important body of French utopian literature. In this chapter I will explore the historical and cartographic underpinning of Paulmier's paradoxical initiative and will then trace briefly the largely occult influence of Paulmier's *Mémoires*.

To situate the Abbé, we must look earlier. At the end of the fifteenth century Pope Alexander VI had permitted Spain and Portugal, through the *Intercœtera* Bull of 1493 and the Treaty of Tordesillas of 1494, to carve up the New World between them. During the sixteenth century, although the French were caught up in debilitating wars of religion and had little energy to devote to conquest, they nevertheless felt excluded from the Iberian partitioning of the world. As François 1er allegedly expressed it in 1541 to the Spanish Ambassador from the court of Charles-Quint:

> the sun shines for me as for the others. I should like to see the clause in Adam's will which excludes me from the apportioning of the world.[2]

---

1 Paulmier, writing in French, uses both the singular and the plural form.

2 '... le soleil luit pour moi comme pour les autres. Je voudrais bien voir la clause du testament d'Adam qui m'exclut du partage du monde' [my translation].

In 1531, in defiance of the pope, the French king had sent Jacques Cartier in search of a west passage to China and Japan; he discovered Canada.[3] In 1555 the Protestant Admiral, Gaspard de Coligny, sent the Chevalier de Villegagnon to set up a French colony, called 'la France Antarctique', in Brazil. De Coligny's aim was to secure havens for the persecuted Protestants, and thus he became the first Frenchman to advocate the founding of colonies overseas to counter the Iberian land grab. The colony established by Villegagnon grouped together both Protestants and Catholics on an island in the Bay of Rio de Janeiro. It was, however, racked by internal dissensions and taken over by the Portuguese in 1559, a year after Villegagnon had returned to France.

During the sixteenth century interest in the idea of *Terra Australis* was already taking shape in France, together with thoughts of possible conquest.[4] One of the earliest maps depicting *Terra Australis* was French. In his cordiform map of 1531 (Figure 6.1), the cartographer and professor of mathematics, Oronce Fine, hypothesized the existence and position of the as yet unknown land 'recently discovered but not yet explored', as he described it, referring to the voyage of Magellan and the hypothesis that Tierra del Fuego was the tip of an Antarctic continent. A proposal to explore the *Terres australes* was made later in the century to Gaspard de Coligny by the D'Albaigne brothers, Francisque and André, from Lucca, with maps drawn by the navigator Bartolomeu Velho. Coligny gave favourable consideration to these initiatives but events leading up to the St Bartholomew's day massacre in 1572 and the death of Coligny in the massacre and that of one of the d'Albaigne brothers put paid to the project.[5]

---

3 Cartier discovered Canada in 1534 and the first French settlement was established by Samuel Champlain in Quebec in 1606.

4 In 1529 the Parmentier brothers had travelled as far as Sumatra and Indonesia and perhaps further, but both died during this voyage and the remaining records are sketchy; the navigator Alfonce from Saintonge had suggested in his *Cosmographie* (1545) that a Great South Land existed which he called Jave la Grande; from the 1540s to the 1560s the Dieppe school of cartographers, whose principal figures were Desceliers, Rotz and Desliens, published a series of maps, thought to be influenced by information from Portuguese navigators, depicting a large land mass that they called Jave la Grande, beneath Indonesia. See Helen Wallis, 'Visions of Terra Australis in the Middle Ages and Renaissance', in William Eisler and Bernard Smith (eds), *Terra Australis: The Furthest Shore* (Sydney, 1988), pp. 35–8.

5 E.T. Hamy, 'Francisque et Andre d'Albaigne: cosmographes lucquois au service de la France', *Bulletin de Géographie Historique et Descriptive* (Paris, 1894): 405–33; 'Nouveaux documents sur les frères d'Albaigne et sur le projet de voyage et de découvertes présenté à la cour de France', *Bulletin de Géographie Historique et Descriptive* (Paris, 1899): 101–10; and 'Documents relatifs à un projet d'expéditions lointaines présentés à la cour de France en 1570', *Bulletin de Géographie Historique et Descriptive* (Paris, 1903): 266–73.

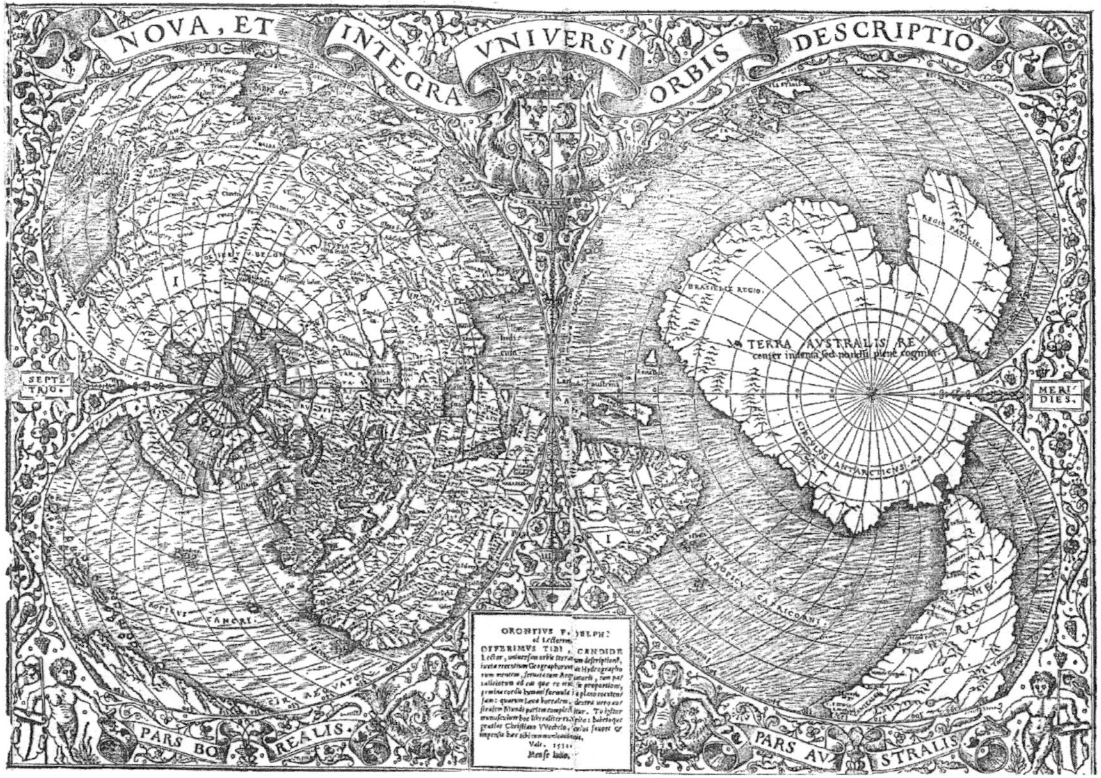

Figure 6.1 Oronce Fine, 'Nova, et integra universi orbis descriptio' (1531). Bibliothèque nationale de France, CP, Rés. Ge D 6440. Reproduced with permission

At the end of the sixteenth century, the Protestant Lancelot Voisin de la Popelinière, in his overtly geopolitical book *Les Trois Mondes* published in 1582,[6] critically described and analysed Spain and Portugal's appropriation of the New World, exploring the reasons for the failure of Villegagnon's utopian project and reproaching the French for their laziness regarding exploration and conquest. La Popelinière used the traditional arguments for the existence of *Terra Australis* and included a copy of Ortelius's world map of 1570 (Figure 6.2). The depiction of *Terra Australis* on this map, adapted from Mercator's world map of the previous year which also incorporated the imaginary sites of Marco Polo's discoveries, would provide the principal model for *Terra Australis Incognita* for the following two centuries.

La Popelinière proposed that the French go in search of the unknown land which, at best, would provide a possible refuge for Protestants and, at worst, a place to which undesirable social elements in France could be banished. In May 1589 La Popelinière himself departed from La Rochelle in search of the *Terres australes* but unfortunately was forced to abandon his project on arriving at Cap Blanc in Africa. La Popelinière's suggestions were not followed by other initiatives at this time, but some 60 years later, his ideas were revived and developed by Jean Paulmier.

In the first half of the seventeenth century, in the context of the strong counter-Reformation push for renewing and expanding the sphere of the French Catholic Church, there was an immense interest in missionary endeavour and in missionaries' accounts of their adventures in far-flung lands.[7] In 1622 the papacy, in an attempt to centralize missionary endeavour and to wrest back power from Spain and Portugal, had founded the *Congregatio de propaganda fide*. This stimulated France, as 'the eldest daughter of the Church', to seek to take her place amongst the nations searching for new lands to conquer and to convert to Christianity. In 1644 the Séminaire des Missions Etrangères was established in Paris and it was from this group that Paulmier, who travelled regularly between Lisieux and Paris, initially sought support for his project, distributing the first draft amongst the members in 1654. This draft was followed by a considerably enlarged and modified version in 1659.

Paulmier's proposal was published in an unauthorized edition in 1663, and then again in an authorized edition, with an introductory explanatory note, errata and slight modifications, in 1664. The title sets out Paulmier's programme

---

6 Henri-Lancelot Voisin, sieur de la Popelinière, *L'Amiral de France, et par occasion, de celuy des autres notions, tant vieiles [sic] que nouvelles* (Paris, 1584); Henri-Lancelot Voisin, sieur de la Popelinière, *Les Trois Mondes*, ed. Anne-Marie Beaulieu (Genève, 1997).

7 Of the 450 accounts of travel to distant countries published between 1600 and 1661, more than a third were written by missionaries.

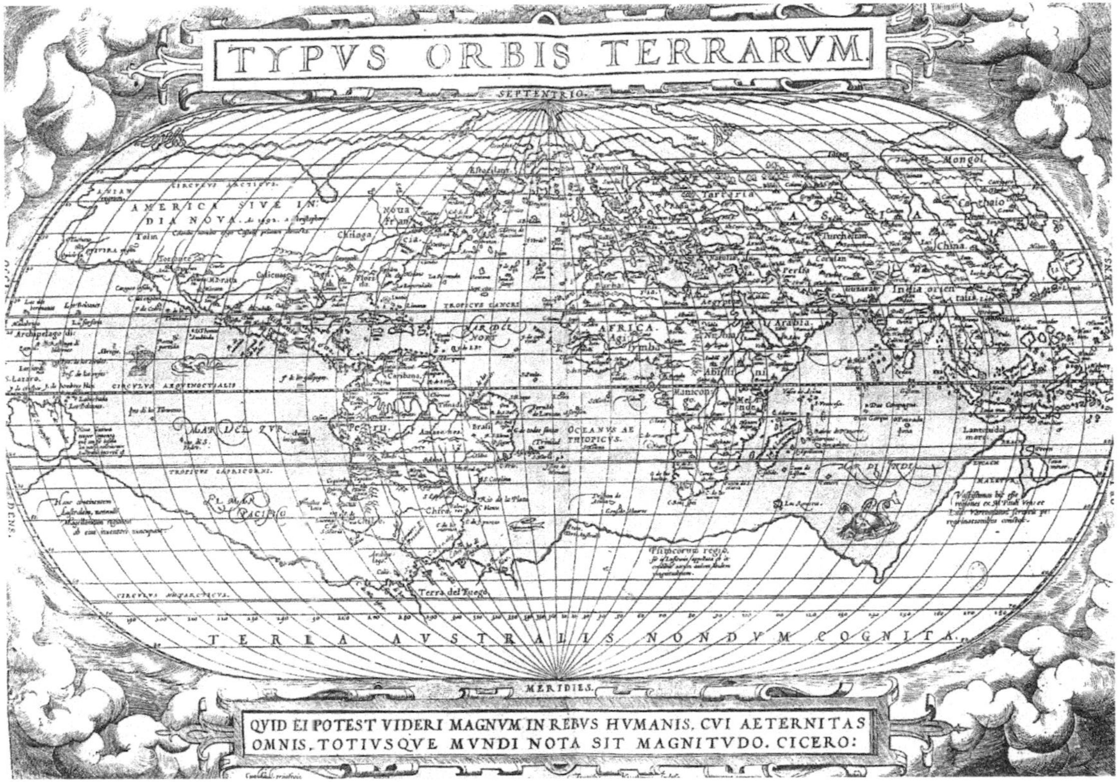

Figure 6.2 Abraham Ortelius, 'Typus Orbis Terrarum', in *Theatrum Orbis Terrarum* (Antwerp, 1570). Bibliothèque nationale de France, CP, Ge DD 4796 (134). Reproduced with permission

and its parameters: *Mémoires pour l'establissement d'une mission chrestienne dans le troisième monde, Autrement appelé, La Terre Australe, Meridionale, Antartique [sic], & Inconnuë*.[8] Through the different drafts Paulmier refined and developed his argument for the establishment of a mission. Initially, citing the work of Bartolomé de las Casas and Francisco di Vitoria,[9] he opposed the activities of missionaries to those of colonists, criticizing the excesses and cruelty of the latter in the strongest terms. Subsequently, however, his anti-colonial stance became more muted. The reality of the situation was that, although it was up to the pope to give permission for the mission, it was for the temporal power, the king of France, to bear the cost. Inevitably the king hoped for some financial return for his investment by way of trade, and for this some colonial framework was necessary.

In his first draft of 1654 Paulmier expounded the case for the existence of the *Terres australes*, hypothesizing their nature from his extensive study of the work of geographers and cartographers. He then argued France's obligation to evangelize this unknown part of the globe, stating that a French navigator and merchant, Gonneville, had discovered the *Terres australes* in 1504:

> Ferdinand Magellan is given the glory of having first recognized them; this brave exploit took place in 1520. And yet before Magellan Captain de Gonneville, a Frenchman, was carried there in 1503 by a storm when he was going to the East Indies.[10]

In his next, expanded manuscript draft of his proposal in 1659, Paulmier reinforced his argument by giving more details concerning Gonneville,

[8] L'abbé Jean Paulmier, *Mémoires touchant l'établissement d'une mission chrestienne dans le troisième Monde, Autrement appellé, La Terre Australe, Meridionale, Antartique, & Inconnuë. Presentez à Nostre S. Pere le Pape Alexandre VII. Par un Ecclesiastique Originaire de cette mesme Terre* (Paris, 1664). Three of the adjectives refer to the southern situation of the land: '*Australe*' – the south wind; '*Méridionale*' – south area; '*Antartique*' – on the other side of the earth from the constellation of the Great Bear (Gk. Arktos-bear).

[9] Bartolomé de las Casas, *La Destruction des Indes* (1552), traduction de Jacques de Miggrode (1579), ed. A. Milhou and J.-P. Duviols (Paris, 1995); Francisco di Vitoria, *Leçons sur les Indiens et le droit de guerre* (1532) ed. Maurice Barbier (Geneva, 1966).

[10] 'On donne à Ferdinand Magellan Portugais la gloire de les avoir le premier reconn*uës*; ce brave exploict est de l'année M.D.X.X. Et touttesfois avant Magellan le Cap. de Gonneville françois en 1503 y fut porté par la tempeste lors quil alloit aux Indes Orientales' [my translation], Scottish Catholic Archives, Temp. dep. 2306 MM3/6, fol. 10v.

Nowhere previously, as far as I know, has the explorer Gonneville ever been mentioned. It is also noteworthy that this is the only mention of Gonneville in the 1654 manuscript and that the subsequent drafts expand his name, adding 'Binot Paulmier de'.

and claimed for the first time to be a descendant of a native of the unknown land, Essomeric, son of the chieftain Arosca, brought back to France by the navigator. Gonneville, according to Paulmier, had discovered the *Terres australes* inadvertently when seeking to travel to the Spice Islands in his ship *L'Espoir*. Blown off course in a violent storm south of the Cape of Good Hope, he had been shipwrecked in an unknown land in which he had spent six months. On leaving the unknown land Gonneville returned to France without visiting the Spice Islands, and was attacked by pirates in the English Channel, losing all his cargo and ship's records. Paulmier in his *Mémoires* claims that Gonneville lodged a summary account of his voyage, the *Relation*, at the Admiralty of Rouen.[11] In the 1659 draft of his proposal, and in the subsequent printings, Paulmier included an extract from the *Relation*, describing the unknown land and its inhabitants. Paulmier names the land as being the *Terres australes*, whereas Gonneville had not offered any suggestions as to the identity of the land in his *Relation*.

Paulmier's *Mémoires* of 1664 is a work of 215 pages containing 12 chapters. In his dedication to the pope, the Abbé foregrounds the French claim to the evangelization of *Terra Australis*, underlining its legitimacy in terms of Gonneville's voyage and his own ancestry. Then follows a well-reasoned argument to support his case. In the preliminary chapters Paulmier makes the case for the existence of the *Terres australes* by summarizing and analysing the writings of geographers, navigators and philosophers, as well as by exploring a wide variety of contemporary and earlier cartographic representations.

The extract the Abbé includes from Gonneville's account is a central part of his argument, justifying the feasibility of the proposed mission. It presents the unknown land in glowing terms: the Indians are friendly, food is abundant and the social order is guaranteed by the firm governance of the king. Paulmier deduces that the climate is ideal for European settlement:

> The Austral lands, like Asia, Africa and America, can congratulate themselves on enjoying perpetual spring in some places: on having lands which bear a double harvest in a single year and such fertile land that it requires fewer weeks than the

---

[11] This document has never been found, but there are what purport to be two seventeenth-century copies of it in Parisian libraries, one in the Bibliothèque de l'Arsenal (Ms. 3221) and the other in the Bibliothèque nationale de France (Ms. NAF 7454). A transcription of the Arsenal document is to be found as an appendix to my critical edition of the Abbé Paulmier's *Mémoires touchant l'établissement d'une mission chrestienne dans le troisième Monde, Autrement appellé, La Terre Australe, Meridionale, Antartique, & Inconnuë. Presentez à Nostre S. Pere le Pape Alexandre VII. Par un Ecclesiastique Originaire de cette mesme Terre*, established, presented and annotated by Margaret Sankey (Paris, 2006).

number of months we need for the production of several fruits, plants, grains and seeds.[12]

It is clear that, in his utopian presentation of the unknown land,[13] Paulmier has been influenced by his reading of the Spaniard Pedro Fernández de Quirós's petitions to Philip III in which the explorer requests permission to return in search of the land glimpsed during his voyage of 1606. Paulmier's acquaintance, Etienne de Flacourt, had translated the eighth petition from the Spanish into French and Paulmier included this translation in the documents supporting his proposal.[14]

In the *Mémoires* the Abbé goes on to discuss the reasons for his proposal in the context of the previous missionary endeavours of the French and other European nations, such as Spain and Portugal. Criticizing the latter two, he proposes a new model for evangelization, more in keeping with the Christian gospel, and argues that since it was a French explorer who discovered the *Terre australe*, it is the duty of the French to relocate and evangelize it. Next he sets out his blueprint for mounting the expedition to the unknown land and discusses the organization of the mission, suggesting that a society be created to raise money for the enterprise.

To further his case, Paulmier not only petitioned the pope but also wrote a long letter to the king's historiographer, François Du Chesne,[15] claiming that Gonneville had discovered the *Terres australes* 20 years before Magellan, and requesting the historian to note this prior discovery in his history of France. To accompany the letter he compiled a dossier of documents and maps, entitled *Recueil de Plusieurs Relations tant anciennes que modernes, Voyages, Navigations et Mémoires*, including extracts from the work of geographers and explorers.[16]

---

12 'Les Terres Australes se peuvent vanter aussi bien que l'Asie, L'Afrique, & l'Amérique, de jouir en quelques lieux d'un Printemps perpetuel: d'avoir des terres qui portent une double moisson en une seule année: Et un sol si heureux, qu'il employe moins de semaines, que le nostre ne consomme de mois en la production de plusieurs fruits, plantes, grains, & semences' (Paulmier, *Mémoires* ..., p. 55 [ed. Sankey, p. 182]).

13 See Margaret Sankey, 'Madagascar et la terre de Gonneville: tribulations d'un mythe des origines', *Ports et voyages dans le sud-ouest de l'océan Indien, XVII–XXe siècles, Revue Historique des Mascareignes*, 5 (2004): 147–56.

14 *Requeste presentée a Philippes III Roy d'Espagne, par Pierre Ferdinand de Quir, Portugais, pour tascher d'obliger sa Majesté catholique a envoyer des Colonies dans les Terres Australes, contenant une sommaire description d'une partie de ces Terres, descouvertes par Ledit de Quir et ceux de sa flotte. Traduite par Monsieur de Flacourt, l'un des directeurs dela Compagnie françoise d'orient, et Commandant de la Colonie françoise de Madagascar* [BnF, Ms. NAF 7454], transcribed in Abbé Paulmier, *Mémoires*, ed. Sankey, pp. 289–302.

15 BnF, Ms. NAF 7454, transcribed in Abbé Paulmier, *Mémoires*, ed. Sankey, pp. 335–45.

16 *Recueil de plusieurs Relations tant anciennes que modernes, Voyages, Navigations, et Mémoires, c'est a dire, La cinquième partie de L'Univers dépeinte dans toutes les Mappes-monde, sous*

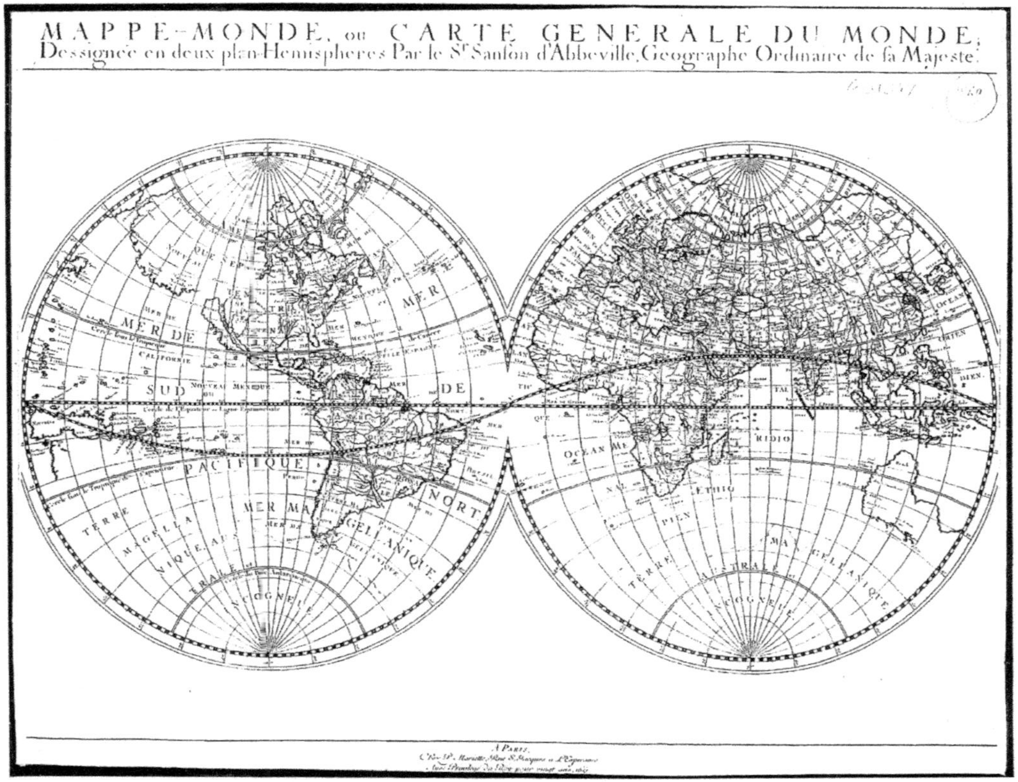

Figure 6.3 Sanson d'Abbeville, 'Mappe Monde, ou Carte générale du Monde, dessignée en deux plan Hemispheres' (1651). Bibliothèque nationale de France, CP, Ge D 12287. Reproduced with permission

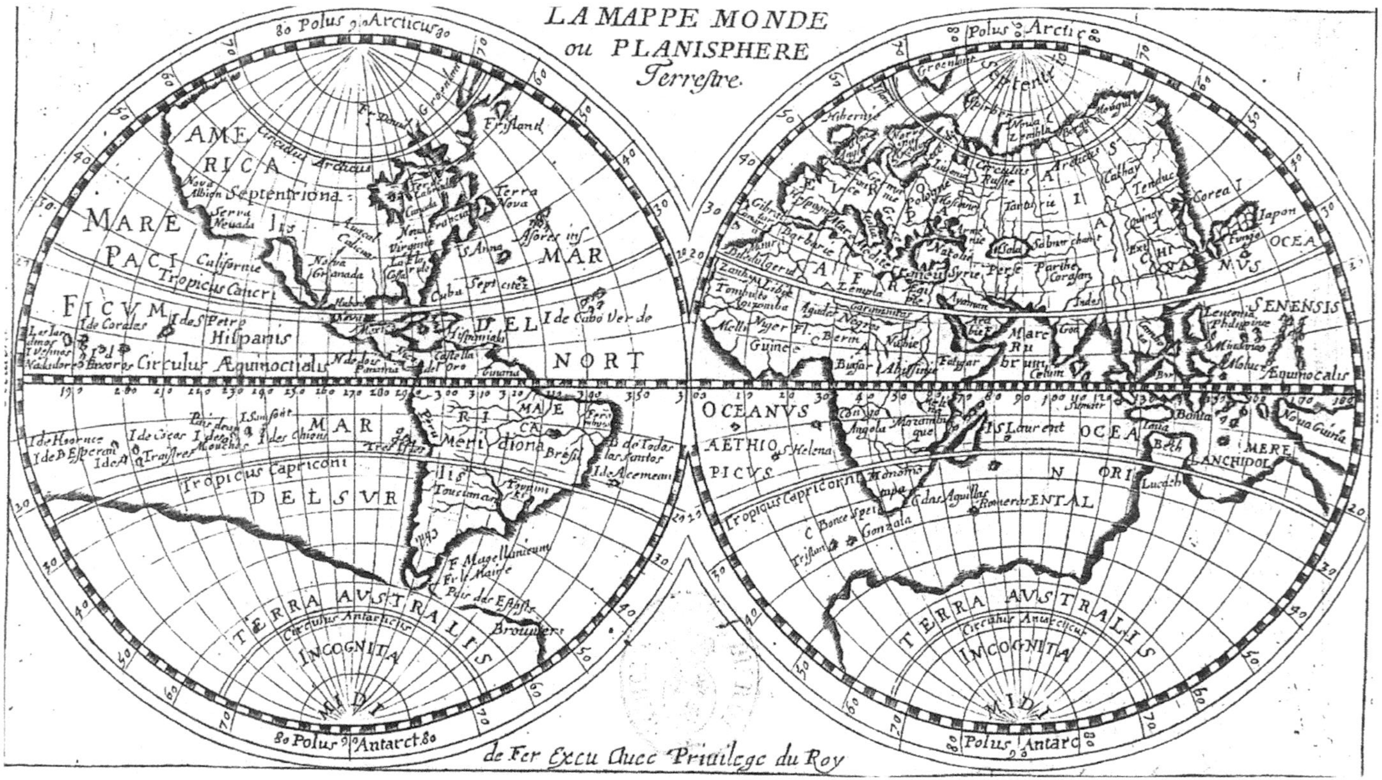

Figure 6.4 Antoine De Fer, 'Typus orbis terrarum. Description de la Terre Universelle, la Mappemonde ou Planisphère Terrestre', undated. Bibliothèque nationale de France, CP, Ge FF 1201 (3). Reproduced with permission

Maps were an essential part of Paulmier's demonstration in the *Mémoires* of the existence of the *Terres australes* and he chose them eclectically, using a variety of sometimes conflicting world maps to frame his argument.[17] The 1659 manuscript was prefaced by Pierre Sanson's world map of 1651 (Figure 6.3), and the second printed issue of 1664 by Antoine de Fer's undated map (Figure 6.4). The *Recueil* contains copies of several other maps and mentions many more. Paulmier speaks of three different kinds of cartographical representation: the planisphere; maps in two parts divided by the meridian; and maps divided by the equator. The maps of which he provides copies are those of Mercator, Hondius (both 1602) and Bertius (1628), all with similar representations of *Terra Australis* (Figure 6.5). On the map of Mercator, redrawn by Hondius, can be seen, beneath the Cape of Good Hope, the *Psittacorum regio*, the Land of Parrots, mistakenly displaced from the earlier maps of Brazil.[18] To the left, and below the Island of Tristan da Cunha, is to be found the Promontory of *Terra Australis*. As well as these maps, Paulmier mentions, without reproducing them, two world maps drawn by the Visscher (Piscator) family, one from 1638–39 engraved by Claes Janszoon Visscher, and another from 1657, probably drawn by his son Nicolaus. These maps represent respectively the old and new world views, demonstrating at the same time how old representations survived alongside the new. On the 1639 map (Figure 6.6), beneath the Moluccas, is to be found the country 'Beach', from Marco Polo's account of his travels. This belongs to the Ortelian *Terra Australis* cartographic tradition, exemplified in Ortelius's map and largely based in the imaginary. The map does, however, depict the Eastern coast of the Gulf of Carpentaria, discovered in 1606 by the Dutch explorer, Willem Janszoon, who believed it to be part of New Guinea.

The 1657 Visscher map (Figure 6.7) signals a new cartographic departure, incorporating Tasman's discoveries of 1642, depicting and naming the coasts of New Holland, which is not identified as such. On the map are marked the Terre de Witz, Houtmans Albohis (sic), Eendracht Land, Dirck Hergotz (sic) Ree, the lands of Nuitz, Edel, Leuwin and Van Diemen. The outline of the mythic *Terra Australis Incognita* is still retained, although backgrounded and partly obscured

le tiltre de Terres Australes Incogneuës pour Monsieur du Chesne Conseiller du Roy, Historiographe de France, en Advocat en tous les Conseils de sa Majesté 1659. [BnF, Ms. NAF 7454], transcribed in Abbé Paulmier, *Mémoires*, ed. Sankey, pp. 317–34.

17 Nancy Bouzrara and Tom Conley have demonstrated the close relationship between literary writing and cartography in the early modern period in France in 'Cartography and Literature in Early Modern France', in David Woodward (ed.), *The History of Cartography* (Chicago, IL, 2007), vol. 3, part 1, pp. 427–37.

18 See W.A.R. Richardson, 'Mercator's Southern Continent: Its Origin, Influence and Gradual Demise', *Terrae Incognitae: Journal for the History of Discoveries*, 25(1993): 67–98.

Figure 6.5 Gerard Mercator/Jodocus Hondius, 'Orbis terrae novissima descriptio' (1602). Bibliothèque nationale de France, CP, Ge B 1514. Reproduced with permission

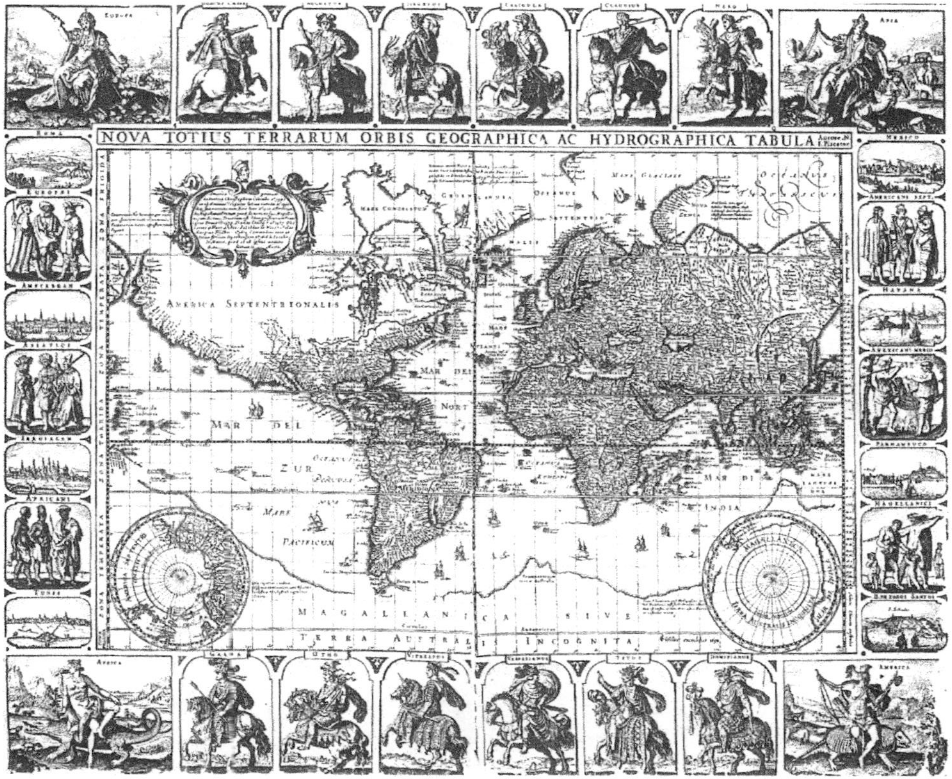

Figure 6.6 Claes Janszoon Visscher, 'Nova totius terrarum orbis geographica ac hydrographica tabula' (1639). Bibliothèque nationale de France, CP, Ge D 7916. Reproduced with permission

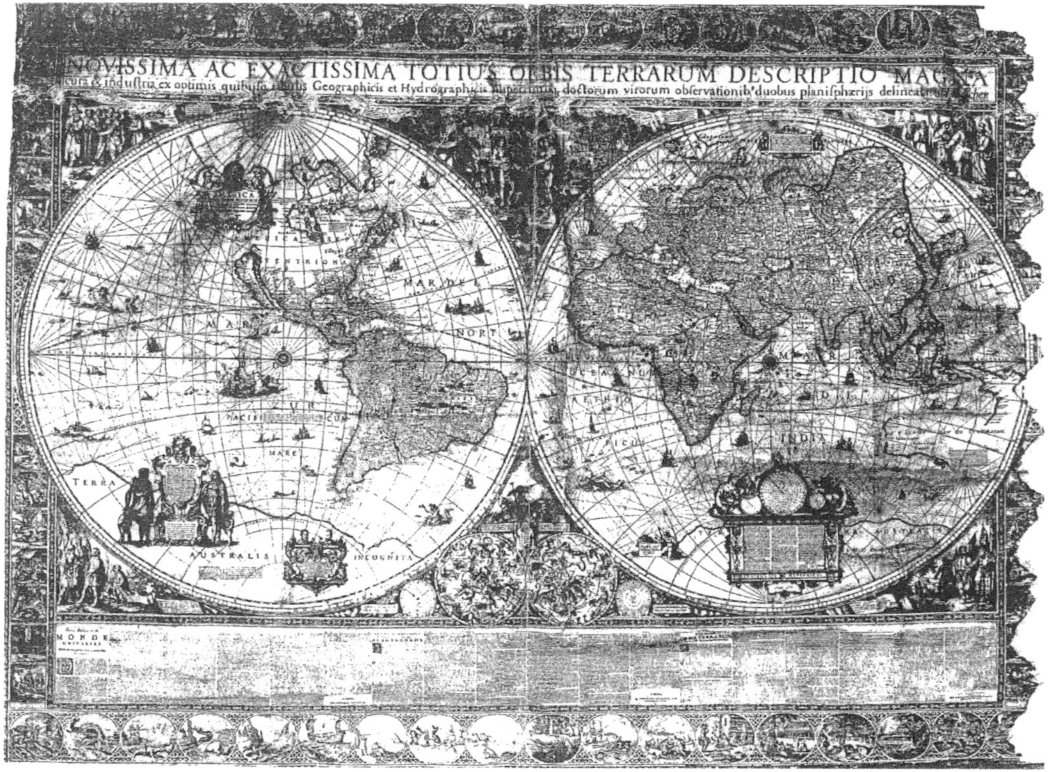

Figure 6.7 Nicolaus Visscher, ‘Novissima ac exactissima totius orbis terrarum descriptio’ (1657). Bibliothèque nationale de France, CP, Ge A 522. Reproduced with permission

by three cartouches appearing on it. In other Visscher world maps of the same period the outline of *Terra Australis* has completely disappeared, and the same is true of another map mentioned by Paulmier in the *Recueil*: the famous 1648 map of Joan Blaeu, dedicated to Gaspar de Bracamonte, Count of Peñaranda. This map is the first in which *Hollandia nova* is identified as a country and Tasman's discoveries of 1642–44 are incorporated. The outline of *Terra Australis Incognita* has completely disappeared, but the name is retained, written in tiny characters round the South Pole (Figure 6.8).

Given that Paulmier was cognizant of all the maps previously mentioned, the choice he made of maps to preface his published *Mémoires* is, on the face of it, curious. In the Sanson map of 1651 (see Figure 6.3) with which Paulmier prefaced the 1658 manuscript of his *Mémoires*, the Dutch discoveries in New Holland have been amalgamated and assimilated into the Ortelian model. Thus Beach (from Marco Polo) is the name given to the tip of Cape York, and the coasts discovered by the Dutch (Nuitz, Edel, Leuwin) are also depicted. The outlines of the *Terres australes* are evident but paler, as though perhaps an attempt had been made to erase them from the engraved plate. This dynamic of dark and light could reflect the cartographer's ambivalence about the existence of *Terra Australis Incognita*, or simply the wish for scientific accuracy. One can ask why then, in this context, Paulmier would choose such a seemingly ambiguous map and why, six years later, in 1664, he would preface the second issue of his *Mémoires* with De Fer's map (see Figure 6.4), which totally ignores the Dutch discoveries and returns to the Ortelian model.

The answers to these questions lie, I think, both in the changing cartographic conventions of the time and in the secretive and suspicious atmosphere surrounding the dissemination of new discoveries. Around the middle of the seventeenth century, as we have seen above in the case of the Visscher maps, the chronological sequence of maps representing *Terra Australis* highlights the confusion operating between the mapping of the real, and imaginary cartography. Mapmaking was becoming increasingly scientific, and the distinction between the drawing of known lands and the hypothesizing of unknown ones was being marked in various ways, from the lighter outline used for the undiscovered *Terra Australis* (which we have seen in the case of Sanson's map), to the name without the outline (Blaeu), to their total disappearance.

The ambivalence of mapmakers is overlaid also by the suspicions harboured by some nations that others were hiding their discoveries for political reasons, to protect their claim over newly discovered lands. In this context Blaeu's map, obliterating *Terra Australis*, would have been read by Paulmier as a ploy to put the French or others off pursuing the search for *Terra Australis*. It is indeed Paulmier's suspicion. In his *Recueil* he remarks:

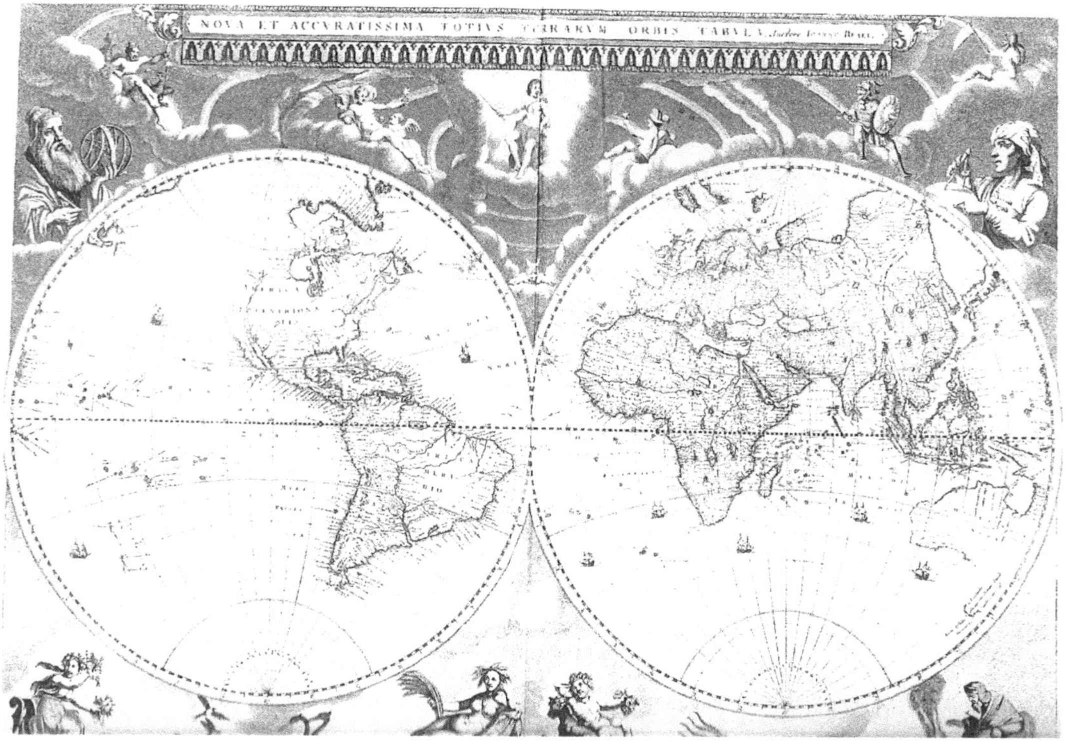

Figure 6.8 Joan Blaeu, 'Nova et accuratissima totius terrarum orbis tabula' (1648), in *Geographia, quae est cosmographiae blauianae pars prima*, vol. 1 (Amsterdam, 1662). Bibliothèque nationale de France, CP, Ge DD 4796 (19). Reproduced with permission

> It is said that the Dutch Ambassador, who in the present year 1659 is in Paris, has many maps and memoranda about the Southern Lands. However it is said that he has received an order from the Director of the Dutch East India Company not to give them to the public.[19]

Thus in the slippage between the politics of map-making and that of map-reading, the *Terre australe* signifier still remained rich in possibilities. Confusion in chronological sequencing of new discoveries could, from Paulmier's standpoint, reflect both cartographic uncertainty and Dutch duplicity. For him the *Terre australe* had yet to be discovered, and New Holland might or might not be a part of it but certainly did not replace it. The reasons for Paulmier's belief lie in his interpretation and presentation of Gonneville's sixteenth-century voyage.

According to the Abbé, the country discovered by Gonneville was merely one region of the vast unknown southern continent and was situated in quite a different part of the traditional *Terra Australis* from New Holland. He identifies the unknown land discovered by the ship-wrecked Gonneville as lying beneath South Africa:

> The country discovered by Captain Gonneville, whose voyage has to some extent given birth to our project, must be situated further on towards the South-East of there [the Land of Parrots].[20]

Thus Gonneville's land, as conceived by Paulmier, was situated between the meridians of 60 and 80 (in relation to the meridian of the Canary Islands) and was to be found in the Indian Ocean, in the area below the islands of Madagascar and Mauritius. The first French map which depicts Gonneville's land is that of Louis de Mayerne Turquet (Figure 6.9), dating from 1661. Calling it the Païs d'Arosca, after the name of the king mentioned in Gonneville's account, Mayerne de Turquet situated it to the right of the Cape of Good Hope. Later in the century, another map (Figure 6.10) was published by Pierre Du Val showing Gonneville's supposed discoveries. In this map Duval equated the Promontory of *Terra Australis*, Arosca's Land and the Land of Parrots but situated them to

---

19 My translation. 'On dit que M L'Ambassadeur de Hollande qui est en la presente année 1659 a Paris a beaucoup de Carthes, et memoires touchant les *Terres australes*; Et mesme je scay qu'il en a faict graver quelques Unes, ensemble des planches des fruicts et plantes de ces pays.'

'Toutes fois on dit qu'il a en ordre de Mrs de la Compagnie Hollandaise de l'Est Inde de ne les donner au public.' [BnF, Ms. NAF 7454], transcribed in Abbé Paulmier, *Mémoires*, ed. Sankey, p. 331.

20 'Il faut mettre plus de là [la Terre des Perroquets] et en tirant au Sud-Est, le pays de la découverte du Capitaine de Gonneville, dont le voyage ayant en quelque façon donné naissance à nostre projet …' (Paulmier, *Mémoires* …, p. 9 [ed. Sankey, p. 156]).

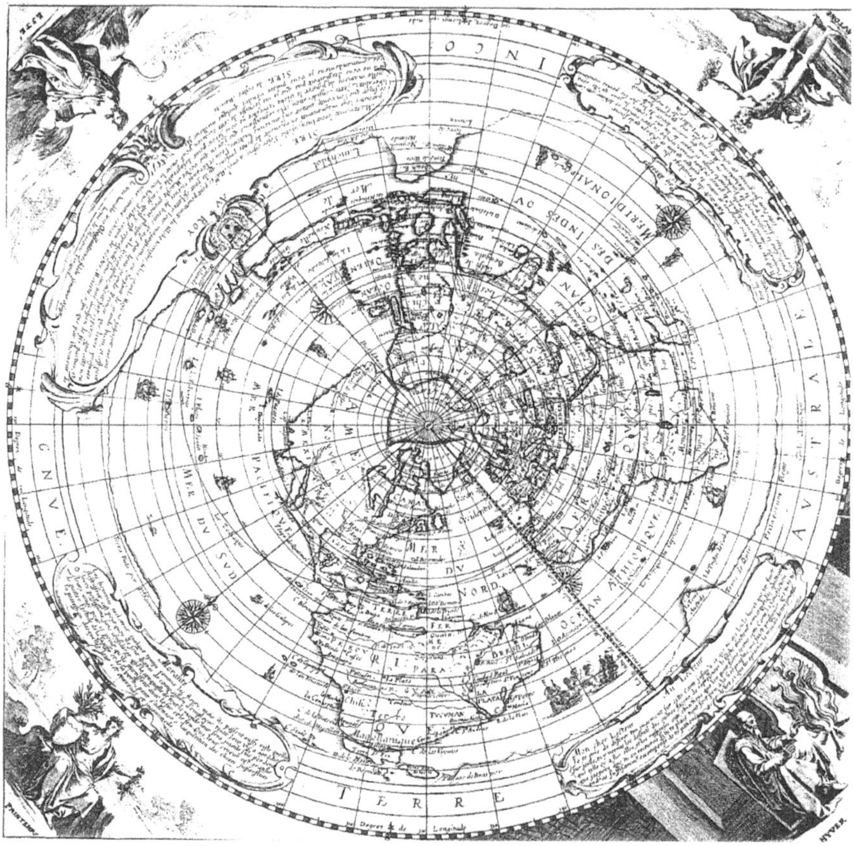

Figure 6.9 Louis de Mayerne Turquet, 'La Nouvelle manière de représenter le globe terrestre en laquelle il est entièrement réduict dans un cercle, sans aucune division de ses parties' (1669). Bibliothèque nationale de France, CP, Ge CC 1270 (5). Reproduced with permission

the left of the Cape of Good Hope. We read on the map: 'Arosca's Land where Gonneville landed in 1504, and brought back Essonier son of King Arosca to Normandy.'[21] These maps were to be followed by others in the course of the next century, locating Gonneville's land either slightly to the west or to the east of the Cape of Good Hope.[22]

[21] 'Terre d'Arosca où en 1504 aborda le nommé Gonneville qui en ramena en Normandie Essonier fils du Roy Arosca.'

[22] See M. Sankey, 'Est ou Ouest: le mythe des terres australes en France aux XVIIe et XVIIIe siècles', in Kumari Issur and Vinesh Hookoomsing (eds), *L'Océan Indien dans les littératures francophones: pays réels, pays rêvés, pays révélés* (Paris and Réduit), pp. 13–26.

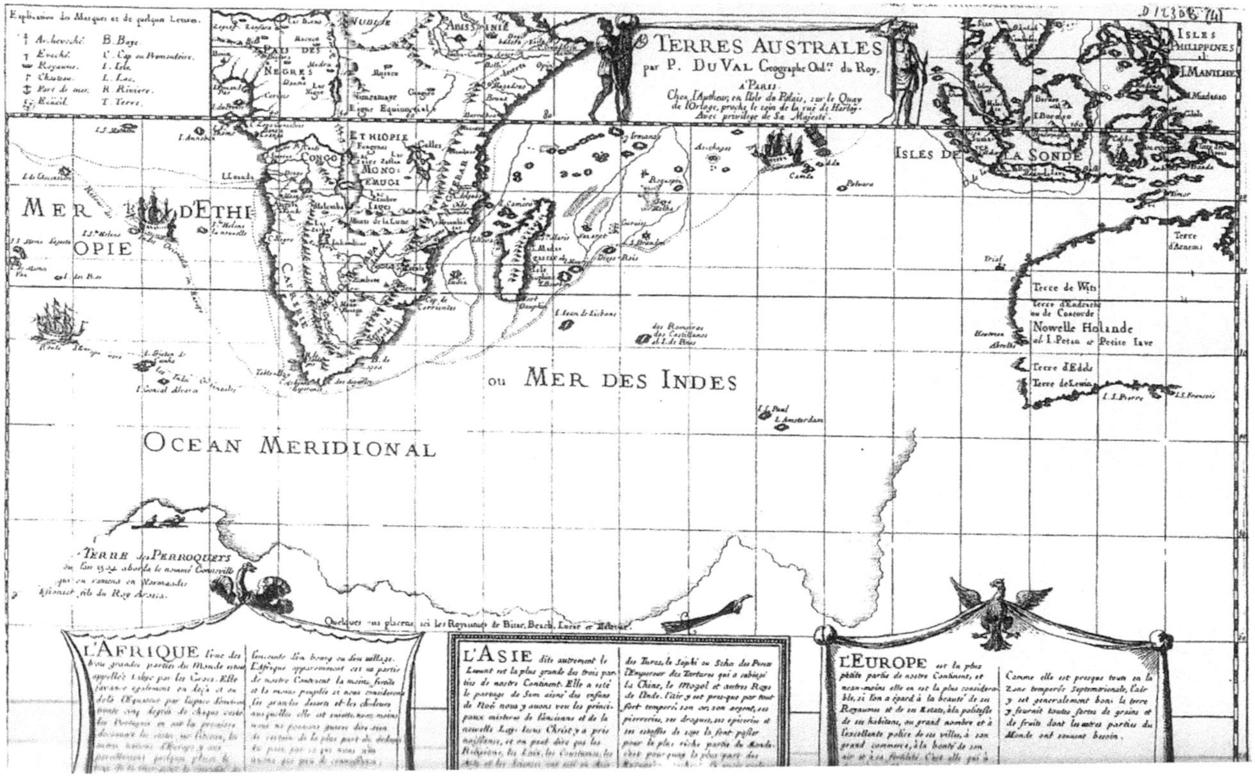

Figure 6.10 Pierre Du Val, *Carte universelle du monde* (late seventeenth century). Bibliothèque nationale de France, CP, Ge D 12308 (4). Reproduced with permission

Paulmier's suggestion that the French search for Gonneville's land received warm support from his friends at the Missions étrangères, and his request was received favourably by the pope. Likewise, Louis XIV agreed to fund and equip the expedition, and costings were made. Plans were also drawn up for the establishment of a Compagnie des *Terres australes*. The project was never realized, however, for political reasons within the Catholic Church. The pope was willing to confer upon Paulmier the title of Bishop of the *Terres australes*, decreeing that the bishopric would include the already established French mission in Madagascar, under the administration of the Lazarist Order.[23] Indignation on the part of the Lazarists, who had long requested a bishopric for Madagascar and who had powerful friends in Rome, caused Paulmier's plan to be quietly dropped. The last known traces of correspondence concerning the proposal date from 1670.

This was not, however, the last to be heard of Paulmier's proposal. During the following century it continued to exercise a powerful effect on the French imaginary, not least because of its utopian aspects. Several voyagers during the late seventeenth century, such as Beaujeu, Voutron and Sainte-Marie, planned to seek Gonneville's land, but the political will or finance was lacking. Writers of utopian literature at the end of the seventeenth century, most associated with Protestantism, such as Gabriel de Foigny, Denis Vairasse and Simon Tyssot de Patot, all mentioned Gonneville's voyage, mediated through Paulmier's *Mémoires*, in their representations of *Terra Australis*.[24]

Paulmier's localization of the *Terres australes* became anchored in French consciousness and was also the focal point for French voyages of discovery to the *Terres australes* in the eighteenth century.[25] The explorer Jean-Baptiste Lozier de

---

23 The missionary order established by Saint Vincent de Paul.

24 Gabriel de Foigny, *La Terre australe connue* (1676), ed. Pierre Ronzeaud (Paris, 1990); Denis Vairasse, *Histoire des Sévarambes, Peuples qui habitent une partie du troisième continent, communément appelé la terre australe. Contenant un compte exact du Gouvernement, des Mœurs, de la Religion, du langage de cette Nation, jusques aujourd'huy inconnuë des Peuples de l'Europe.* Traduit de l'anglois, 4 tomes (2 vols, Paris, 1677–79; repr. Genève, 1979); [Tyssot de Patot], *Voyages et avantures de Jaques Massé* (Bourdeaux (La Haye), 1710). For a discussion of Paulmier's influence on these writers see M. Sankey, 'L'Abbé Paulmier méconnu: le mythe et l'histoire des *Terres australes* en France au dix-septième et dix-huitième siècles', *Studies in Eighteenth-Century French Fiction in Honour of Angus Martin*, ed. Wallace Kirsop, *Australian Journal of French Studies*, 38.1 (2001): 54–68. For a discussion of literature concerning the utopian aspects of *Terra Australis* see David Fausett, *Writing the New World: Imaginary Voyages and Utopias of the Great Southern Land* (Syracuse, NY, 1993). Fausett does not, however, make the link with Paulmier's *Mémoires*.

25 This question is explored in a paper given at the symposium, *Discovery and Empire: The French in the South Seas*, at the University of Adelaide on 8 July 2009 and is to be published in the *Proceedings*.

Bouvet undertook his search for the *Terres australes* in 1738–39, sent by the East India Company. Having read the extracts of Gonneville's *Relation* in Paulmier's *Mémoires*, he searched in the region below the Cape of Good Hope indicated by the Abbé, but encountered nothing but a tip of land shrouded in ice and snow that he named Cap de la Circoncision, which later exploration showed to be part of an island. The cartographer Philippe Buache illustrated Lozier de Bouvet's attempt in a series of fanciful maps, culminating in the fictional, fertile *Terres australes* at the South Pole, surrounded by a rampart of ice. The 'Terre de Gonneville' is marked on this map at top right-hand corner of the larger landmass (Figure 6.11).

Inspired by Maupertuis and encouraged by Buffon, Charles de Brosses published in 1756 his *Histoire des navigations aux Terres australes* to encourage the French to search anew for the *Terres australes*.[26] As proof of the existence of the undiscovered lands he included Paulmier's extract from Gonneville's *Relation*. Subsequently Bougainville in 1768, Surville in 1768–69, Marion du Fresne in 1771 and Kerguelen in 1771 and 1773 would search for Gonneville's land without success. During this period Captain James Cook was setting out on his momentous second voyage,[27] which would finally bring to an end the French search for the mythic *Terres australes*.

But where was Gonneville's land, if not somewhere in the area beneath the Cape of Good Hope, as Paulmier had believed? The nineteenth-century geographer Armand d'Avezac, distorting Gonneville's *Relation* (of which he was using the recently discovered seventeenth-century copy of the complete version, rather than extracts), concluded that the country that Gonneville had discovered was Brazil. But was it? Textual analysis of the timelines of Gonneville's voyage and inconsistencies in his account would suggest not.[28] Others suggested that Gonneville's land might be Maryland in America, or Madagascar.

---

26 Charles de Brosses, *Histoire des Navigations aux Terres Australes contenant ce que l'on sçait des mœurs & productions des Contrées découvertes jusqu'à ce jour: & où il est traité de l'utilité d'y faire de plus amples découvertes, & des moyens d'y former un établissement* (Paris, 1856; Amsterdam and New York, 1967). Paolo Carile characterizes de Brosses's book as being 'une sorte de chant de cygne du mythe du continent austral', in Sylviane Leoni et René Ouellet (eds), *Mythes et géographies des mers du Sud* (Dijon, 2006), p. 27.

27 1772–75. English interest in the search for *Terra Australis* had been stimulated both by the publication of John Callender's *Terra Australis Cognita, or, Voyages to the Terra Australis or Southern Hemisphere* (Edinburgh, 1766), which was essentially a translation of De Brosses's work, and by Alexander Dalrymple's *An Account of the discoveries made in the South Pacifick Ocean previous to 1764* (London, 1767), which argued for the mounting of an English voyage of exploration to the southern hemisphere. Dalrymple hoped to be given the charge of such an expedition, but instead Cook was appointed to undertake what became his first voyage, 1768–71.

28 See Abbé Paulmier, *Mémoires*, ed. Sankey, pp. 112–16.

Figure 6.11 Philippe Buache, 'Carte des Terres Australes' (1754). Bibliothèque nationale de France, CP, Ge AF Pf 34 (74). Reproduced with permission

But other, more serious questions have been raised about Gonneville and his voyage. Why is Gonneville nowhere mentioned in history or literature before the Abbé Paulmier and why can one find no other documentation concerning him or the native of the *Terres australes* brought to France in 1504? Armand Rainaud[29] mentions a world map of Louis de Mayerne Turquet, dated 1648, on which is indicated Arosca's Land (see Figure 6.9). If this were the case, this document would pre-date Paulmier's redactions and possibly provide external corroboration for Gonneville's existence. However, analysis of the five maps dated 1648 in the Bibliothèque nationale in Paris, shows that only one is in fact from that year, since it bears the coat of arms of Louis XIII.[30] Significantly, the Land of Arosca does not appear on it, whereas it does on the four other maps which each bear the coat of arms of Louis XIV and are therefore posterior to 1661.

Given the lack of external corroboration for Gonneville's existence, Jacques Lévêque de Pontharouart has suggested that Gonneville's account is perhaps a seventeenth-century forgery and that Jean Paulmier himself was the forger.[31] Some of Lévêque's arguments are specious but a nagging suspicion remains and a growing body of evidence suggests that all is not as historical tradition would have it appear and that there is indeed reason to question Paulmier's presentation of events. This is a serious matter as Gonneville is hailed in both France and Brazil as the first Frenchman to have visited the South American country.[32]

Be that as it may, the real and imagined spaces in the southern oceans, foregrounded in Paulmier's vision, provided a powerful incentive for dreams of French conquest during the eighteenth century. The idea of Gonneville's land kept the dream alive and a search for this lost paradise yielded for France a meagre harvest of cold and windswept possessions in the southern oceans: the Kerguelen and Crozet Islands. And in 1892 France officially took possession of the islands of St Paul and Amsterdam. The Cap de la Circoncision, discovered by Bouvier de Lozier, is part of Bouvet Island, which now belongs to Norway.

29 Armand Rainaud, *Le Continent australe: hypothèses et découvertes* (Paris, 1893).

30 BnF, CP [Ge.D.12282].

31 Jacques Lévêque de Pontharouart, *Paulmier de Gonneville: son voyage imaginaire* (Beauval en Caux, 2000).

32 For example, 2009 was celebrated in France as *l'année du Brésil* and events took place in both France and Brazil centred on the figure of Gonneville.

Chapter 7

Ceremonial Encounters: Spanish Perceptions of the South Pacific, 1567–1794

Mercedes Maroto Camino[1]

On arrival in today's Santo in Vanuatu in 1606, Pedro Fernández de Quirós performed a series of rituals of possession that bewildered his crew and must have astonished the island's inhabitants. Quirós declared that he had found the mythical *Terra Australis Incognita* and named it *Austrialia del Espiritu Santo*, a name which was coined in honour of the Holy Spirit ('espíritu santo') and the Spanish Habsburg ('Austrian') king, Philip III (r. 1598–1621). The king had agreed to sponsor Quirós's expedition following ten years of persistent efforts to canvass support for a second voyage to the South Pacific after sailing as pilot in the failed expedition led by Álvaro de Mendaña to the Santa Cruz Islands in 1595 (Figure 7.1).[2] He was finally put in command of this Spanish Pacific voyage, which departed from Peru's harbour of El Callao in Lima in 1605. The fleet comprised two ships, the *San Pedro y San Pablo* and the *San Pedrico*, and a launch, *Los Tres Reyes*. Quirós was in charge of the *San Pedro y San Pablo*, with his declared enemy, Diego de Prado y Tovar, as second in command. Luis Váez de Torres, after whom Torres Strait and Islands are named, was captain of the *San Pedrico* and Pedro Bernal de Cermeño of *Los Tres Reyes*. Neither the Solomons

1 I am very grateful to Glyn Williams for his kind help with the last image of this article. Thanks also to the Museo Naval and Museo América in Madrid for kindly allowing me to use their material. Lastly, I wish to acknowledge the organizers of the *Perceptions of Terra Australis* conference in Perth (Australia) and the editors of this volume, especially Anne Scott, for their patience and support.

2 Regarding Quirós' efforts to canvass support from 1595 to 1606, Ian Cameron remarks: 'The Viceroy of Peru referred Quirós to the court of Spain; the court of Spain referred him to the Papacy; the Papacy referred him back to Philip III; Philip handed him over to the Council of The Indies, and the Council of The Indies passed him on to the Council of State' (Ian Cameron, *Lodestone and Evening Star: The Epic Voyages of Discovery 1493 B.C.–1896 A.D.* (New York, 1966), p. 163).

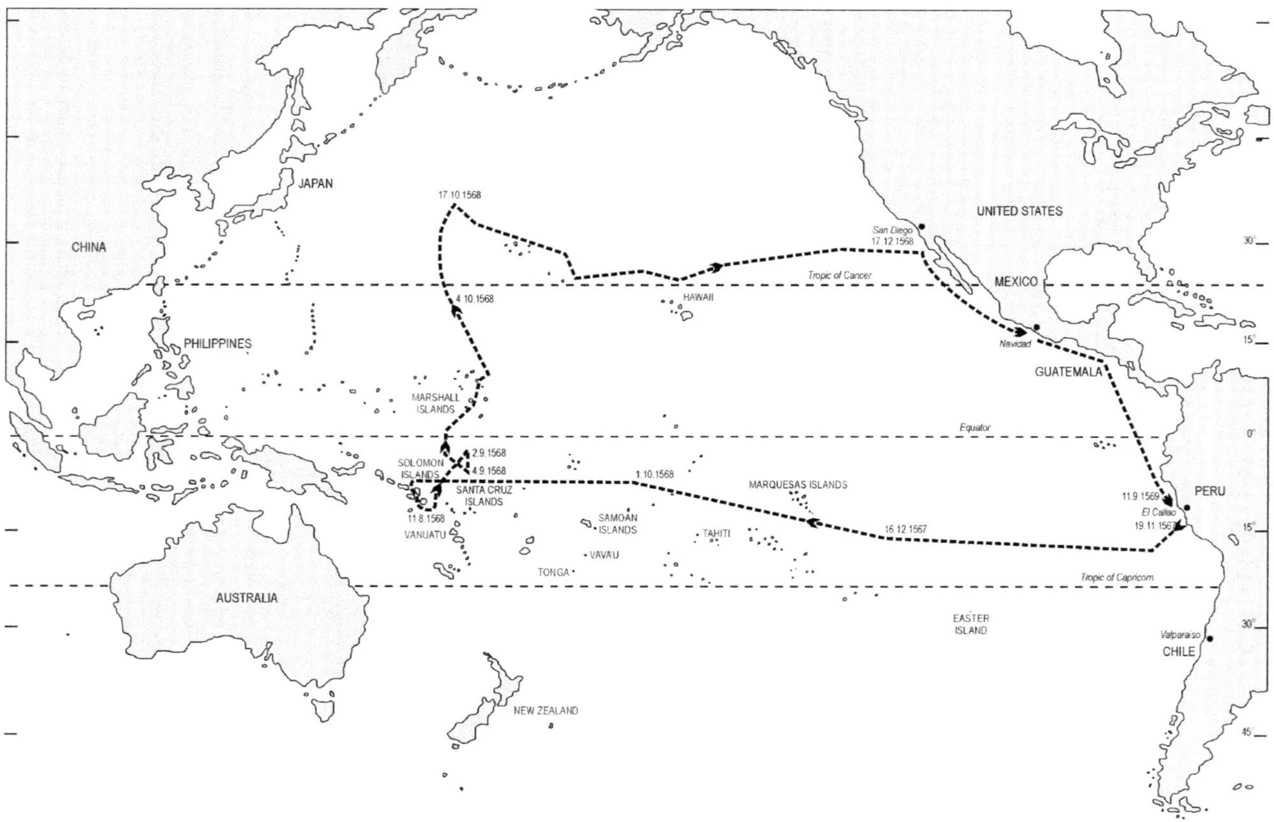

Figure 7.1 Itinerary of Álvaro de Mendaña's 1595 voyage to the Santa Cruz archipelago. Reproduced with permission

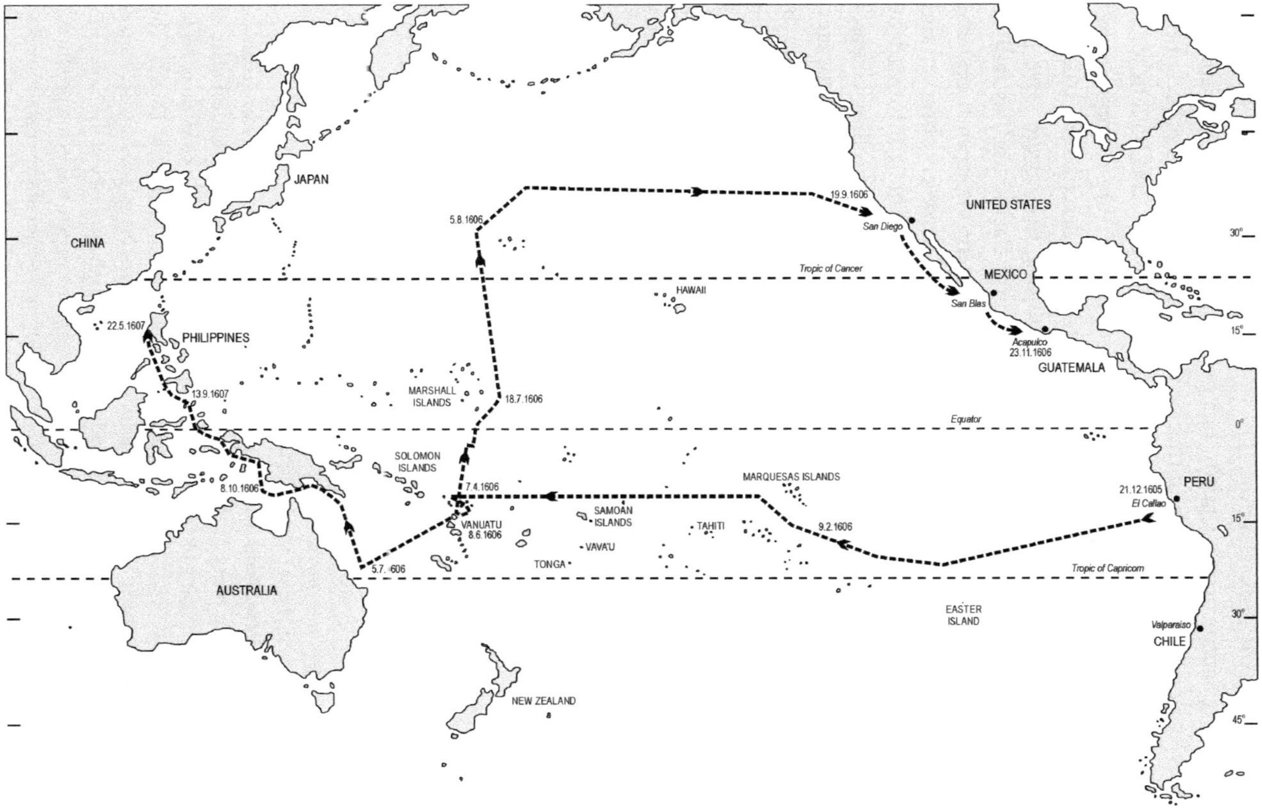

Figure 7.2 Itinerary of Pedro Fernández de Quirós's 1606 voyage to Vanuatu's Santo. Reproduced with permission

nor the Santa Cruz archipelago was found on this journey, even though Quirós landed on some Pacific islands, including Taumaco, and on Vanuatu's Santo, his *Austrialia del Espíritu Santo* (Figure 7.2).

On reaching Santo's beach, Quirós, in a gesture reminiscent of Christopher Columbus, knelt and kissed the land that he thought was part and parcel of the Southern Continent, hailing it in the following words: 'Ah! Land, sought for so long, pretended by so many and so intensely desired by me!'[3] Quirós represented Oceania as a Terrestrial Paradise, using feminine terms to describe the land and its people. The ceremonial practices performed during his voyage were part and parcel of the Spanish apprehension of the South Pacific and, as will be seen below, remained of paramount importance up to the eighteenth century. There is, however, a substantial rearrangement of roles between observer and the observed from the earlier to the later voyages. This shift is informed by contemporary notions of gender and class and exemplifies some telling changes regarding the assimilation of the other between the earlier and the later voyages.

Quirós's perception of the Pacific occupied a privileged position for nearly 200 years, inflecting eighteenth-century discourses. This was largely due to the translation and widespread dissemination of one of his memorials, known as Memorial Eight. His perspective of a bountiful land went on to become the hallmark of the apprehension of the Pacific, remaining popular when James Cook embarked on his momentous voyages (1768–71, 1772–75 and 1776–79), as well as during the Spanish four-year voyage led by Alessandro Malaspina between 1789 and 1794.[4] Besides many written accounts, this journey included artists who produced a large number of drawings of indigenous people and the local landscape. These are relevant to the perception of *Terra Australis* insofar as the journey sojourned in Port Jackson, New Zealand and the Vava'u archipelago of Tonga.[5] Furthermore, in terms of design and objectives, Malaspina's expedition

---

[3] Pedro Fernández Quirós, 'Viajes de Quirós', in Justo Zaragoza (ed.), *Historia del descubrimiento de las regiones austriales hecho por el general Pedro Fernández de Quirós* (3 vols, Madrid, 1876. Facsimile in 1 vol.; Madrid, 2000), pp. 115–308.

[4] According to Andrew David, 'the aim of the expedition was to produce hydrographical charts for the most remote parts of the Americas and to investigate the political state of Spanish possessions with regard to Spain and foreign countries and to record their trade, natural resources and defence, rather than discovering new lands' (Andrew David, 'The Voyage of Alejandro Malaspina 1789–1794', *Annual General Meeting of The Hakluyt Society 7 July 1999* (London, 2000), pp. 3–34, 3). David offers a thorough summary of the journey in this article, which was written while the Hakluyt edition of this voyage was being produced.

[5] In fact, as Donald Cutter remarks, the 'most enduring legacies of the expedition was the copious artistic archive amassed during the voyage, as the artists drew both places and people ... The many drawings give a visual dimension to the anthropological details, and help to identify the scenes described in the journals. Since further details were often added in the

followed the patterns established by the earlier 'scientific' expeditions of Cook or Louis de Bougainville.[6]

The parameters embedded in Quirós's words thus connect perceptions of *Terra Australis* inherent in narratives and drawings of sixteenth- and eighteenth-century Spanish voyages. These cultural products contributed to creating or, to use my own terminology, producing the South Pacific or South Seas.[7] Differences notwithstanding, both the earlier and later representations relied on an apprehension of landscape (including seascapes and beaches), indigenous peoples and women that was determined by European social relations.[8] This way of perceiving the universe also reveals the relationships of power inherent in observation, description and interpretation. It is a mode of observing that can be linked to Michel Foucault's description of the *panopticon*, in which observers gaze upon and construct their subjects, which is integral to the development of ethnography within the colonial setting.[9]

finished versions, the drawings generally are most reliable in their original state than when efforts were made to meet artistic expectations' (Donald Cutter, 'Introduction', in Andrew David, Felipe Fernández-Armesto, Carlos Novi and Glyndwr Williams (eds), *The Malaspina Expedition, 1789–1794: Journal of the Voyage by Alejandro Malaspina*, vol. 1 (London and Madrid, 2001–2004), pp. xxix–lxxvii, lx).

6 Bernard Smith explains the contradictions within which the journey operated: 'One of the most important voyages modelled on those of Cook was the Spanish one under the command of Alejandro Malaspina which set out from Cadiz in 1789. Despite their excellent work of collecting and describing, completed on the voyage itself, those on the expedition came home to Spain to confront a social and political situation highly unfavourable to the encouragement of the arts and sciences. The momentum of the French Revolution, with all its attendant excesses, threatened the values upon which the dynamics of Enlightenment science were based' (Bernard Smith, *Imagining the Pacific: In the Wake of the Cook Voyages* (New Haven, CT and London, 1992), p. 47).

7 Mercedes Maroto Camino, *Producing the Pacific: Maps and Narratives of Spanish Exploration (1567–1606)*, Portada Hispánica, 18 (Amsterdam and New York, 2005), pp. 15–16.

8 In this context, it is salutary to remember that the notion of landscape, as W.J.T. Mitchell categorically puts it, is nothing but 'the dreamwork of imperialism'. W.J.T. Mitchell, *Picture Theory: Essays on Verbal and Visual Representation* (London and Chicago, IL, 1994), p. 10.

9 For Foucault, domination, subjection and exploitation are inherent to the exercise of power that runs through the different social strata (Michel Foucault, 'The Subject and Power', in Hubert Dreyfus and Paul Rabinow (eds), *Michel Foucault: Beyond Structuralism and Hermeneutics* (Chicago, IL, 1982), pp. 208–26). According to James Clifford, 'Ethnographic work has indeed been enmeshed in a world of enduring and changing power inequalities, and it continues to be implicated. It enacts power relations. But its function within these relations is complex, often ambivalent, potentially counter-hegemonic ... The effect of domination in such spatial/temporal deployments ... [referring to Orientalism but not limited to it] is that they confer on the other a discrete identity, while also providing the knowing observer with

Through his account, Quirós contributed uniquely to the genealogy of the Pacific as a plentiful land in terms of flora and fauna. This myth was, in turn, extended to embrace the sexual availability of its women, especially Polynesian women. Indeed, the perception of 'free love', which was especially popular in narratives and paintings from the eighteenth and nineteenth centuries, has only recently been seriously questioned by Serge Tcherkézoff.[10] Interestingly, this notion ran hand-in-hand with the eighteenth-century view of upper-class women as weak and delicate creatures in need of protection and enclosure. This idealization of women was heavily inflected by class differences, and it is a view that was translated to the Pacific into a feminization of indigenous men, who were deemed to be objects of vision within the paradigm of observation characteristic of the Enlightenment project, as will be seen below.

The creation and appropriation of the Pacific in the European mind is explicitly conveyed by Quirós's words quoted above, which were part of the rituals of possession performed on arrival in his *Austrialia*. Quirós's words were accompanied by several ceremonies that elaborated on those performed by Columbus on reaching the Americas and which were modelled on French traditions. These performances began with Quirós creating war ministers, choosing the office holders and establishing the Order of the Knights of the Holy Spirit. Following this, Quirós took formal possession of the island in the name of the Holy Trinity, the pope and His Catholic Majesty Philip III. He then gave everyone a blue cross to wear on their breast, a decision which was hotly contested by the Franciscan Friar Martin de Munilla, who believed that displaying such insignia contradicted the charter of his order. Munilla's description of these events is both interesting and ironic:

> Afterwards, as the General stood on the bank of a cool stream, he sent for certain men of standing and appointed them officials for the administration of justice and for government, such as *alcaldes ordinarios*, officials of the Hermandad, *Regidores* and other similar offices. It was a marvellous thing to see such a diversity of knights, for truly nothing like it has ever been seen since the world began, because here there were sailor-knights, grummet-knights, ship's page-knights, mulatto-knights and negro-knights and Indian-knights and knights who were knight-knights.[11]

---

a standpoint from which to see without being seen, to read without interruption' (Stephen Tyler, 'Post-modern Ethnography: From Document of the Occult to Occult Document', in James Clifford and George E. Marcus (eds), *Writing Culture: The Poetics and Politics of Ethnography* (Berkeley and Los Angeles, CA and London, 1986), pp. 122–40, 9, 12.

[10] See Serge Tcherkézoff, *Tahiti-1768: Jeunes filles en pleurs* (Papeete, 2004).

[11] Fray Martín de Munilla, 'La Austrialia del Espíritu Santo: The Journal of Fray Martín de Munilla O.F.M. and Other Documents Relating to The Voyage of Pedro Fernández de

It was after arguing with Munilla, telling him 'things that cannot be put down on paper', that Quirós knelt and kissed the land, hoisted a flag and took possession of the island again, this time in the name of the Holy Trinity, the Catholic Church, Saint Francis and his Order, John of God and his Order, the Holy Spirit and, last but not least, His Majesty the King of Spain. Three masses were said, with everyone taking communion and this was followed by a fiesta in which two slaves were freed only, however, to return to slavery thereafter. In relation to this latest gesture, one interpreter of the events, Juan de Iturbe, wrote:

> Quiros wished to establish an Order and to call it the Order of the Holy Ghost. And so at Whitsun he gave crosses of blue taffeta to all the crew of the fleet, who wore them on their breasts ... even two negro cooks were rewarded by such largesse, great liberality and munificence for their gallantry and courage. Besides, on that day, he granted them their liberty, although they did not belong to him, and what is more they afterwards continued in the self-same state of slavery.[12]

During the following days, Quirós also planted maize as a sign of possession and sent some soldiers to reconnoitre the island, while further festivities took place. These included the celebration of Corpus Christi, where the Spaniards heard another three masses and staged an elaborate procession, parading the religious images of Saint Peter and Saint Paul. The procession was led by a soldier carrying a heavy cross made of orangewood, followed by one friar, an array of various groups playing music and other clusters of people carrying the royal banner. All sang hymns, shooting volleys in the air before a fourth mass was said. It would be interesting to know what the indigenous boy renamed Pedro thought about the event. Pedro took part in the procession dressed up in taffeta and seems to have been well pleased with the festivities, as well as proud of the cross he bore on his chest; he was both a participant and an observer of the spectacle in front of him, occupying a privileged position in this regard. As will be seen in relation to Malaspina's voyage, this paradigm would become a hallmark of European scientific observation from the eighteenth century onwards.

For reasons that are subject to some debate, Quirós's fleet did not seem as happy as Pedro with his command, and they left *Austrialia* without attempting to corroborate whether it was the Southern Continent. They proceeded to Mexico and failed to meet with Torres at their appointed rendezvous. Confused

---

Quirós to the South Sea (1605–1606) and the Franciscan Missionary Plan (1617–1627)', in Celsus Kelly O.F.M. (ed.), *La Austrialia del Espíritu Santo* (Cambridge, 1966), pp. 134–270, 222–3.

[12] Juan de Iturbe, 'Sumario Breve', in Kelly (ed.), *La Austrialia del Espíritu Santo*, pp. 273–93, 286.

as to the reasons for Quirós's absence, Torres led his ship through the strait that today bears his name between the southern coast of New Guinea and the north of Australia, proving the insularity of New Guinea and sighting Australia. Torres' deed was, however, largely unknown until, in the eighteenth century, Alexander Dalrymple found in Manila a letter by Torres recounting the voyage. Dalrymple published it, and proposed that Torres' name be given to the strait and islands that today bear it. As Brett Hilder notes:

> That such great discoveries were made only to be filed away so successfully in the archives that the world has been largely ignorant of them to this day is a very poor reward for Torres and his men, who received neither thanks nor repayment from the Spanish Crown. Quirós, on the other hand, thanks to his genius for exaggeration and propaganda, achieved great renown and lasting fame for his meagre discoveries, which were still inspiring exploration at the time of Cook.[13]

In Mexico, Quirós had to contend with Torres' report and, after addressing rumours about his inability to lead the journey, he renewed his relentless memorials to the king, seeking sponsorship for a further voyage to the South Pacific. This voyage was finally granted in 1614, but failed to materialize only because Quirós died when he was *en route* to achieving his aim. Thus, poetic justice met with Quirós, who ended as he had lived, on the way to his promised land. It would take more than 150 years for other Europeans to find Quirós's desired land, and for the Spaniards to venture into the South Pacific again. Between these voyages, the Southern Continent sought by Quirós remained as elusive to geographers and explorers as it had been since the classical era.[14]

The last ocean to be explored, the Pacific, was gradually constructed in European maps and minds from a resilient mythology that was only laid to rest by Cook in the last third of the eighteenth century. Prior to Cook's voyages, the South Pacific remained one of the least known parts of the world. The fabled Islands of Solomon, discovered in 1567 by Álvaro de Mendaña, found their way into maps but remained intangible to later explorers until well into the eighteenth century, when Louis de Bougainville completed their discovery (1766–68).[15]

---

13 Brett Hilder, *The Voyage of Torres: The Discovery of the Southern Coastline of New Guinea and Torres Strait by Captain Luis Báez de Torres in 1606* (St Lucia, 1980), p. 10.

14 The South Seas were a landscape created by a series of myths, adventures and misadventures from its conception in the minds of Classical scholars. On this topic, see Camino, *Producing the Pacific*, p. 28, and the chapters by Alfred Hiatt, Bill Richardson and Margaret Sankey in this collection.

15 Álvaro de Mendaña is believed to have been born in a village in the province of León in 1542 and went to Peru when he was 20, after his uncle, Lope García de Castro, was named

For two and a half centuries after Ferdinand Magellan's circumnavigation (1519–21) mapmakers laid down *Terra Australis Incognita* on their maps with no evidence for its existence.

Sixteenth-century explorers, including the first world circumnavigator, were mesmerized by the idea of this Southern Continent, which was based on the legacy of classical scholars, as well as that of Marco Polo. From Columbus's voyage in 1492 to Magellan's journey in 1519, the mapping of the world underwent a revolution of unprecedented proportions. It was during this time that the Pacific entered the European imagination and became the target for future explorations. The name Pacific is attributed to Magellan, when he was deceived by the peaceful appearance of the ocean after the tumultuous waters of the strait that today bears his name.[16] From this time onwards, the Pacific Ocean started to appear on maps produced in various European countries. Indeed, these maps and the narratives about Magellan's voyage can be said to have 'produced' the Pacific, for it was not until this journey was popularized that the name and limits he had given to the ocean became part of European knowledge.

After the return of Magellan's ship, *Victoria*, official cartographers working for the master chart, the *padrón general*, drew maps to illustrate the voyage's geographical discoveries. This *padrón* was compiled from the diverse charts, accounts, official interviews, logbooks and similar documents, which were delivered to the House of Trade (*Casa de la Contratación*) in Seville. There they produced master charts, which were meant to be secret and devoted to the exclusive use of Spanish authorities for future expeditions or for military purposes, although much information leaked in one way or another and found its way into maps of the time.

From Columbus onwards, then, the Spanish perception of the Pacific was that of a sea bound by the Southern Continent. This view was already widespread by the middle of the sixteenth century, following a number of voyages which also included that of Miguel de Legazpi to the Philippines in 1565. It was a vision that prevailed with few alterations until the eighteenth century, even after the Spanish incursions into the Pacific that led to the discovery of the Philippines in 1565, the Solomons in 1567, the Santa Cruz archipelago in 1595 and Vanuatu in 1606.

President of Lima's Audiencia (High Court). There, a distant relation of Mendaña, Pedro Sarmiento de Gamboa, influenced Mendaña's decision to solicit support for an expedition to the fabled Southern Continent. This landmass, Mendaña thought, could be found west of Peru, where King Solomon's ships took the gold with which his famous temple was built in Jerusalem.

16 The first map to use the name Pacific supposedly given to the sea by Magellan is Sebastian Münster's (1546).

Quirós's voyage to Vanuatu was the third voyage of Pacific exploration that the Spanish crown sponsored. The first of these was Álvaro de Mendaña's 1567 journey to the Solomon Islands, which departed from the harbour of Callao in Lima (Figure 7.3). Mendaña always thought that he would find *Ophir*, the land of the Queen of Sheba and, like Quirós after him, was not deterred by the failure of an enterprise that cost not just money but many lives. From his arrival back in Peru in 1568, Mendaña dedicated much time and energy to canvassing support for a further expedition to the Solomons. However, the new viceroy, Don Francisco de Toledo, was not sympathetic towards his aspirations and Mendaña appealed directly to the king, Philip II. After four years of missives, Mendaña went to the court in person and was received by the king, who gave Mendaña the *capitulaciones* (sailing contract), following which he returned to Panama in 1577. Once there, he again encountered opposition from Viceroy Toledo and the expedition was finally approved when Don García Hurtado de Mendoza, Marquis of Cañete replaced Toledo as new viceroy.

Hurtado de Mendoza and his wife, doña Teresa de Castro, stimulated the social life of the town and were often surrounded by many noble locals, among them Isabel de Barreto. Since Isabel was a good friend of Teresa's and Álvaro de Mendaña was close to the viceroy, they met and, in spite of the fact that Mendaña was twice as old as Isabel, married soon after, in May of 1586. With funds obtained from his wife, Mendaña commanded the second journey to the Pacific in 1595, with a group of men and women ready to settle on the islands (see Figure 7.1). Quirós was the pilot of this journey, which failed to find the Solomons, landing instead in the Santa Cruz archipelago, where Mendaña died. The expedition then returned to the Philippines under the command of Isabel Barreto, who thus became the first ever female Admiral.[17] After failing to find the lost ship *Santa Isabel* (also known as *Almiranta*), and with a decimated

[17] To date, the only biography of Isabel, written in 1943 by a self-appointed biographer, Manuel Bosch-Barrett, presents a rather unsympathetic view of his subject matter. Manuel Bosch-Barrett, *Doña Isabel Barreto: Adelantada de las islas Salomón* (Barcelona, 1943). Likewise, Robert Graves' novel, *The Isles of Unwisdom* (London, 1950), portrays Isabel as an attractive, ambitious and manipulative woman, cast in a quasi-Victorian mould. According to Bosch-Barrett, Isabel was born into a Galician family from the northwest Spanish town of Pontevedra who had migrated to Peru when she was a little girl. Her father, Don Francisco Barreto, had been governor in the Portuguese Indies and was obsessed with exploring the region of Monomotapa on the Abyssinian coast where, it was believed, the famed Queen of Sheba collected the gold for King Solomon. Francisco Barreto, however, died in an expedition to this area in 1574. In one of history's ironies, Isabel would be hailed as the Queen of Sheba upon her arrival in the Philippines in 1596, returning from the frustrated attempt to rediscover the Isles of Solomon in which her husband and two of her brothers lost their lives.

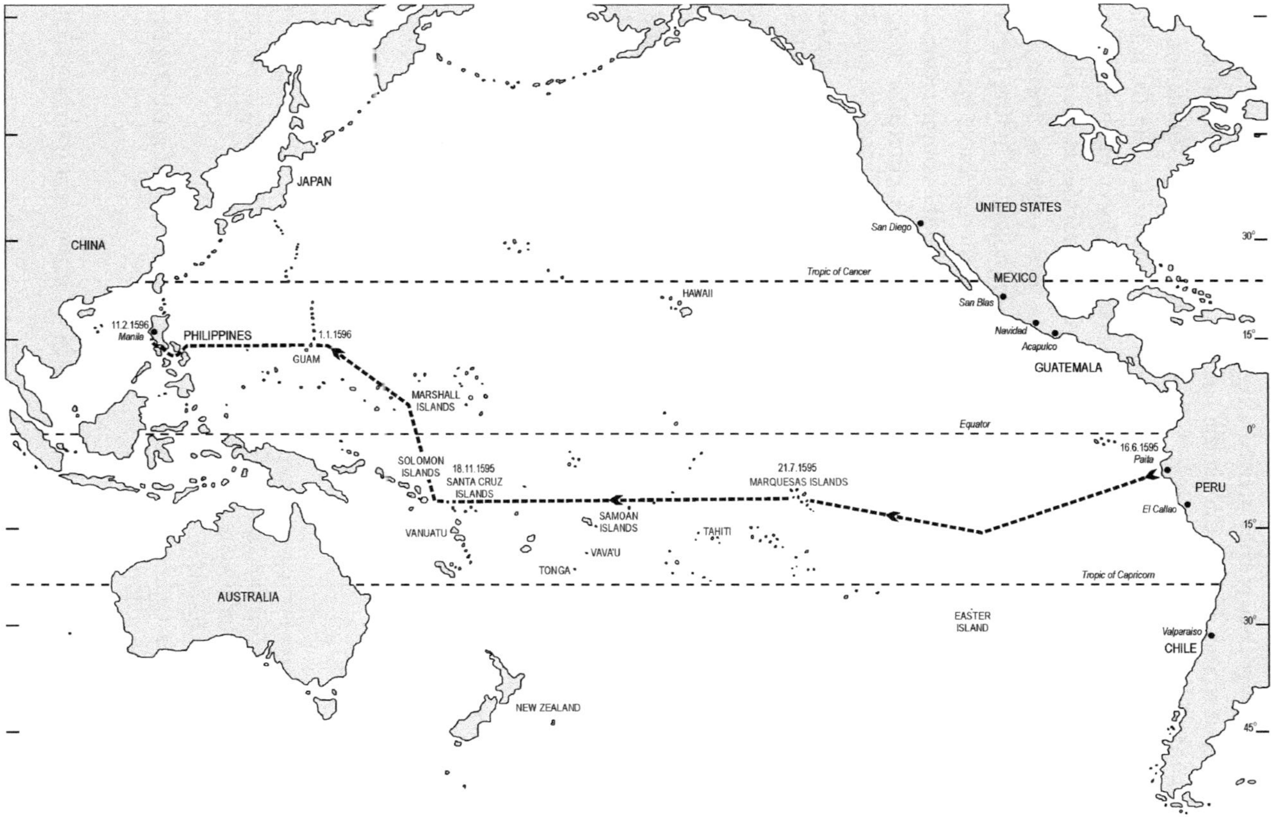

Figure 7.3 Itinerary of Álvaro de Mendaña's 1567 voyage to the Solomon Islands. Reproduced with permission

crew, Isabel decided to head for the Philippines, under the guidance of the pilot, Quirós. Many more people on the ship died from malaria *en route* to the then Spanish archipelago, where they arrived after some mutinous attempts against Isabel were put to rest. In fact, during the last days of the journey, Isabel seems to have contemplated her own death.[18] They finally arrived in Manila on 11 February 1596 where the locals, believing the survivors had come from the Solomons, came out into the streets and called Isabel the Queen of Sheba.[19]

The survivors of this journey were received by the authorities in Manila and by local nobles who considered it an honour to have them as guests. There were few Spanish women in Manila and all the widows on board, including Isabel, soon found replacements for their lost husbands or partners.[20] Isabel married a Spanish nobleman, Fernando de Castro, in November and the couple returned to Mexico where they requested to continue Mendaña's enterprise from Acapulco in 1596.[21] Fernando went back to the Philippines in 1598, by which time Isabel already had a child. There, he asked for licence to return to Spain for eight years and to resume the travels to the Solomon Isles in 1602. When Isabel and Fernando heard, in 1603, that Quirós had been given permission and finances for the aforementioned journey to Vanuatu, they protested their own rights, which they did again after the journey, in 1608, to no avail.[22] Nothing more is known of them so that their fate, like that of Isabel's first husband, Mendaña, became linked to the mythical Islands of Solomon. These islands would only be 'rediscovered' in the eighteenth century, by which time people doubted their existence. In fact, the locations of their landings, as well as the fate of the *Almiranta* and other contemporary shipwrecks, still remain the subject of many hypotheses.[23]

After Quirós's 1606 voyage, the Spanish crown did not sponsor any more voyages until the eighteenth century. Nevertheless, many of the ideas about *Terra Australis Incognita*, including the notion of the 'noble savage', can be traced to some

---

18 'The governess in her retreat seemed to be preparing for death. She had a Book of Hours in her hands, her eyes looking up to heaven and she was praying, and was as afflicted and tearful as everybody else' (Quirós, 'Viajes de Quirós', pp. 253–4).

19 Bosch-Barrett, *Doña Isabel Barreto*, p. 169.

20 Pedro Fernández Quirós, *Descubrimiento de las regiones austriales*, ed. Roberto Ferrando (Madrid, 1986), p. 170.

21 Quirós, 'Viajes de Quirós', p. 287.

22 Celsus Kelly, O.F.M. and Gerard Bushell, O.F.M. (eds), *Austrialia Franciscana*, vol. 6 (6 vols, Madrid, 1973), p. 200.

23 In *The Lost Caravel* (Sydney, 1975) and *The Lost Caravel Re-explored* (Canberra, 1988), Robert Langdon attributes the ethnic differences among Pacific peoples to one of those shipwrecks, that of García de Loaysa's *San Lesmes* in 1525.

of the mystified views that he expounded.[24] Quirós himself, as his contemporaries noted, was imbued with ideas from the early conquest and settlement of the Americas more than one century after Christopher Columbus's momentous arrival in 1492. Indeed, Quirós was rightly compared with Columbus and he continually stressed the supposedly innate innocence of indigenous peoples, very much along the lines of Bartolomé de las Casas.[25] This view contrasted with the presentation of indigenous peoples as possessing inhuman brutality, exemplified by the practice of human sacrifice or cannibalism.[26]

One obvious difference between the mindset of eighteenth-century explorers and their predecessors is that, throughout the settlement of the Americas and as a result of the subjugation of the great Aztec and Inca empires, early Spanish voyagers, by and large, believed in a providential view of the world and of humanity. Unlike their 'enlightened' followers, sixteenth-century explorers could justify colonization by claiming that they were instruments of the divine will to extend Christianity's kingdom. This was the case even when the difficulties of incorporating the discovered worlds into a universal Christian paradigm became immediately apparent to many, and the contradictions inherent in that outlook remained contested and unresolved.

The time when gold and God infused travel and exploration was all but fading away by the end of the eighteenth century, with commercial exchanges and other forms of economic domination gradually gaining ascendance. This was the case initially for the Netherlands, followed by England and, later on, for France and even Spain.[27] Charles III of Spain, who reigned between 1759

---

24 Glyn Williams shows that the growth and spread of publications about the Pacific and the desire to believe in an ideal place on earth infused eighteenth-century explorations, spreading the idea of the 'noble savage' at the time: 'Although the idea of the "noble savage" can be traced back at least to Montaigne's writings, it was during the eighteenth century that it became a cult' (Williams in *The Malaspina Expedition*, ed. David et al., p. 102).

25 Las Casas, who was born in Seville in 1484, became a Dominican friar and an outspoken advocate of the rights of Amerindians throughout his life. He engaged in theological debates with Juan Ginés de Sepúlveda, which culminated in the great debate of 1550 in Valladolid, where Las Casas defended the humanity and rights of Amerindians with passion. The publication of his *Brevísima relación de la destrucción de las Indias* (*A Brief Account of the Destruction of the Indies*) (1552), galvanised public opinion against the methods used by conquistadors and *hacendados* to subdue the Amerindian population.

26 'A Europe newly convinced of the innate sinfulness of man, and increasingly conscious of the need for a powerful state organization to restrain the forces of disorder had little inclination to idealize the virtues of primitive societies' by the seventeenth century (John H. Elliott, *The Old World and the New 1492–1650* (Cambridge, 1970), p. 103).

27 Pagden observes that the French, whose imperialism had initially been informed by Catholicism, eventually, 'shifted their ultimate objectives from the cultivation of souls to the

and 1788, worked closely with ministers clearly steeped in French revolutionary ideology and appreciated the differences between the past empire and his own project.[28] It was during this epoch that Spain underwent a centralizing impulse comparable to that which had taken place in the sixteenth century and, under its Bourbon rulers, tightened the commercial rules to be followed in its colonies. This, as Peter Bakewell observes, has led historians to see the later part of the eighteenth century as the only time when 'Spain's American territories became true colonies in a modern sense of the term: overseas possessions whose own interests were subservient to those of the metropolis'.[29] These differences are apparent in the voyages that Charles III sponsored, the leading men of which, like their European counterparts, embraced the paradigm of scientific observation, even if Christianization remained both the intended and the actual goal of many.

Following upon the new era of Pacific exploration, which was heralded by the important eighteenth-century voyages of George Anson (1740–44), Samuel Wallis (1766–68), Louis Antoine de Bougainville (1766–69) and, especially, James Cook (1768–71, 1772–75 and 1776–79), the Spaniards took a renewed interest in this part of the globe, which they had until then considered their own 'lake'. Commercial and scientific aspirations were undoubtedly reflected in the way the expeditions to the South Pacific from colonial Latin America were mounted and in the composition of the people on board the ships. The changes in the understanding of the world and the way travellers and explorers approached it are noticeable in the use of scientists in the voyages. In this regard, the voyages of Cook and Bougainville established a model that was readily followed thereafter. For the first time, key scientists, often from different nationalities, travelled on the ships. These included the famous Joseph Banks, Daniel Solander, Johann Reinhold Forster and George Forster on Cook's expeditions or Tadeo Haenke and Luis Née in the Spanish-sponsored voyage led by the Italian Alessandro Malaspina (1789–94).[30]

cultivation of land and the opportunities for trade ... By the second half of the eighteenth century all three empires, even that of Castile, came to be seen by their respective mother countries as predominantly commercial enterprises' (Anthony Pagden, *Lords of All the World: Ideologies of Empire in Spain, Britain and France c.1500–c.1800* (New Haven, CT and London, 1995), pp. 37 and 73).

28 Ibid., p. 194.

29 Peter Bakewell, 'Spanish America: Empire and Its Outcome', in John H. Elliott (ed.), *The Hispanic World: Civilization and Empire: Europe and the Americas: Past and Present* (London, 1991), pp. 65–84, 79. In fact, Bakewell goes so far as to label the last years of Ferdinand VI's reign (1746–59) and that of Charles III (1759–88), as 'a second reconquest of Spanish America', which he contrasts with the long period from 1600 to 1750, in which the empire underwent a continuous process of decentralization (pp. 76–9).

30 Besides navigators like Juan Antonio Gutiérrez de la Concha, 'The scientific staff chosen by Malaspina consisted of the Spanish army officer Antonio Pineda as chief of natural

With these scientists travelled artists, whose work included maps, bird's-eye views and artistic representations of what they saw, including flora and fauna, landscapes and the peoples encountered.[31]

Eighteenth-century Spanish voyages of Pacific exploration have not been very widely documented. During this time, four significant Spanish explorations took place besides the better-known scientific journey of Malaspina, on which the last section of this chapter will concentrate.[32] These five voyages, four of which were sponsored by the Spanish crown and organized by the Viceroy of Peru, Manuel de Amat, departed from the Peruvian port of El Callao in Lima, with the first one directed to today's Rapa Nui or Easter Island in 1770. This expedition was commanded by Felipe González de Haedo, with Juan Hervé as pilot, who drafted some remarkable bird's-eye views of the island. The next two Spanish-sponsored voyages were directed to Tahiti, and took place in 1771–72 and 1774–75 respectively. The first voyage was led by Domingo de Boenechea, who also commanded the second one, in which he died on the island, being succeeded by Tomás de Gayangos for the return journey.[33]

history, with French-born Luis Neé, a naturalized Spaniard, as his assistant. The landscape artist José Guío completed the scientific team. At the King's suggestion the services of the brilliant Bohemian-born natural scientist Tadeo Haenke was also obtained' (David, 'The Voyage of Alejandro Malaspina', p. 5). Pineda died of heat exhaustion on the voyage, in the Philippines, where he was buried, in March 1972. The artists, Juan Ravenet and Fernando Brambila, joined the expedition in Mexico, in October 1792.

31 It is worth remembering in this context Denis Cosgrove's now-classic formulation about the landscape: '[T]he landscape idea represents a way of seeing – a way in which some Europeans have represented to themselves and to others the world about them and their relationships with it, and through which they have commented on social relations. Landscape is a way of seeing that has its own history, but a history that can be understood only as part of a wider history of economy and society; that has its own assumptions and consequences, but assumptions and consequences whose origins and implications extend well beyond the use and perception of land; that had its own techniques of expression, but techniques which it shares with other areas of cultural practice. The landscape idea emerged as a dimension of European elite consciousness at an identifiable period in the evolution of European societies: it was refined and elaborated over a long period during which it expressed and supported a range of political, social and moral assumptions and became accepted as a significant aspect of taste' (Denis Cosgrove, *Social Formation and Symbolic Landscape* (Madison, WI, 1998), p. 1).

32 María Dolores Higueras grouped the materials related to this journey, cataloguing them in three volumes and also edited Malaspina's journal (1985). Quotations from Malaspina's journal in this section are taken from the translation and edition from the Hakluyt Society (2004).

33 For a description of the cross-cultural encounters in these and the earlier voyages of Mendaña and Quirós, see Mercedes Maroto Camino, *Exploring the Explorers: Spaniards in Oceania 1519–1794* (Manchester and New York, 2008).

The last two voyages of Spanish exploration to the Pacific reached the Vava'u group of islands in the Tongan archipelago. The earliest of these was led in 1780–81 by Francisco Mourelle de la Rúa, the first European ever to land in those islands. The same group of islands was visited in the most 'enlightened' eighteenth-century Spanish voyage to the Pacific: the scientific journey led by Malaspina in 1789–94. Malaspina's fleet consisted of two ships, *Descubierta* and *Atrevida*, which were commanded by Malaspina and José de Bustamante respectively. As part of the four-year voyage, they travelled from the North Pacific to the Philippines, New Zealand, Vava'u and Australia, landing in Port Jackson and visiting Sydney, Parramatta and Botany Bay in 1793. Malaspina's fleet arrived in the Pacific from Alaska and Mindanao in 1793. Initially, they sailed to New Zealand's Dusky Sound but because of the weather, they had to anchor off nearby Doubtful Sound. As the bad weather continued, they abandoned their task, though they left some names on the New Zealand landscape, which are still preserved (Figure 7.4).[34]

Even though there were periods in Malaspina's voyage during which imperial goals remained important, the segment of the long voyage at Port Jackson and in Vava'u, was mostly devoted to inspection and those forms of scientific exploration associated with the Enlightenment. Prominent here was investigation of flora and fauna, as well as the lifestyle of the indigenous peoples, although they seem not to have had much direct contact with Australian Aborigines, who nonetheless found themselves as 'ethnographic sitters', as well as anonymous workers on ships.[35] Similarly, Australian convicts made a surreptitious appearance in one of the drawings, from which their chains were eloquently removed in the version given to the Australian Governor, Philip Grose, to avoid offending him.[36]

---

[34] Cf. John Robson, 'Somebody Knows What!: The Cartographic Results of the Visits by Vancouver and Malaspina to New Zealand in the 1790s', *New Zealand Map Society Journal*, 15 (2002): 35–47. Cutter believes Malaspina's aims to have been twofold, and to have been involved in espionage: 'Although when Malaspina's ships reached Port Jackson the visiting Spaniards might easily have been viewed as agents of an alien power, on the surface they were received most civilly in acknowledgment of the expedition's role in the advancement of science ... the visit coincided with a brief period when Great Britain and Spain were allies ... there is little doubt but that he was engaging in what at a later date might be called espionage' (Cutter, 'Introduction', p. lxxi).

[35] 'He described their physical appearance in some detail, but took much of his comment on their customs from Cook's report of his contact with them in 1770, agreeing with him that they were "the most miserable and least advanced nation which exists on earth" ... without agriculture and industry, and without any product which would attest their rationality' (Cutter 'Introduction', p. lxxii).

[36] See Peter Barber, 'Malaspina and George III, Brambila, and Watling: Three Discovered Drawings of Sydney and Parramatta by Fernando Brambila', *Australian Journal*

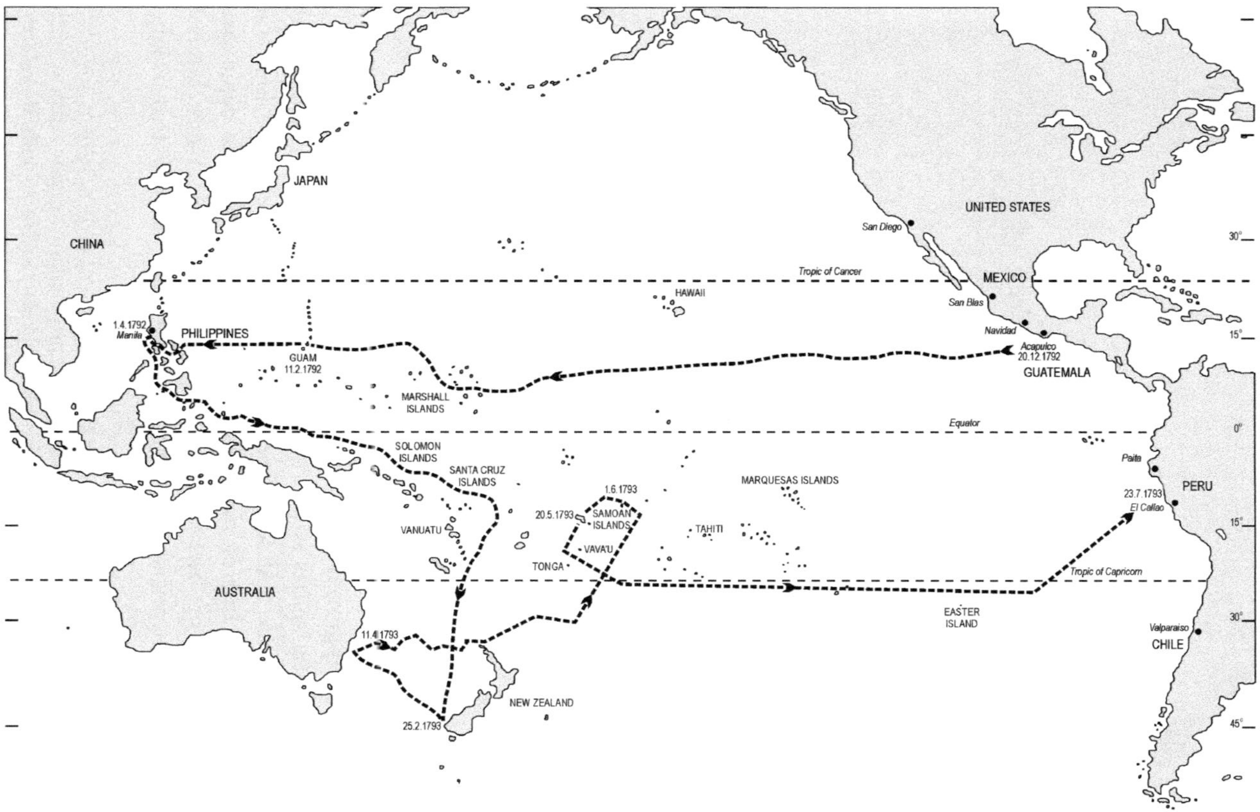

Figure 7.4 Itinerary of Alessandro Malaspina's 1789–94 voyage. Reproduced with permission

Figure 7.5 Luis Planes, *Baile de los hombres en Vavao* (Men's Dance in Vava'u). Madrid, Museo Naval, MS 1724 (12). Reproduced with permission

While some draughtsmen accompanied the expeditions to Easter Island and Tahiti that were more or less contemporaneous with those of Cook, Malaspina's journey mirrored quite closely Cook's and Bougainville's with the inclusion of botanists, as well as artists, such as Felipe Bauzá, Luis Planes and the Italians Fernando Brambila (Ferdinando Brambilla) and Juan Ravenet, who joined the expedition in 1792. Their drawings help scholars identify the scenes in the journals, giving also a visual dimension of the socio-cultural parameters embedded in their approach to cultures different from their own.

The artists of Malaspina's expedition produced a large number of drawings and sketches of people and scenes of Vava'u, including coastal profiles and ceremonial gatherings. Interestingly, in some of these 'ethnographic' depictions the explorers' position of observers is clearly highlighted. These images thus not only provide interesting social data, but are also a salutary reminder that, as John

of Art, 11 (1993): 31–55, 88. Differences between original and revised versions also offer some curiosities. 'Occasionally, different versions were prepared by the same artist, as when Brambila during the ships stay at the new British settlement in New South Wales drew two versions of a scene. One, which was sent to Spain, showed some of the convicts as a chain gang; the other, presented to the governor, Philip Grose, omitted the chains' (Cutter, 'Introduction', pp. lx–lxi).

Figure 7.6 Luis Planes, *Baile de las mujeres en Vavao* (Women's Dance in Vava'u). Madrid, Museo Naval, MS 1724 (13). Reproduced with permission

Berger remarks, 'We never look at just one thing; we are always looking at the relation between things and ourselves' and that 'we situate ourselves in' relation to what we see.[37] This can certainly be said of the scenes presented, where we share the position with the explorers who view them. Furthermore, here the men are strategically located in the foreground, providing, as it were, a point of entry into the view, representing themselves in a privileged position as mediators between the subjects who view the images and the objects of observation. This feature is clearly embedded through the location of Europeans in these ceremonies performed in their honour, where they stand between the painter and the viewer to whom these are addressed. Needless to say this viewer is not just there observing the scene but also somewhere in Europe, assessing the scene thus mediated (Figures 7.5 and 7.6).

The drawings of these scenes share the dominant belief taking hold of intellectuals throughout the latter part of the eighteenth century, that the world ought to be observed carefully in its minutiae and that this observation would reveal the hidden rules underpinning it. Interestingly, the same tenets that applied to the observation of nature were extended to geography and non-European cultures, all visible from the gaze that Mary Louise Pratt has incisively

37 John Berger, *Ways of Seeing* (London, 1972), pp. 5, 11.

labelled 'imperial eye'. This 'imperial eye' cast its gaze upon a world which might be apprehended in a panorama-like view and could be subsequently possessed and controlled by human forces. The typical, masculine observer, whom Pratt has labelled 'seeing-man', would be a paradigmatic explorer-cum-scientist white male whose personal gaze presented the world from a point of view that was assumed to be universal.[38]

By comparison with Quirós's ceremonials of possession, eighteenth-century rituals emphasize the attempt to understand the other, while still considering oneself and European mores as superior. In this, the explorers were attempting to show respect for the local customs, which they observed as much as they took part in them. They could thus be said to have taken part in a Western 'gazing act' of 'participant observation'. As the name suggests, participant observation requires that one take part in the cultures described, while attempting to accept them on their own terms. Nonetheless, as Clifford remarks, observers cannot but be outsiders who must, of necessity, remain beyond the cultures under scrutiny.[39]

The ceremonies surrounding this journey were mostly staged to honour the visitors, who spent ten days on the island. They were almost always performed by the indigenous people, although the explorers corresponded with the display of military drills. On the day after they arrived, on 21 May 1793, the visitors were met with a kava ceremony, which was followed by the chanting of 'some twenty men ... to the beat of split or hollow canes'. These men were inside a circle 'divided into three sections for men, women, and youths', which made their welcoming performance rather harmonious, as Malaspina was duly appreciated.[40]

The kava ceremony that preceded this celebration was not as satisfying to the explorer, though Malaspina tried to show respect for the local customs. This event took place after the scientist Haenke and the draughtsman Bauzá returned from a short trip to explore the island, when many of the locals gathered to share kava, the preparation of which, as had happened with the previous voyager, Mourelle, disgusted Malaspina. In spite of their repugnance at the fact that a

---

38 Instead, Pratt reminds us that he was nothing if not a 'traveller which deploys an imperious and imperial way of perceiving and viewing from above, offering a visual topography from heights which suggest a masculine and imperial possession of landscape and peoples' (Mary Louise Pratt, *Imperial Eyes: Travel Writing and Transculturation* (London and New York, 1992), p. 7).

39 James Clifford, *The Predicament of Culture: Twentieth-century Ethnography, Literature, and Art* (Cambridge, MA and London, 1988), p. 93. The standard method of fieldwork, which was first set down by Malinowski in the Trobriand Islands, requires the presence of the ethnographer, who observes and participates in the culture, learning and speaking the language. Bronislaw Malinowski, *Argonauts of the Western Pacific: An Account of Native Enterprise and Adventure in the Archipelagoes of Melanesian New Guinea* (London, 1922), p. 25.

40 *The Malaspina Expedition*, ed. David et al., p. 107.

woman chewed the root to produce fermentation, the Spaniards took the drink 'with the required ceremony' to show their appreciation of indigenous culture.[41] The voyagers sat with Vava'u's chief, Vuna, accompanied by some local women, spending the afternoon 'in the greatest concord and cheerfulness' while the chief's nephew, Feileua, and an older boy who seemed to tutor him, Latu, acted as go-betweens. Food and female company, as well as meetings, were part of the encounter with Malaspina highlighting the kindness of the women of those 'happy climes'.[42]

The following day, 22 May, the Spaniards were honoured by further 'amusements', which were announced to them by Feileua. On this occasion, Feileua indicated that the festivities would take place in the afternoon, when a 'chorus of young women' sat at the centre of a group. The Spaniards corresponded with gifts, which were followed by 'a large meal of roots sweetened with sugar'. Then, the men took their turn, with 30 of them performing 'to the beat of the canes'. After 'pyramids of foodstuffs' were consumed, around a dozen women performed, although this time 'nervously', as they were obliged to do it by Vuna against their will[43] (see Figures 7.5 and 7.6).

The Spaniards eventually corresponded with some ceremonies of their own, which the indigenous people admired. On 25 May, five days after they had been on the island, people sat again in circles, the explorers made a display with their weapons, first marching and then firing some volleys in different directions. The islanders then reciprocated by playing instruments, while a choir of 32 men sang, following which some 60 men staged a battle. This was followed by more dances, firstly those 'performed by at least forty men' and then 'fifty women' whose skills were admired as 'even finer than those we had already seen', as they showed 'modesty and shyness'.[44] Their movements inspired Malaspina to compare them with 'a vivid representation of the Golden Age', which was by then a widely-used analogy in many contemporary descriptions of Pacific life.[45]

---

41 Ibid., p. 107.

42 Ibid., p. 113. 'It would be difficult, without being accused of the exaggeration so often found in traveller's tales, to describe the attractiveness with which fortune has endowed the fair sex in these happy climes, whose charms we still had only a very imperfect notion.' Curiously, however, there seems to have been a good degree of restraint on the part of the sailors, which was often enforced by those in command, using strict punishments for those who sought to have intercourse with local women. Again here, both in terms of respect and admiration for local women, the journeys of Mourelle and Malaspina exemplify the eighteenth-century approach.

43 *The Malaspina Expedition*, ed. David et al., pp. 113–18.

44 Ibid., pp. 128, 129.

45 Ibid., p. 129. Malaspina likened their performance to those taking place in the temples of 'Cnidus and Amathus', which was Aphrodite's sanctuary in Cyprus.

These encounters provide early examples of 'participant observation' when Malaspina and his officers watched the women and the men dancing, or when they took part in the kava ceremony. Like gift-giving, ceremonies were often a way to display degrees of authority and of cementing closer relationships between visitors and hosts. It is interesting to note in this context that, as seen in the images of the festivities, the alternation between male and female is matched evenly, and both are equally located within the explorers' field of vision. Effectively, this places the indigenous peoples in the position of women *vis-à-vis* the explorers or the scientists who view and describe them. This feature is corroborated by the way the Spaniards represented two named islanders, Feileua and Latu, in individual portraits (Figures 7.7 and 7.8).

The explorers had practical reasons for appreciating the friendship of Vuna and of his nephew, Feileua, a boy around eight to ten years old, because, whenever these two were on board, any items stolen were quickly restored to them. Indeed, Feileua became good friends with the Spanish and, as a sign of his friendship, exchanged names with José Espinosa and acquired Spanish dress. Both Feileua and Latu, the older boy who accompanied him, appear in portraits by Juan Ravenet, which are significant in the depiction of their gentleness and quasi-feminine appearance (see Figures 7.7 and 7.8). This can indeed be appreciated when compared with the representation of one of Vuna's *fatafegis* (wives) (Figure 7.9).

In Ravenet's portrait, Latu holds the loincloth opened for the viewer, so that the central and first point of access to the painting is his groin. His demeanour, however, sharply contradicts his supposed masculinity, as the gentleness of his 'unveiling' is matched by the expression in his face. Along those lines, Ravenet's drawing of Feileua is likewise feminized, with the countenance, his arms folded together and general demeanour clearly feminizing the boy.

The intricate connection between art, science and power that was at the centre of these voyages is apparent in the drawings thus produced. In fact, the extent to which hegemonic relations underline these eighteenth-century products can be clearly appreciated in the scenes presented. These drawings freeze reality at a specific moment, removing it from the flux of time and change, and rendering it the property of the observer.[46] They therefore locate the other in a temporal space distant from the speaker's and sum up the different notions of reality that both cultures embraced, as well the role played by class and gender in such understandings.[47]

---

[46] Cosgrove, *Social Formation and Symbolic Landscape*, p. 22.

[47] On the temporal distance created by the ethnographer or observer and their subjects, see Johannes Fabian, *Time and the Other: How Anthropology Makes Its Object* (New York, 1983).

Figure 7.7 Juan Ravenet, *Fayleua, hijo de Vuna, Principe Eredero de las Yslas de los Amigos* (Feileua, son of Vuna, heir to the Friendly Islands). Madrid, Museo Naval, carp. I (34). Reproduced with permission

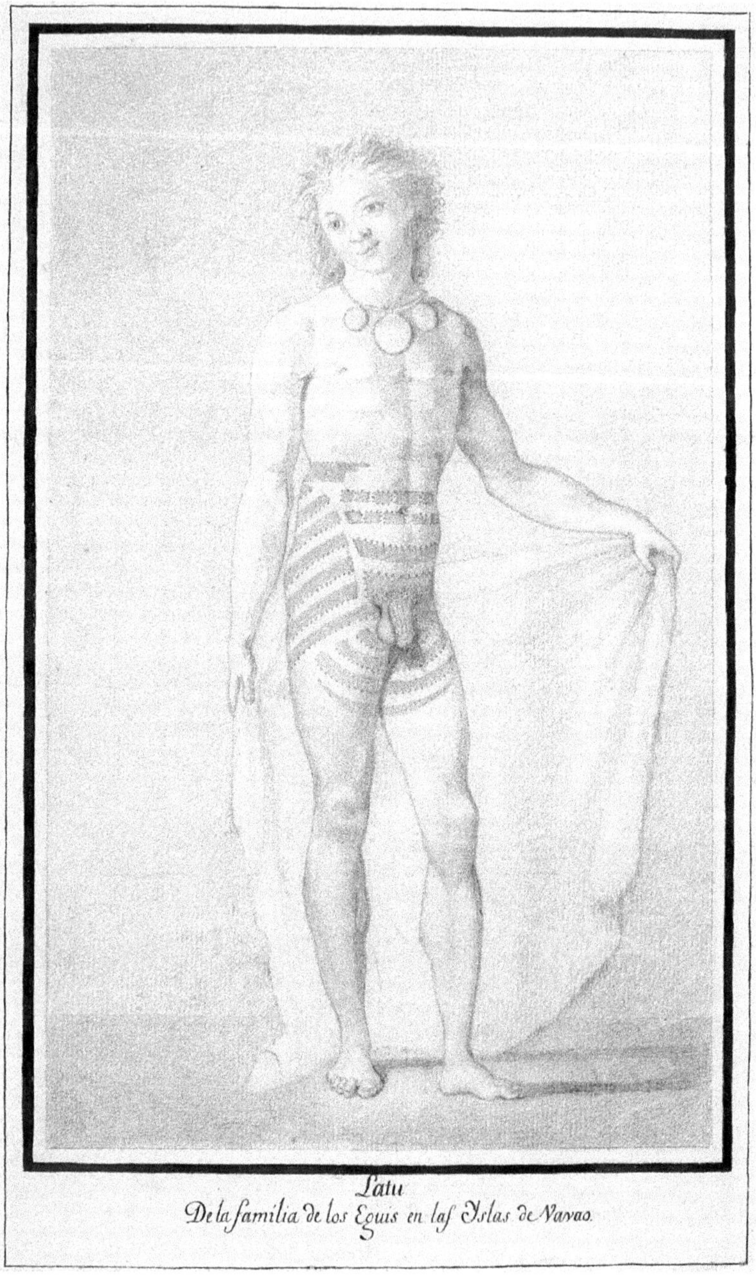

Figure 7.8 Juan Ravenet, *Latu, De la familia de los Eguis en las Yslas de Vavao* (Latu, from the chiefs' family in the Vavao Islands). Madrid, Museo Naval, carp. I (37). Reproduced with permission

Figure 7.9 Juan Ravenet, *Fatafegui, De la familia de los Eguis en las Yslas de Vavao* (Fatefegi, from the chiefs' family in the Vava'u Islands). Madrid, Museo Naval, carp. I (38). Reproduced with permission

Figure 7.10 Juan Ravenet, *Señorita de Panamá en una hamaca*. Madrid, Museo de América. Reproduced with permission

These notions are corroborated by one of the most unusual drawings produced on this journey. Wishing to be given some utensils, Vuna insisted on offering women in exchange 'for our use', a method which Malaspina classed as the weapon Vuna thought to be more effective with them.[48] At this point, and to counter the offer of local women, the Spaniards asked Ravenet to produce a likeness of an ideal European woman, which he did, drawing 'a woman graced with all the personal charms usually admired in our Europe ... dressed in imitation of the ladies of Panamá and reclining negligently in a hammock' (Figure 7.10).[49]

The Spaniards showed this portrait to Vuna, telling him that the wives of the officers are just as beautiful but would not go with them because of their delicate constitution.[50] On seeing the portrait, Vuna, assuming the woman to be real, proclaimed his wish to meet and possess her. In exchange, Vuna offered as many Vava'u women as the visitors might wish to have. Malaspina interpreted the scene with a summary of the relationship between painting and original,

[48] *The Malaspina Expedition*, ed. David et al., p. 122.

[49] Ibid., p. 123.

[50] Ibid. The women 'did not accompany us because we considered the hardships of life at sea to be far too great for their delicate constitutions.'

stressing how Vuna would prefer the 'imaginary subject of our tableau' to local women, which the Spaniards would desire 'with good reason'.[51]

The reference to this woman's beauty places the artist and those contemplating the portrait in the role of observers, very much reminiscent of Albrecht Dürer's *Draughtsman Doing Perspective of a Reclining Woman*. The position of both women further suggests not only that they are objects of display but that, as I have stressed in an earlier reading of Dürer's drawing, they are a 'kind of landscape'.[52] This woman, clearly bounded by her class and not physically active, is presented as the epitome of beauty for contemporary Europeans. Dressed in apparel that appears to be soft taffeta or muslins, she is slightly orientalized.[53] Although presented as European, desirable and belonging to the privileged classes, her presentation and clothing make her resemble a woman in a harem, while her placement on a hammock locates her firmly as a colonial subject of the Americas.

The significance of the hammock and its prominent position as regards colonialism has been scrutinized in relation to the Americas by Tom Conley, who observes the recurrence of the artefact in the frontispiece of Abraham Ortelius's Atlas, *Teatrum Orbis Terrarum* (1570) and the paradigmatic representation of the arrival in the Americas in Jan van der Straet's 'Vespucci Discovering America'. For Conley, 'the weave of the hammock [is] one of the principal attributes of Amerindian culture ... [and] can be seen among "effects" brought back to Europe from the New World'.[54] The portrait is thus an example of a way of defining the self by opposition to the other, in this case the Spaniards by opposition both to their own women and to indigenous people like Vuna.

Differences notwithstanding, eighteenth-century explorers, like their predecessors, relied on a paradigm that associated the landscape with the indigenous peoples of the places encountered, feminizing their embodiment. Throughout these examples, to represent is to assume the position of the observer in a duality in which indigenous peoples, women and the landscape are observed and denied agency. Any claim of realism that these drawings present is in fact

---

51 Ibid.

52 On this drawing's objectification of women, see Mercedes Maroto Camino, 'A Waxen World: Early Modern Women and Geographical (Un)Awareness', *Parergon*, 16.2 (1999): 101–32.

53 Edward Said's classic definition of Orientalism is that of 'a Western style for dominating, restructuring, and having authority over the Orient' (Edward Said, *Orientalism* (London and Harmondsworth, 1978), p. 3).

54 In this way, Conley concludes, 'The conquest of the New World is engineered by a process of gridding that is common to both reason and mapping ... [so that] Ortelius' hammock [may be seen] as an abstracted figure that folds together or conflates into one form both the cause and effect of conquest' (Tom Conley, 'Pierre Boaistuau's Cosmographic Stage: Theater, Text and Map', *Renaissance Drama*, 23 (1992): 59–86, 76.

ideological.[55] It is, to borrow John Berger's classic formulation about Western art (1972), a 'way of seeing' or a 'gazing act' performed by the explorers and the intended viewers of the scenes. The way this gazing act underlies both the earlier and later voyages is brought home by Michel de Certeau's now-classic reading of Jan Van der Straet's representation of Vespucci's arrival in the Americas:

> Amerigo Vespucci the voyager arrives from the sea. A crusader standing erect, his body in armor, he bears the European weapons of meaning. Behind him are the vessels that will bring back to the European West the spoils of a paradise. Before him is the Indian 'America,' a nude woman reclining in her hammock, an unnamed presence of difference, a body which awakens within a space of exotic fauna and flora ... This erotic and warlike scene has an almost mythic value. It represents the beginning of a new function of writing in the West ... This is *writing that conquers* ... For the moment of a rupture between a subject and an object of the operation, between a *will to write* and a *written body* (or a body to be written), this writing fabricates Western history.[56]

As Certeau mentions, from the early voyages, colonialism was premised on a division between those who could write and those who did not.[57] Like Vespucci, his eighteenth-century counterparts carried with them the symbols of European power, including their scientific artefacts and their ships. Although the cruciform staff is nowhere to be seen, Vespucci's navigational astrolabe became an Arnold chronometer, while the banner bearing the Southern Cross was replaced by an ensign with the insignia of the royal or private sponsors. Also, Vespucci's large sword was removed from sight, although the manoeuvres and the displays of military skills would certainly give the locals a perception of the visitors' might.

From the three early modern discourses of mastery: religious truth, scientific truth and military power, only the first remains in the background by the eighteenth century, while the second and the third are, if anything, intensified. Whereas religious truth had become increasingly questionable, the claim of

---

55 Cosgrove, *Social Formation and Symbolic Landscape*, p. 26.

56 Michel de Certeau, *The Writing of History*, trans. Tom Conley (New York, 1988), pp. xxv–xxvi.

57 As Stephen Tyler posits: 'To represent means to have a kind of magical power over appearances, to be able to bring into presence what is absent, and that is why writing, the most powerful means of representation, was called "*grammarye*," a magical act. The true historical significance of writing is that it has increased our capacity to create totalistic illusions with which to have power over things or over others as if they were things. The whole ideology of representational signification is an ideology of power' (Stephen Tyler, 'Post-modern Ethnography', p. 131).

reality of scientific truth, often backed by military power, remained a cornerstone of European thought. As this chapter proposes, this is exemplified by the Spanish perceptions of *Terra Australis* and its inhabitants, which presupposed a notion of the universe in which nature, indigenous peoples and women are subjected by the 'universal male' that 'produced' them. This universal, 'disembodied' viewer, as Daniel Cosgrove reminds us, was 'endowed with a will to power' and was 'thus the sovereign subject of history'.[58] This man objectified, mystified or idealized imagined women who, like the new lands and peoples encountered, was left behind when the ships departed.

[58] 'Humanist values were promoted as a universal and progressive achievement, to be adopted with time and "development" by all peoples, an element in the construction of a modern global identity. Central to this progressive narrative of human achievement has been the figure of the individual European male, conceived as a universal subject, exercising rational self-consciousness within a largely disembodied mind, and endowed with a will to power: thus the sovereign subject of history' (Cosgrove, *Social Formation and Symbolic Landscape*, p. xvii).

Chapter 8

Naming and Shaming: The Baudin Expedition and the Politics of Nomenclature in the *Terres Australes*

Jean Fornasiero and John West-Sooby

When the Baudin expedition set out from Le Havre, on 19 October 1800, on a voyage of discovery to the *Terres australes* or Southern Lands, one of the principal objectives set down in its instructions – indeed its prime objective – was geographical. And the expedition's mission in this regard was an ambitious one, especially given the time allotted to the task: in the space of 15 months, from the end of March 1801 to the end of June 1802, the French commander was to visit Tasmania and Bass Strait before undertaking a detailed survey of extensive sections of the southern, western and northern coastlines of mainland Australia, with the western part of the island of New Guinea thrown in for good measure.[1] This was, in effect, a mission to complete the contours of the Southern Lands, and, in passing, to dispel lingering myths about *Terra Australis* while at the same time resolving more recent speculations about a divided land mass – speculations that had been aroused by the starkly contrasting shores of New Holland and New South Wales. While circumstances would conspire to prevent the planned charting of the Gulf of Carpentaria and adjacent sections of the northern coastline, Baudin's hydrographers did succeed in producing detailed charts of many other parts of the continent, and in particular of the south and west coasts, where they

[1] This geographical objective is announced in the opening sentence of the 'Plan of Itinerary for Citizen Baudin', drawn up by Count Fleurieu: 'The aim of the Government ... has been to have examined in detail the *south-west, west, north-west* and *north* coasts of *New Holland*, some of which are still entirely unknown, while others are known only imperfectly.' The 'Itinerary' has been translated by Christine Cornell and incorporated in her translation of Baudin's sea-log, which was recently updated and re-issued as *The Journal of Post Captain Nicolas Baudin Commander-in-Chief of the Corvettes* Géographe *and* Naturaliste (Adelaide, 2004 [1974] (see p. 1)). It should be noted that, in all the French documents of the period, the 'south-west' coast designates the entire length of continental Australia's southern coastline, including that section which at the time was still unknown.

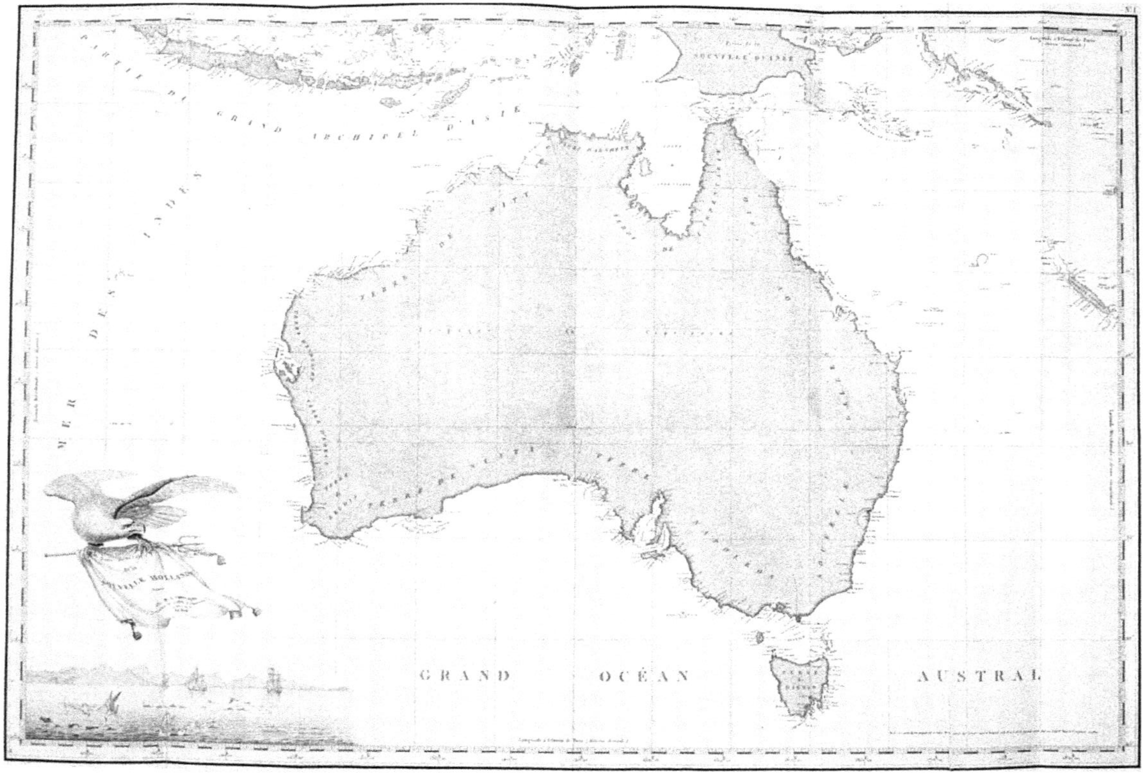

Figure 8.1 Louis Freycinet, 'Carte Générale de la Nouvelle Hollande' (1808), published in the *Atlas* of the *Voyage de découvertes aux Terres Australes* (Paris, 1811). Reproduced with permission

also made a number of important discoveries. As a result, Louis Freycinet was subsequently able to fulfil his government's wishes by publishing in 1811 the first complete map of the *Terres australes*, thereby making a first bid to establish the modern cartographic identity of Australia (Figure 8.1).[2]

This was a major achievement. However, like so many of the expedition's accomplishments, the ground-breaking work it had undertaken in charting and naming the often dangerous coasts of a little-known continent was destined to become overshadowed by controversy. Instead of bringing honour and glory to the French nation, the publication of the maps from the Baudin expedition would initially lead to embarrassment and shame. The controversy in this case centred on the matter of nomenclature, and more specifically on the plethora of French names that dotted the expedition's maps of the previously uncharted south coast.[3] Given the long-standing convention according to which naming rights belonged to the discoverer, the British were aggrieved at what they took to be a blatant and perfidious dismissal of their own discoveries in that region, particularly those of Matthew Flinders, whom the French expeditioners had encountered in 1802 and of whose prior claims they were well aware. The fact that the French labelled this vast stretch of land 'Terre Napoléon' and that many of the names on their maps of its coastline had Bonapartist resonances only added insult to injury, given the context of war and the later imprisonment of Flinders by the French in Mauritius (Figure 8.2).

The main aspects of this particular controversy have been discussed at some length by historians of the Baudin expedition, with the result that many of the

---

[2] The publication history of the official account of Baudin's voyage is somewhat complex. The first volume of the historical narrative, written by the expedition's chief zoologist François Péron, was published in 1807 under the title *Voyage de découvertes aux Terres Australes, exécuté par ordre de Sa Majesté l'Empereur et Roi sur les corvettes le* Géographe, *le* Naturaliste, *et la goëlette le* Casuarina, *pendant les années 1800, 1801, 1802, 1803 et 1804* (Paris, Imprimerie Impériale). It was accompanied by an *Atlas* comprising two parts: the first part, published likewise in 1807, contained various engravings (portraits of native inhabitants, drawings of animals, coastal profiles); while it had been intended that two maps, of Australia and 'Terre Napoléon', should be included in this *Atlas*, their publication was delayed and they were eventually produced, along with a number of other maps, in a second, companion *Atlas*, published in 1811. Volume II of the historical narrative, some chapters of which Louis Freycinet compiled following the death of Péron in 1810, was published in 1816. In the interim, Freycinet had published his volume of navigational and hydrographic observations – *Navigation et géographie* (1815) – which was preceded by the publication of a new, companion atlas of maps (dated 1812 but not actually published until 1814). A second edition of the historical narrative, revised and enlarged, was published in 1824, also with an atlas.

[3] Péron indicated many of these in his narrative, but a great many more were revealed on the detailed maps contained in Louis Freycinet's *Navigation et géographie* atlas (see previous note).

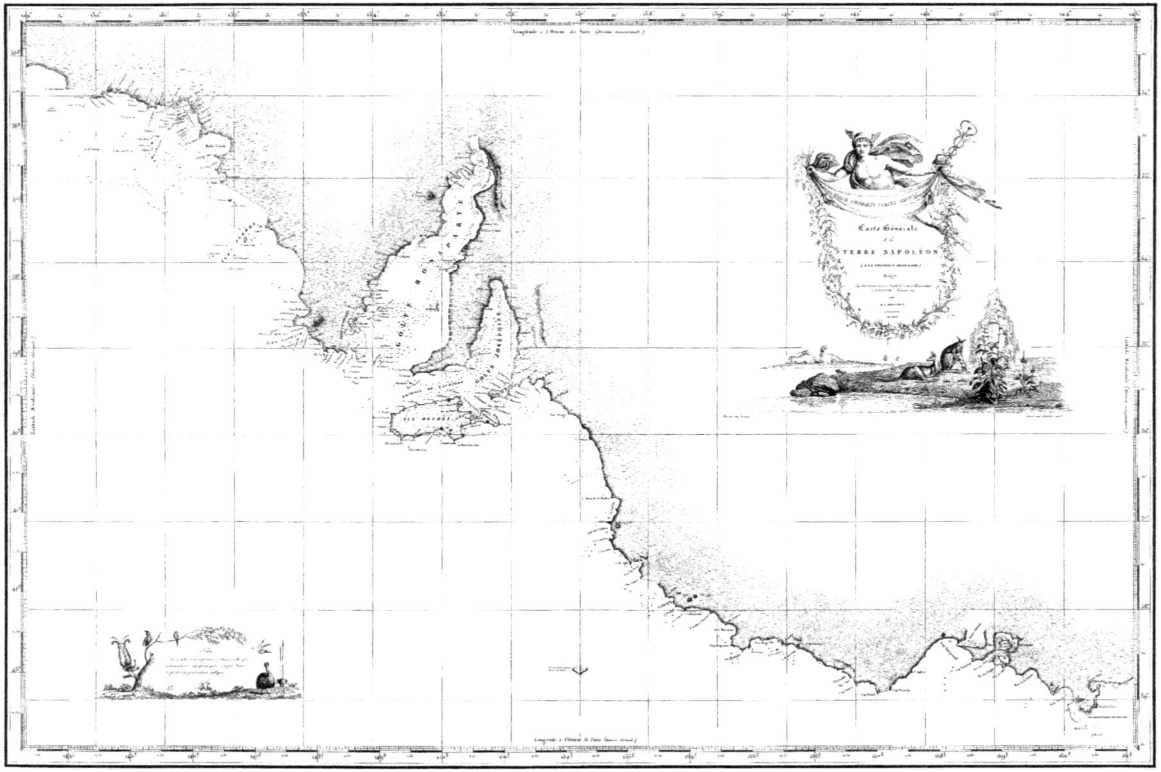

Figure 8.2 Louis Freycinet, Map of 'Terre Napoléon' (1808), published in the *Atlas* of the *Voyage de découvertes aux Terres Australes* (Paris, 1811). Reproduced with permission

original accusations of French treachery or plagiarism no longer hold.[4] The maps of the Baudin expedition can now be viewed for their own qualities, and as a legitimate first representation of the Australian continent, rather than as a tainted product.[5] Various negotiations over the years have also restored some balance between the competing French and British claims for naming rights on the 'unknown coast'.[6] However, before the matter of nomenclature can be dismissed entirely, there remains one major question to resolve: that of the responsibility for the controversial names on the French maps of 1811. These names were not, in fact, those originally bestowed on the southern Australian coastline by Nicolas Baudin, and which appear both in his journal and on the manuscript charts prepared under his supervision. Baudin's names were systematically changed by those who were charged with compiling the official narrative and maps following the death of the commander – his travelling companions and bitter enemies, François Péron and Louis Freycinet, who deliberately sought to shame Baudin by not naming him at all, either in their narrative or on the maps. This means not only that the controversial names bear no relation to the principles and practices of the commander, and hence to the original intentions of his mission, but also that, as hydrographer Greg Eccleston has recently proposed, there might well be a case to be made for replacing Péron's and Freycinet's names on Australian maps with those of Baudin, in view of the priority of his nomenclature.[7]

In order to pass judgment on the objectives and outcomes of the Baudin expedition, particularly in the hydrographic domain, we therefore need to be in a position to distinguish between the names that were originally bestowed on features of the Australian coastline and those that were adopted for political reasons at a later date. Initial research by historians has cast some light on both

---

4 Ernest Scott was the first to cast light on this episode, in *Terre Napoléon: A History of French Exploration and Projects in Australia* (London, 1910), chapter 5, and *The Life of Captain Matthew Flinders, R.N.* (Sydney, 1914), chapters 22–6. Subsequent studies by Geoffrey Ingleton, *Matthew Flinders, Navigator and Chartmaker* (Guildford, Surrey, and Australia, 1986), and Frank Horner, *The French Reconnaissance: Baudin in Australia 1801–1803* (Carlton, 1987), have reviewed, confirmed and amplified Scott's findings.

5 Ingleton, *Matthew Flinders*, p. 295, notes that Freycinet's maps have 'a special place in the history of Australian cartography' – and not merely for their beauty. He declares that the complete Australian map drawn by Freycinet is the first 'to be based entirely on hydrographic knowledge from both French and English sources'.

6 Most notably, the visit of Count Alphonse de Fleurieu to Australia in 1911 was to lead to the restoration of several French names to the South Australian coast. See Horner, *The French Reconnaissance*, p. 16.

7 Gregory C. Eccleston, 'The Neglect of Baudin's Manuscript Charts of the Victorian Coastline', *The Globe*, 66 (2010): 27–58.

the existence and the content of Baudin's manuscript charts[8] and established some of the differences between Baudin's nomenclature and that of Péron and Freycinet.[9] Useful comparisons have also been made with the names adopted by their rival, Matthew Flinders.[10] The comparative tables that Greg Eccleston has drawn up for the two sets of French names given to features of the Victorian coastline likewise make a useful contribution to our knowledge on this subject.[11] Nevertheless, further systematic study is required. Our approach here is to focus on the historical context in which the substitution of names took place by re-examining the full range of archival and secondary documents available. We intend firstly to determine the precise role played by the various protagonists, as well as the personal and political motivations that influenced their naming practices. A close examination of the manuscript charts and coastal profiles, along with the various textual records, will then allow us to compare the nomenclature adopted by Baudin himself during the voyage with the naming principles that guided Péron and Freycinet. And lastly, the analysis of Baudin's nomenclature will provide us with some insights regarding the way in which the French commander seems to have sought, through the names he conferred on features of the Australian coastline during the course of the voyage, to express his more personal views – not only his frustrations and resentments with respect to the failures and shortcomings of his fellow travellers, but also his response to the unfamiliar landscape which lay before him.

---

8 In his *Matthew Flinders*, pp. 294–8, Ingleton gives a detailed material description of the Baudin expedition's original manuscript charts. Leslie Marchant published a number of these charts relating to the Western Australian coast in *France Australe* (Perth, 1998).

9 British commentators were the first to attribute to Baudin a seaman-like respect for the names bestowed by other navigators and to contrast this with Péron's controversial choice of Napoleonic nomenclature. In the review article of the first volume of Péron's narrative (*Quarterly Review*, 4.7 (August 1810): 42–60), published anonymously, but attributed to Sir John Barrow, the author says of Baudin: 'we cannot think so ill of him as to suppose that he would lend his authority, in so wanton and unjustifiable a manner, "to pluck the laurels from a brother's brow"' (p. 52). In *Matthew Flinders*, p. 280, Ingleton comments on Baudin's scrupulous policy with regard to published English place names.

10 H.M. Cooper utilizes both Flinders' and Baudin's original charts in his self-published study of *The Unknown Coast (A Supplement)* (Adelaide, 1955). Anthony Brown and Christine Cornell list the different names conferred by the Flinders and Baudin expeditions in 'Legacy of the Encounter: British and French Place Names of the South Australian Coast', *South Australian Geographical Journal*, 99 (2000): 10–15, while Jean Fornasiero, Peter Monteath and John West-Sooby analyse differences in nomenclature and naming practices in 'Hydrographic Reputations: Matthew Flinders, Nicolas Baudin and Australia's Unknown Coast', *Hydrographic Journal*, 121 (July, 2006): 17–23.

11 Eccleston, 'The Neglect of Baudin's Manuscript Charts', pp. 45–8.

The details of the 'Terre Napoléon' controversy are well enough known, but it is worth recapitulating the main points here.[12] When Baudin began his survey of the 'unknown south coast', on 27 March 1802, he was unaware that the section of it corresponding roughly to the southern coast of present-day Victoria had recently been examined and charted, albeit cursorily in parts, by James Grant, Francis Barrallier and John Murray. In fact, the English had visited this coast three times during the previous 16 months: Grant had been the first to examine it, in December 1800, while sailing to Port Jackson from England through the recently discovered Bass Strait; he then returned, with Barrallier, in March to May 1801, at the behest of Governor King, who had not been satisfied by the results of the initial survey; still unsatisfied after this second survey, King sent John Murray to the same section of coast in November 1801, leading to his discovery and exploration, in February 1802, of Port Phillip Bay. Baudin, of course, could not have been aware of these surveys: as he coasted his way westwards from Wilson's Promontory, he believed he was entering entirely uncharted waters. At the same time, Matthew Flinders, as we know, was conducting his own survey of the same stretch of coast, but from the other direction. When the two navigators crossed paths in Encounter Bay, on 8 April 1802, and exchanged information on their explorations, the precise limits of their respective discoveries along the unknown coast were set: even though certain details remained to be finalized on the map, such as the southern coast of Kangaroo Island, the dots had essentially been joined and the integrity of the continent of New Holland had finally been demonstrated. Flinders could therefore lay claim to the discovery of the southern coastline from Nuyts Archipelago to Encounter Bay (including Kangaroo Island and the two gulfs above it), while Baudin's discoveries were limited to the section of coastline extending westwards from Cape Northumberland, or possibly Cape Banks, to Encounter Bay[13] – although, as Trevor Lipscombe has recently pointed out, the French expedition should also be credited with the first charting of certain portions of the Victorian coast that Grant, Barrallier and Murray had been unable to examine.[14] Both Flinders and Baudin would have been disappointed at the sudden realization that they would not be able to claim sole right of discovery over the unknown coast, but Baudin would have been the more despondent of the

12 For a full account, see Horner, *The French Reconnaissance*, pp. 8–16.

13 Flinders, in his *Voyage to Terra Australis*, makes no claim for Grant's discoveries west of Cape Northumberland, though Grant did sight and name Cape Banks. On the French map, Cape Northumberland is represented as 'Cap Bélidor' and Cape Banks appears to correspond to 'Cap Buffon'.

14 Trevor Lipscombe, 'Two Continents or One? The Baudin Expedition's Unacknowledged Achievements on the Coast of Victoria', *Victorian Historical Journal*, 78.1 (2007): 23–41.

two, given the barrenness of the shores he had discovered.[15] His disappointment would be further compounded when he visited Port Jackson two months later and learned of the extent of the discoveries made by Grant, Barrallier and Murray.

The encounter between Baudin and Flinders thus served to set a western limit to Baudin's discoveries, and the subsequent revelation, in Sydney, of the earlier English explorations likewise made the eastern limit of those discoveries abundantly clear to the French commander and his fellow expeditioners. This did not prevent Péron and Freycinet, however, from applying the name 'Terre Napoléon' to the entire stretch of coastline collectively discovered and charted by Flinders, Baudin, Grant, Murray and Barrallier. Péron uses this term as a chapter title in the first volume of his official account of the French expedition, published in 1807, and retains in his narrative of this part of the voyage the French names conferred on all parts of the coast; that same chapter also opens with the patently false claim that, 'the English not having explored further south than Western Port, the whole section of coast between this latter point and Nuyts' Land was consequently still unknown at the time we arrived on these shores'.[16] As for Freycinet's maps of Australia and the unknown coast, published in 1811, the names he inscribes on them mirror exactly the terminology used by Péron. Both men therefore left themselves wide open to charges of misrepresentation and deceit.

The English were quick to notice this anomaly, with various newspaper articles expressing indignation at this slight on the part of the French.[17] Flinders, who, since his release from detention on Mauritius in 1810, had been finalizing his own narrative and charts for publication, was understandably affronted when he learned of this appropriation of his discoveries. Although he generously presumed that Péron's misrepresentation was the result of political pressures and

15 While he found the coast of present-day Victoria to be of a pleasant aspect, the whole stretch of what is now known as the Coorong, which would soon be revealed to be the limit of his discoveries, struck him and the other Frenchmen as featureless and uninteresting. In his journal, Baudin frequently mentions the endless line of arid and sterile sand-dunes. See *Journal of Post Captain Nicolas Baudin*, pp. 376–9.

16 *Voyage de découvertes aux Terres Australes*, vol. 1 (1807), p. 316 (our translation). Péron begins this paragraph by setting out the state of knowledge of this coast 'at the time of our departure from Europe', but he was well aware that the discoveries made by the English along this coast had greatly reduced the length of it that was unknown when Baudin 'arrived on these shores'.

17 It was in August 1810, for example, that the *Quarterly Review* published its review article of the first volume of Péron's narrative (see above, note 9). Matthew Flinders, in his private journal, also mentions that he was consulted by John Weyland, who was seeking his opinion on Péron's *Voyage* for a review he was preparing to publish. See *Matthew Flinders Private Journal*, edited and with an introduction by Anthony J. Brown and Gillian Dooley (Adelaide, 2005), pp. 345–6.

that it 'smote him to the heart' to make such deceitful claims, he nevertheless went to great pains in his *Voyage to Terra Australis* to point out that neither Péron nor Freycinet could claim to be ignorant of the British claims to prior discovery along this coast: 'M. Péron was present afterwards at Port Jackson, when I showed one of my charts of this coast to Captain Baudin, and pointed out the limits of his discovery; and ... the officers of the *Géographe* always spoke of them as belonging to the *Investigator*.'[18]

Thanks to the work of Ernest Scott, Flinders' suspicion that Péron was only acting under intense pressure from the French authorities has been definitively dispelled.[19] Scott also debunked other myths that had sprung up as a result of this controversy by demonstrating that the French did not plagiarize Flinders' charts, nor did the French Government attempt to hasten publication of the French maps in order to stake a territorial claim over the region. Flinders' continued detention on Mauritius was likewise not the result of a plot to prevent the English navigator from publishing his work before the French had published theirs. Indeed, the French authorities were so little interested in publishing the expedition maps that Freycinet was obliged to harass them about their inaction.[20] The arguments Scott presented in defence of the integrity of the French maps, however, did not prevent him from casting doubt on the character of Louis Freycinet himself,[21] whom he associated with the actions of Péron on the question of the controversial Napoleonic nomenclature. It is of interest, then, to investigate further the role played by Freycinet in this matter in order to determine the level of responsibility he may have shared, despite his numerous disclaimers, with his fellow author, Péron.

Whenever the 'Terre Napoléon' controversy is discussed, it is Péron who is portrayed as the prime offender while Louis Freycinet generally comes across as the embarrassed and unfortunate colleague who is left with the awkward task of cleaning up the mess created by the zoologist's rashness and indiscretion.[22] This is largely due to the fact that the initial 'offence', in a public sense at least, was committed under Péron's name, with the publication in 1807 of the first volume

---

[18] Matthew Flinders, *A Voyage to Terra Australis; Undertaken for the Purpose of Completing the Discovery of that Vast Country, and Prosecuted in the Years 1801, 1802 and 1803, in His Majesty's Ship, the* Investigator (London, 1814), vol. 1, p. 192.

[19] See Scott, *Terre Napoléon*, chapter 5, and Scott, *The Life of Matthew Flinders*, chapters 22–6.

[20] Scott, *The Life of Matthew Flinders*, p. 280.

[21] Ibid., p. 281.

[22] Horner, for example, without going so far as to disculpate Freycinet, nevertheless describes him in this context as a 'victim' of unclear lines of authority or as being left to deal with a series of 'embarrassments' (*The French Reconnaissance*, pp. 11, 12).

of the historical account of the voyage. Freycinet appeared to confirm Péron's responsibility for the controversy when he came to write the Preface to the second volume of the *Voyage*, which he had been left to complete following Péron's death in 1810 and which was not published until 1816. While attempting to remain loyal to his now deceased colleague, Freycinet nevertheless asserted that 'Péron had conceived the project of all the names', adding that 'this project had been officially adopted'.[23] Given their well-attested indifference to publishing the records and maps of the Baudin expedition, however, it remains unclear how committed the authorities actually were to the choice of nomenclature.[24]

There is room, therefore, to question Freycinet's portrayal of the roles played by Péron and himself in this contentious episode. It is certainly plausible that the idea of adopting Napoleonic nomenclature for the unknown coast was initially that of Péron – his earlier participation in the revolutionary army had been a formative experience for him, and some of his other writings reveal a strong patriotic spirit.[25] There also exist numerous records of his attempts to secure the patronage of the Empress Joséphine upon the expedition's return.[26] It is well established, then, that Péron had Bonapartist leanings. Nevertheless, it is highly unlikely that Freycinet played little or no part in the choice of nomenclature for 'Terre Napoléon' and that he simply (and reluctantly) acquiesced to Péron's decisions in this regard. As important a figure as Péron had become, it is difficult to believe that he would have been able to impose his will on a senior naval officer against that officer's better judgment. Ultimate responsibility for the maps and for the presentation of the geographical and navigational results of the voyage rested with Freycinet, as he himself pointed out in the Preface to his *Navigation et géographie* volume, published in 1815.[27] In that same Preface, Freycinet showed that he was not afraid to associate himself with the iconography and

---

23 Péron and Freycinet, *Voyage de découvertes aux Terres Australes*, vol. 2 (Paris, 1816), p. viii (our translation).

24 The delay in the publication of the second volume of the official account is one sign of the government's apathy with respect to the Baudin expedition; the difficulties Freycinet experienced in obtaining funding to engrave and publish his maps are another.

25 His confidential report on Port Jackson, entitled *Mémoire sur les établissements anglais à la Nouvelle Hollande, à la Terre de Diémen et dans les archipels du grand Océan Pacifique*, is noteworthy for its frequently bellicose tone and its compliments directed at Napoleon. The manuscript of this document is held at the Muséum d'histoire naturelle, Le Havre (Collection Lesueur, dossier 12). A transcription of it was published in French by Roger Martin in the *Revue de l'Institut Napoléon*, no. 176 (1998).

26 See Edward Duyker, *François Péron, An Impetuous Life* (Carlton, 2006), p. 212.

27 Freycinet notes in this Preface that, whereas Péron was given responsibility for everything relating to the natural sciences and the historical narrative of the expedition, the 'nautical and geographical details were naturally the domain of the navy; the Minister of

narrative of the expedition. He had no qualms, for example, about taking full ownership of his 1812 (1814) atlas, calling it 'my atlas', and noting that 'The first volume of the history of our voyage, prepared by Péron, has been the occasion for Mr Flinders to attack our nomenclature'.[28] Freycinet's use of first person plural possessives, particularly in the second instance, marks a frank acceptance of the shared nature of their work. This stands in contrast to his 1816 Preface to the second volume of the *Voyage*, where he attributes the idea for the names solely to Péron. Freycinet's attitude was clearly evolving as the political pressure surrounding 'Terre Napoléon' intensified, particularly after the publication of Flinders' narrative and maps.

It is therefore difficult to escape the conclusion that Freycinet was a compliant, and possibly even a willing participant in the 'Terre Napoléon' project. It is definitely not Péron's hand that has modified the names on Baudin's charts and coastal profiles of the region; nor is it Péron who has almost systematically replaced the commander's notation 'Côte du Sud Ouest' by 'Terre Napoléon' (Figure 8.3). While the handwriting does not appear to be that of Freycinet,[29] it is reasonable to assume that the changes could not have been made by one of his subordinates without his supervision and approval.[30] Furthermore, the inconsistencies and contradictions in Freycinet's various attempts at justifying the French nomenclature, from his Preface to the 1815 *Navigation et géographie* volume through to the Preface he compiled for the second edition of the *Voyage de découvertes aux Terres Australes*, published in 1824, all betray an embarrassment that is as personal as it is collective. Finally, and perhaps crucially, given the well-established hostility of the two Freycinet brothers towards their commander and the determination expressed by Louis to expunge Baudin from the expedition records,[31] there is every likelihood that Louis derived great pleasure from seeing

that department entrusted me with their account' (our translation). See Freycinet, *Voyage de découvertes aux Terres Australes*, vol. 3, *Navigation et géographie* (Paris, 1815), p. vi.

28 Ibid. (our emphasis, our translation).

29 Ingleton, *Matthew Flinders*, p. 297.

30 The handwriting of the person who has effected the deletions and substitutions of names on Baudin's original charts does not appear to be the same as that to be found in Freycinet's journal. However, whether or not Freycinet actually deleted Baudin's names himself or instructed another to do so, his ownership of the changes is clear from the responsibility he claims for the charts, and hence for any alterations to them, in the various prefaces and introductions to the volumes of the *Voyage*.

31 This hostility and this objective are both mentioned in the correspondence between the two brothers. In a letter to Henri dated 30 October 1806, for example, Louis explains that Baudin's name is to be completely expunged from the official account of the expedition, the first volume of which was being finalized for publication the following year: 'That, my dear Freycinet, is the state of play with respect to the work of our expedition. Moreover,

Figure 8.3 Coastal profiles from Baudin's exploration of the 'unknown' south coast of New Holland. Archives nationales de France, Série Marine 6JJ4. Reproduced with permission

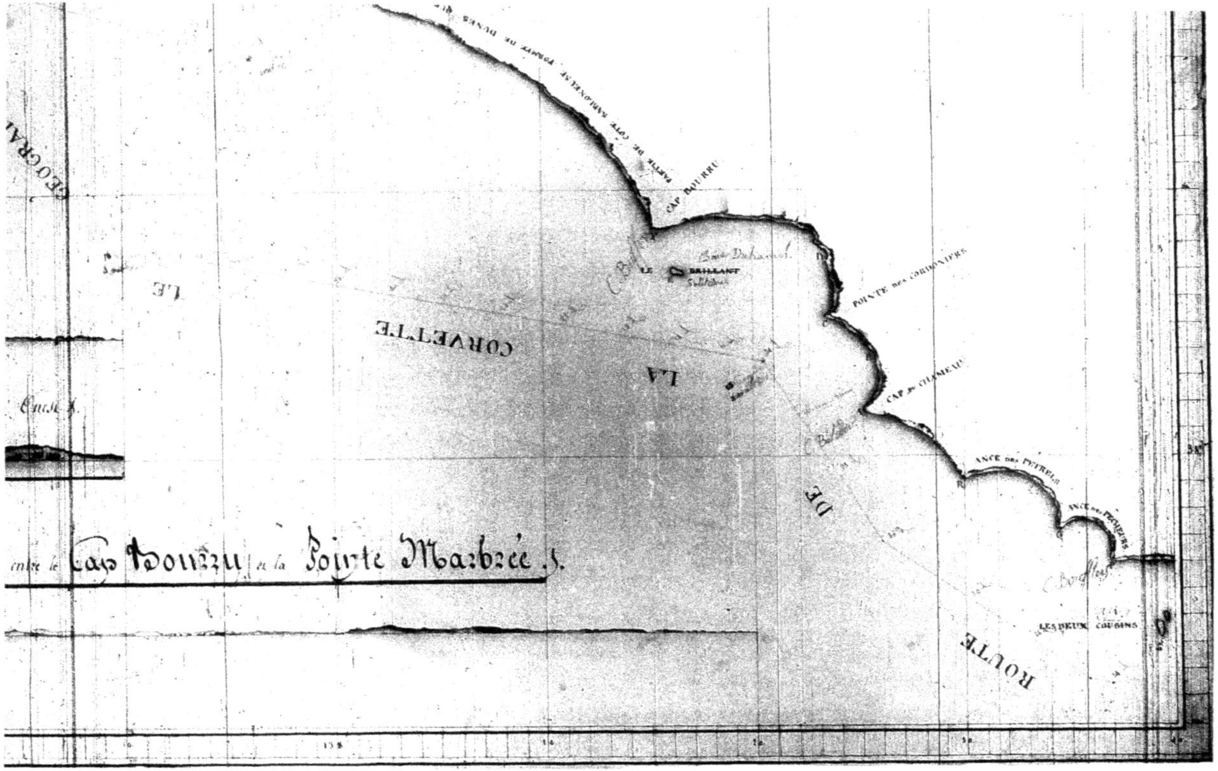

Figure 8.4 Draft French map of a section of the Victorian coast, detail. Archives nationales de France, Série Marine, 6JJ4. Reproduced with permission

Baudin's names replaced with new ones of his own or Péron's invention (Figure 8.4). After all, Baudin had paid scant homage to his companions in his choice of names, whereas under the new system the names of Péron and Freycinet abounded.

Indeed, the names adopted by Freycinet and Péron for the final maps demonstrate not just a penchant for self-aggrandizement or an enthusiasm to pay homage to the Napoleonic régime, but also a significant shift with respect to the nomenclature principles adopted by Baudin.[32] Both sets of names of course pay due homage to patrons and other important government figures, past and present. In Baudin's case, however, such names are relatively few in number and the emphasis is on distinguished members of the French navy rather than on statesmen. Among the few political figures he commemorates are Louis XIV's finance minister, Colbert, after whom a strait was named, and the Councillor of State and friend of the 'philosophes', Malesherbes, whose name was given to a bay. Count Fleurieu, the former navigator and Minister for Marine who had drawn up his itinerary, was honoured with a peninsula, while Baudin named a cape and an island after his patron, Admiral Bruix, under whom he had served in the Mediterranean in 1799.[33] Other naval figures he commemorated include:

---

the name of that despicable commander will not figure in it. His memory and his name, forever blackened, will not dishonour the splendour of the fine work conducted in spite of him and the vexations he gave rise to at every stage' (our translation). The letter, which is presently held in the Woodside Valley Foundation collection, stands in stark contrast to the conventional portrait of Freycinet as more of a diplomat than Péron, on the evidence that he (Freycinet) had restored Baudin's name to the second edition of the *Voyage*. This letter makes clear that the responsibility for obliterating Baudin's name was equally shared by Péron and Freycinet in the project to publish the *Voyage*.

32 The analysis that follows is based on a close examination of the names that figure on the final maps as well as on the draft maps and coastal profiles that were drawn up during the course of the voyage. Baudin's responsibility for the names that are to be found on the draft charts and profiles, a great number of which are crossed out by an unidentified hand, is made clear both by his sea-log, in which he records several of the names he conferred, and by the journals kept by the officers and scientists during the course of the voyage, which likewise note the commander's decisions in this regard. One example from the journal of the botanist Leschenault shows that other members of the expedition occasionally made suggestions, but Baudin always had the final say. The discovery of some admiral shells on an island off Australia's northwest coast, notes Leschenault, 'determined the Commander to give this island the name of *Admirals* Island; someone had suggested giving it the name of *Three Fountains*. This name would have alerted navigators to the fact that they would find water there, an essential thing on a coast where, up till now, we have not found any fresh water' (Archives nationales de France, Série Marine, 5JJ 56, p. 43 – our translation).

33 For details of this service, see Michèle Rivas, 'Un navigateur-naturaliste d'origine poitevine célèbre en Australie, méconnu dans sa patrie: Nicolas Baudin (1754–1803)', *Revue historique du Centre-ouest*, 5 (2006): 77–8.

the duc de Vivonne, one of Louis XIV's highest ranking naval officers who, in retirement, was a great patron of writers; Antoine Poisson(n)ier-Desperrière, who was a member of the Royal Marine Academy; the naval officer and one-time Governor of the Ile de France, La Bourdonnais; and the naval engineer Sané, whose name was given to the cape at the eastern tip of Kangaroo Island (now Cape Willoughby).[34] Baudin also shows great deference to those explorers who preceded him, such as Tasman or Forbin. This sign of respect extends to his contemporaries, as is demonstrated by his persistent use of the name 'Port de Basse' to refer to Western Port. Most of these names were changed by Péron and Freycinet, who apparently did not share Baudin's predilection for celebrating naval history. They also added a vast number of names to those Baudin had already given in commemoration of contemporary political figures, and gave particular attention to the Bonaparte dynasty – the Golfe Bonaparte and the Golfe Joséphine being the most notorious examples, though Péron and Freycinet also honoured several other members of the Bonaparte family (Caroline, Pauline, Louis) and paid homage to a number of Napoleon's victories (Rivoli, Mondovi and others).[35]

Another stark contrast is provided by the commemoration of literary figures. These are few and far between in Baudin's list, Cap Du Bellay on the eastern side of Eyre Peninsula being a rare example. As his personal library demonstrates, Baudin had a keen interest in literature, but the presence in his collection of several guides to literature shows that he was an auto-didact and had no pretensions in that domain.[36] His principal interest was in history, and in maritime history more particularly. Péron and Freycinet, on the other hand, make liberal use of literary names, especially along the coast of present-day South Australia. There is a welcome reference to the irreverent Rabelais, but it is predominantly writers from the seventeenth and eighteenth centuries who are commemorated: Mme de Sévigné, Regnard, Corneille (who replaces Baudin's Du Bellay), Descartes, Raynal, Deshoulières, Scudéry, Bossuet, La Rochefoucauld, Montaigne, Volney, d'Alembert, Fénelon, Montesquieu, Boufflers, to name just some. To this list we could add artists, such as Chardin, or historical figures, such as Bayard or Jeanne Hachette. There is of course every reason for explorers to seek to honour in this

---

34 Baudin initially named this cape 'Cap de la Reprise' to mark the place where he resumed his charting of this section of the unknown coast.

35 Baudin himself was not completely averse to honouring Bonaparte's triumphs – his Cap Marengo is one example, and Cap Lodi, also on the present-day Victorian coast, is another (though, curiously enough, this was changed to Cap Volney – its present name – by Péron and Freycinet to commemorate the contemporary philosopher of that name).

36 For a presentation and analysis of Baudin's library, see Jean Fornasiero and John West-Sooby, 'Baudin's Books', *Australian Journal of French Studies*, 39.2 (2002): 215–49.

way the literary and artistic luminaries of their nation; there are also many other examples of discoverers having recourse to a particular theme in naming the features of a given region (Flinders' reminiscences of his native Lincolnshire, for example, in the area around South Australia's Port Lincoln). However, it is also possible to consider that, in establishing such a marked difference with respect to their commander's nomenclature, Péron and Freycinet were subtly adding to the campaign aimed at portraying Baudin as uncouth and without social graces.

One area of overlap, on the other hand, is the honouring of scientific endeavour. Both sets of names show a similar willingness to commemorate naturalists, mathematicians, astronomers, geographers and other savants: Borda, Saussure, Maupertuis, Delambre, Bossut, Buache de la Neuville, Lacépède, Jussieu, Bouvard, D'Après de Mannevillette, La Condamine, Réaumur and so on. In Baudin's case, the homage he paid to these savants demonstrates that he was certainly not the enemy of science that his detractors, Freycinet and Péron among them, claimed him to be. It is interesting to note that Péron and Freycinet not only retained many of Baudin's names in this category, but also added several more to the list. They were not going to allow themselves to be outdone in this respect by their commander.

Evidence of Baudin's interest in the natural world is provided by his frequent recourse to animals when it comes to naming coastal features. There are innumerable capes, coves, islands and bays named after all manner of birds – albatrosses, snipes, shrikes, petrels, pelicans, larks, eagles, crows, tits, nightingales and cassowaries – as well as creatures of the sea, large and small – jellyfish, sponges, sea urchins, sea elephants, seals, sperm whales and other 'spouters' – and finally land animals of all varieties, native and exotic – snakes, butterflies, camels, dragons, kangaroos and other 'jumpers'. To their credit, Péron and Freycinet did keep a number of these, albeit with some modifications (changing 'Spouters Point' to 'Dolphin Point', for example, or, just as gratuitously, 'Albatross Cove' to 'Cormorant Cove' – present-day Horseshoe Bay in South Australia's Encounter Bay). They did, however, abandon a large number of Baudin's animal names in order to honour people – understandable, perhaps, if the aim is to flatter patrons or pay homage to prominent national figures, but in retrospect the case for changing Baudin's names in order to honour the sixteenth-century royal surgeon, Ambroise Paré, or the fifteenth-century defender of Beauvais, Jeanne Hachette, might be considered less than compelling (Cape Ambroise Paré was Baudin's 'Serpents Point', while 'Cape Jeanne Hachette' was his 'Eagle Point'). The point is no doubt somewhat moot – it is, after all, refreshing to see a woman commemorated in this way, and the 'duty of memory' can always be invoked as a reason for reminding the world of some less well-remembered figures – and Baudin himself might have made a number of similar such modifications if he had survived to supervise the draughting of the final versions of his charts back in Paris. It is nevertheless evident that, while Péron and Freycinet were happy to allow a certain number of animal names to subsist, they did not consider this a noble enough source of nomenclature for such names to proliferate on their maps in the way they did on those of Baudin.

Two final categories of names highlight the distinction that has already begun to emerge between Baudin and his two nemeses. The first of these points further, perhaps, to the mariner in Baudin: he had the sailor's keen eye for identifying distinctive shapes and forms, and was well aware how useful these could be for navigation. He consequently endowed many of the features of Australia's coastline with names that reflected their physical appearance. Some examples include: Three Point Cape, Red Cape, Green Cape, Round Cape, Hollow Cape, Broken Cape, Long Island, The Dome, Granite Point, Low Land Point, Marbled Point, Sandy Point, or, more imaginatively, Camel Cape, the Dragon Islets and Anvil Cape. He was also sensitive to the aesthetic appeal of certain coastal features, hence Cape Pleasant, Cape Belle Vue and Picturesque Point. Baudin's nomenclature drew equally on the moral or on the psychological as it did on the physical. He was not averse to conferring names that reflected his mood, or the general mood on board the ship, as well as particular events. There is no prize for guessing where he drew his inspiration for names such as Dangerous Cape, Encounter Cape, Detour Cape, Cape Surprise, Termination Point or Resumption Point. Others are less obvious, though he does explain some of them in his journal. Protest Cape, writes Baudin, earned its name from a moment of disgruntlement on the part of the officers;[37] Disorder Bay, Evil-doers Point, Grumpy Cape and Cape Ingratitude were presumably inspired by similar such incidents. The depressing nature of long stretches of the unknown coast is surely the source of such names as Dull Bay, Point Desolation, Point Misery or the Point of Sighs. And finally, the Bay of Vain Hope on Kangaroo Island expresses with wonderful clarity Baudin's frustration at Freycinet's delay in returning from his survey of the gulf waters (unsurprisingly, perhaps, this was later changed to Freycinet Bay, though it would ultimately receive the name of D'Estrees Bay).

Generally speaking, then, Baudin's nomenclature is a reflection of the commander himself, that is to say of his interests and his personality. He had great respect for navigators and maritime tradition; he was vitally interested in natural history and sought to show his respect for the great men of science; he was not afraid to show his displeasure and frustration through his choice

[37] *Journal of Post-Captain Baudin*, p. 372 (entry dated 9 germinal, Year 10 [30 March 1802]).

of names, but could equally demonstrate that he had a wry sense of humour; and finally, he had quite a creative imagination. Freycinet and Péron may have shared some or even all of these qualities, but they do not shine through in their nomenclature, which is heavily dominated by proper names – including those of their fellow travellers (and their own), which they sprinkled along the Australian coastline in a much more liberal fashion than Baudin had done. Almost all of the commander's names, for example, in the last two categories we have examined (visual or physical features and shipboard incidents or sentiments) were replaced by names of people on the final charts; and, as colourful, in retrospect, as 'Golfe Joséphine' and 'Golfe Bonaparte' might appear to us today, they do not have the whimsy or the evocative power of Baudin's 'Golfe de la Misanthropie' and 'Golfe de la Mélomanie'. Freycinet and Péron were clearly in the business of honouring by naming – which included honouring their own names – and, in the case of their commander, of shaming by not naming. Baudin, for his part, was just as willing to bestow honour on worthy national figures as were Freycinet and Péron, though he did so with a lighter hand; however, on the odd occasion when he felt moved to record through a place name an instance of shameful behaviour, he did so discreetly, highlighting the incident and maintaining the flavour of shame, but without identifying the protagonist – confidentially shaming without publicly naming.

In this as in other categories of nomenclature, then, the fact that so many of Baudin's creative and imaginative names did not survive is perhaps the greatest shame of all. His was a more varied, less contentious and more creative perception of the 'Terres Australes' than the 'Terre Napoléon' of his companions. He saw the voyage firstly as a human and a collective enterprise and tended to name significant features after events that concerned the entire company, rather than putting forward individuals simply for the sake of it. He was also interested in the actual features of the land he surveyed, attempting to conserve for posterity a record of exactly what he had seen, and also providing recognizable landmarks for fellow navigators. Finally, as we have noted, he was scrupulously respectful of the names already bestowed by his predecessors. In this way, Baudin showed that he was equally in tune with the primary objective of his mission – which, as this issue serves to highlight, was geographical – as he was with the traditions and ethics of navigation. Freycinet himself was as aware of the importance of this aspect of the voyage as his commander had been, declaring that 'the reputation of our expedition depends wholly on the success of our geographical work'.[38] However, in 1811, by the time that this statement was made, he had

---

38 Louis Freycinet, Letter to the Minister of Marine, dated 29 August 1811, Archives nationales de France, BB4 996, cited by Scott, *The Life of Matthew Flinders*, p. 281.

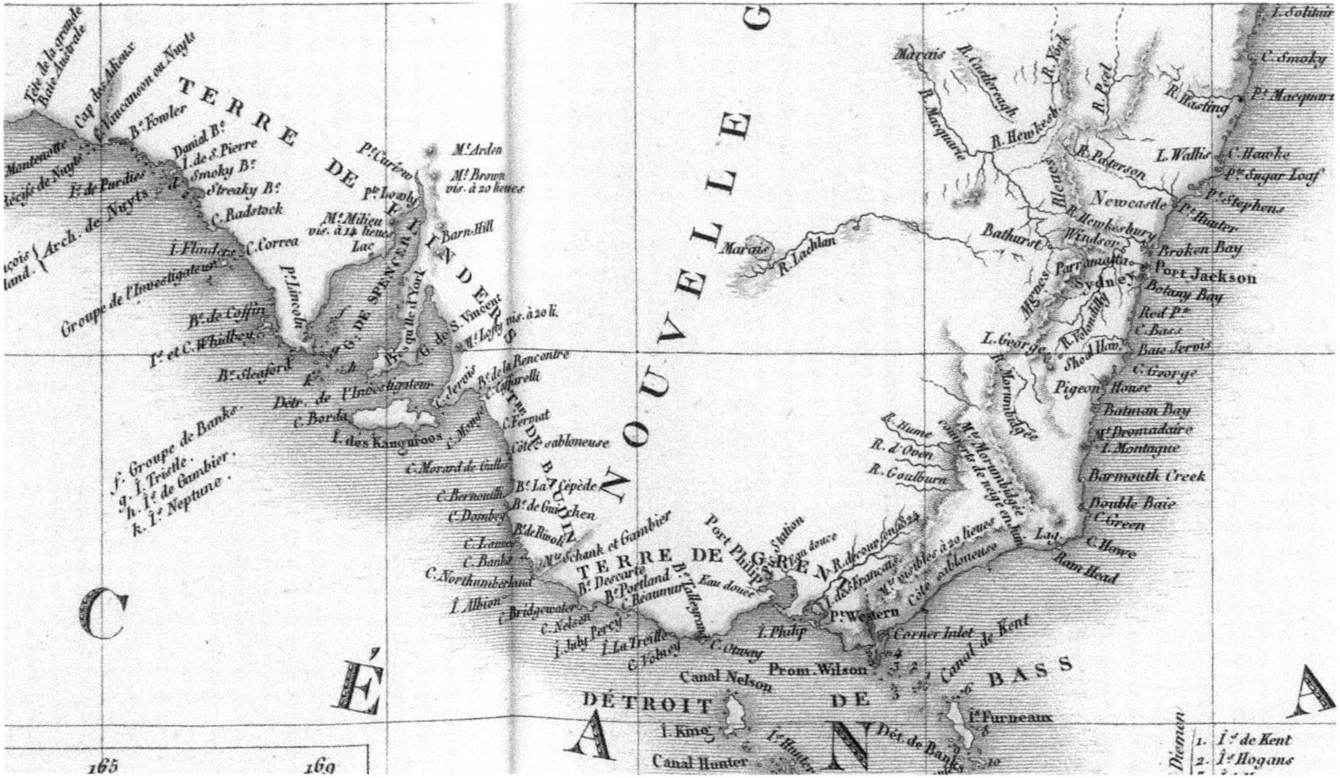

Figure 8.5 Adrien-Hubert Brue, 'Carte de l'Australie' (1826), detail. Reproduced with the kind permission of the David Rumsey Map Collection (www.davidrumsey.com)

already embraced the political nomenclature that retrospectively tarnished that reputation, and had thereby abandoned the ethics that might have helped to protect it.

Cartographers of the nineteenth century were to come some way towards recognizing Baudin's geographical achievements and the pride he took in his role as navigator, by inscribing his name, though without his original nomenclature, on the section of mainland Australia to which he had earned the right of discovery. It is both just and fitting that the 'Terre Napoléon' of Freycinet and Péron was transformed to become 'Terre Baudin' and 'Terre Flinders' on the exquisite 1826 map of Australia drawn by Adrien-Hubert Brue – the celebrated cartographer who had himself made that ill-fated journey to Australia with Baudin as a midshipman (Figure 8.5). This was a worthy, if rare, acknowledgment, in the cartographic record, of Baudin's achievements. There were, however, few such attempts at the time to set the record straight and to counter the campaign of vilification exemplified in the public domain, paradoxically, by the omission of Baudin's name from the official account of his own voyage. As a result, not only were many of the names Baudin conferred on the coastline of southern Australia consigned to oblivion, but, for many years to come, shame rather than glory would be attached to (the absence of) his name. And from this shame, the names of neither Péron nor Freycinet can be dissociated.

Chapter 9

Who Do You Trust? Discrepancies Between the 'Official and Unofficial' Sources Recording Explorers' Perceptions of Places and Their People

Michael McCarthy

Introduction

The Western Australian Museum's examination of the primary sources relating to William Dampier and to Louis and Rose de Freycinet's perceptions of New Holland and its people commenced soon after the tercentennial of Dampier's landing at Shark Bay in 1699. There it was decided that the author would attempt to locate the wrecks of His Majesty's ship *Roebuck* which carried Dampier and the French naval corvette *L'Uranie*, transport of the lovers Louis and Rose de Freycinet to Shark Bay and beyond. These ships lay at Ascension Island and in the Falkland Islands respectively. A considerable amount of research ensued, including the location of many primary sources. After they were located in 2001 the wrecks and their chief characters then became key elements in two major museum exhibitions: *Journeys of Enlightenment* and *Voyages of Grand Discovery*. These presented amongst other things some of the personal journals, published works and expedition artwork.

While initially utilized for the purposes of locating the wrecks, understanding the chief proponents and presenting the exhibits, these sources also revealed circumstances where the official records were found to be at odds with the unofficial accounts. They further provided instances where the official and unofficial artwork depicting the peoples and the places visited also appeared at odds. This chapter will consider some discrepancies between original perceptions recorded by explorers for their own private use and later versions of the same incidents released in the public domain.

William Dampier: 'Pirate and Hydrographer'[1]

William Dampier's appointment as a buccaneer and privateer – and in some eyes a common pirate – to the command of *Roebuck*, a naval vessel, came six years after his return from an extended circumnavigation of the globe. Soon after his arrival home in 1691, Dampier began exhibiting the famous 'Painted [tattooed] Prince' Jeoly who was from Miangis in the Spice Islands near Sulawesi. He and his captive also appeared before Their Majesties William and Mary, providing Dampier with a modicum of respectability in London society.[2]

Dampier's reputation was further enhanced with the publication of an account of his circumnavigation entitled *A New Voyage Round the World*. Appearing in 1697, it was such a literary and maritime sensation that it went through many editions. Overnight Dampier became a celebrity, as the following quotation from the diary of John Evelyn shows:

> I dined with Mr Pepys, where was Captain Dampier, who had been a famous buccaneer, had brought hither the painted prince Job [Jeoly], and printed a relation of his very strange adventure ... He was now going abroad again by the King's encouragement, who furnished a ship of 290 tons. He seemed a more modest man than one would imagine by relation of the crew he had associated with.[3]

As a result of his great fame, Dampier sat for a painting that hangs in the National Portrait Gallery under the title 'Pirate and Hydrographer'. His book, his experience and his knowledge had also greatly impressed Evelyn's other guest Samuel Pepys – the acknowledged 'father' of the British Navy and the British Admiralty. As a result Dampier's proposal to examine the then unexplored east coast of New Holland via Cape Horn was accepted. He had also proved such a keen and expert observer that British mariners following in his wake carried the scientific treatises dealing with oceanography, meteorology magnetism and tides that he published separate to his better-known work. As some indication of their value, naval leaders such as Admirals Pellew, Howe and Nelson and the naval explorers Flinders, Cook and Byron are all known to have consulted them.

---

1 The title earlier appearing on his official portrait in the National Portrait Gallery, London.

2 William Dampier, quoted in Geraldine Barnes, 'Curiosity, Wonder, and William Dampier's Painted Prince', *Journal for Early Modern Cultural Studies*, 6.1 (Spring/Summer, 2006): 31–50.

3 From the diary of John Evelyn, entry dated 16 September 1698. Reproduced in Alex S. George, *William Dampier in New Holland: Australia's First Natural Historian* (Melbourne, 1999), pp. 135–6.

Dampier's influence was also international. His first book was translated into French and Dutch in 1698 and into German in 1702. As a result he became highly regarded as a mariner and chronicler even amongst enemies such as 'the French, the Dutch, and even the Spaniards' (whom he had terrorized as a buccaneer and privateer in his earlier life). It is also reported that Napoleon had a bust of Dampier in his Gallery of Notables.[4] Dampier's 'New Voyage' proved such a source of great and lasting fascination that it has never been out of print. Some of his descriptions also became the inspiration for romanticized and mythological accounts of unknown regions. Defoe's *Robinson Crusoe*, Swift's *Gulliver's Travels* and Coleridge's *Ancient Mariner* are most often mentioned.[5]

Having been provided with *Roebuck* too late in the season to make a safe rounding of the Horn, Dampier was forced to approach his destination from the Cape of Good Hope instead. *En route*, in 1699, he landed at Dirk Hartog Island on the west coast of New Holland and remained for a while at what he named 'Sharks Bay'. In making a natural science collection and in describing the flora and fauna there, and at places further north, Dampier came to be justly described as 'Australia's First Natural Historian'.[6] In that context, Diana and Michael Preston wrote in their biography *A Pirate of Exquisite Mind: The Life of William Dampier*:

> He pioneered what is today known as descriptive botany and zoology, later developed by Joseph Banks, the naturalist who sailed with Captain James Cook – the careful, detailed and objective recording of the world's living things.[7]

In continuing on towards the east coast of New Holland via New Guinea and New Britain Dampier found that his ship was slowly rotting beneath him and he was forced to abandon his plans just short of the east coast. Though successfully retracing their outward voyage back into the Atlantic, *Roebuck* abruptly sank at Ascension Island. Dampier and his crew were later rescued and he returned to Britain with his journals and some of his specimens. His descriptions of the peoples, places, flora, fauna and natural phenomena under the title *A Voyage*

---

4 Leslie R. Marchant, *An Island unto Itself. William Dampier and New Holland* (Perth, 1988), p. 5.

5 Gary C. Williams, 'William Dampier: Science, Exploration and Literary Influence Including his Hydrographic Treatise of 1699', *Proceedings of the California Academy of Science*, series 4, 59.14 (2008): 533–663, 576.

6 See the title of George, *William Dampier in New Holland*.

7 Diana Preston and Michael Preston, *A Pirate of Exquisite Mind: Explorer, Naturalist and Buccaneer: The Life of William Dampier* (New York, 2004), p. 5.

to New Holland were to become yet another best seller, adding further to his international fame.[8]

Dampier's ship remained lost until *Roebuck* was located at Ascension Island by the Western Australian Museum's team in 2001.[9] As indicated above, this success was partly a result of the location of primary documents relating to the ship, its construction, career and loss. Another result of this research – and a common feature of historical archaeology as a whole – was the use of the site as a catalyst and focus for a wider range of studies. This included a re-examination of other primary sources, including those carrying Dampier's descriptions of New Holland and its people.

In the context of the explorer's perceptions of New Holland and its people, of special interest amongst this material were discrepancies between his journals and his published works. From an Australian perspective, the most notable of these appeared in Dampier's influential and very controversial descriptions of the Indigenous people he encountered on the Western Australian coast on his first voyage. This occurred in 1688 while in the company of privateers on board *Cygnet* under Captain John Read. They landed at what is today known as Cape Leveque north of Broome.

> The 4th day of January 1688 we fell in with the land of New Holland ... The inhabitants of this country are the miserablest people in the world. The Hodmadods of Monomatapa, though a nasty people, yet for wealth are gentlemen to these; who have no houses, and skin garments, sheep, poultry, and fruits of the earth, ostrich eggs, etc., as the Hodmadods have: and, setting aside their human shape, they differ but little from brutes. They are tall, straight-bodied, and thin, with small long limbs. They have great heads, round foreheads, and great brows. Their eyelids are always half closed to keep the flies out of their eyes; they being so troublesome here that no fanning will keep them from coming to one's face; and without the assistance of both hands to keep them off they will creep into one's nostrils and mouth too if the lips are not shut very close; so that, from their infancy being thus annoyed with these insects, they do never open their eyes as other people: and therefore they cannot see far, unless they hold up their heads as if they were looking at somewhat over them.
>
> They have great bottle-noses, pretty full lips, and wide mouths. The two fore-teeth of their upper jaw are wanting in all of them, men and women, old and young; whether they draw them out I know not: neither have they any beards. They are long-visaged, and of a very unpleasing aspect, having no one graceful feature in

8 See J.A. Williamson (ed.), *William Dampier: A Voyage to New Holland* (London, 1939).

9 Michael McCarthy, 'HM Ship *Roebuck* (1690–1701): Global Maritime Heritage?', *The International Journal of Nautical Archaeology*, 33.1 (2004): 54–66.

> their faces. Their hair is black, short, and curled like that of the Negroes; and not long and lank like the common Indians. The colour of their skins, both of their faces and the rest of their body, is coal-black like that of the Negroes of Guinea.[10]

Preston and Preston believe that Jonathan Swift's portrayal of the 'Yahoos' may have been based on what they characterize as 'Dampier's vivid and uncompromising portrayal' of the Aborigines of New Holland. It also appears that even pragmatic and experienced explorers were affected by his words. On his voyage along the east coast of New Holland with James Cook, Joseph Banks referred to Dampier's influence when he wrote:

> ... five people who appeared through our glasses to be enormously black; so far did the prejudices which we had built on Dampier's account influence us that we fancied we could see their colour when we could scarce distinguish whether or not they were men.[11]

Modern Indigenous leaders and historical commentators often reference Dampier's words with a justified sense of outrage – most recently the high-profile Dawn Casey, a former Director of the National Museum of Australia. In a lecture entitled 'Being Collected' she noted that 'Dampier's contemptuous and oft-quoted dictum – "the inhabitants of this country are the miserablest people in the world" – was to be echoed as a self-justification by the land-takers two centuries later'.[12] Here she was referring to John Mulvaney's assessment of Dampier's effect on others, with an additional nod towards the controversial 'terra nullius' arguments that arose in the context of Indigenous land title claims in recent times.[13] On the other hand, in being aware of the context of the offending words, Casey also looked beyond them towards broader societal influences affecting Dampier's choice of words.

> William Dampier, who published the first best-selling travel books about Australia around the reign of Queen Anne, set the fashion for an unsympathetic press on Aboriginal people. And although many would argue Dampier presented accurate anthropological observations about Aboriginal people from the Kimberley's' [sic] (tall, slender, slight limbs, heavily-boned eyebrows, eyes troubled by flies), his

---

10 Reproduced in Marchant, *An Island unto Itself*, pp. 152–9.

11 Preston and Preston, *A Pirate*, p. 175.

12 Dawn Casey, 'Being Collected'. Lecture given at the University of Sydney on 28 August 2006, reproduced in full at: http://sydney.edu.au/museums/whatson/webcasts/lecture_beingcollected.shtml [accessed: 30 October 2011].

13 See ibid.

perceptions were clouded by his European values. In an age when trade benefited one country at the expense of others, mercantilist, bourgeois and protestant Christianity, naked heathens lacking goods for trade and who neither farmed nor husbanded animals, epitomised the negation of ordered society.[14]

When she presented this lecture Casey was the Chief Executive Officer of the Western Australian Museum – in the same period that its research and exhibits featuring Dampier and the de Freycinets were in preparation. Perhaps as a result Casey was not as dismissive of Dampier's Australian legacy as were other eastern states scholars such as John Bach who wrote the entry on Dampier for the online edition of the *Australian Dictionary of Biography* thus:

> Dampier's direct contribution to Australian history was slight; indeed, his own impression of the west coast was unfavourable, since it seemed to him to be a long series of reefs and shoals behind which lay sandhills and barren country, apparently without water and inhabited by 'the miserablest People in the World'.[15]

The same social forces alluded to by Casey had been also examined at length first by Leslie Marchant, a West Australian scholar, and most recently by G.C. Williams from California.[16] They are both of the understanding that Dampier, conscious of his own reputation as a buccaneer and privateer (and pirate),[17] was writing with an eye to assuaging the attitudes and expectations of contemporary society. Under the heading 'Dampier's Two Different Descriptions of the Aborigines' Leslie Marchant also noted that while the published version 'implies he was not a dispassionately, objective scientific observer ... [t]he manuscript version of his diary, as in other matters, gives a different account from the published versions ... [and] is more in keeping with his reputation and talents as a natural history observer'.[18] Expanding, Marchant noted that the reference to the Hottentots of the Cape of Good Hope in the context of the description of the Aborigines of New Holland proves that the published account was not based on what Dampier wrote at the time of his landing in Australia, for he did not arrive at the Cape until 1691:

---

14 See ibid.

15 *Australian Dictionary of Biography*, http://adbonline.anu.edu.au/biogs/A010265b.htm [accessed: 30 October 2011].

16 See Marchant, *An Island unto Itself* and Williams, 'William Dampier'.

17 Many scholars sympathetic to Dampier gloss over his period on board the *Batchelor's Delight*, a pirate ship in any estimate.

18 Marchant, *An Island unto Itself*, pp. 72–3.

> The much quoted account of the Aborigines of New Holland therefore obviously was composed years after he saw the Aborigines of New Holland. The manuscript version, kept in diary form in this case, consequently appears to be the best one to use to check comments about the natural history of New Holland and the native peoples he observed there. It was not written for a reading public.[19]

What Dampier wrote in his journal was this:

> ... the people of this country have no houses nor any thing like a house neither have they any sort of grain or pulse, flesh they have not nor any sort of cattle not so much as cat or dog for indeed they have no occasion of such creatures unless to eat them for of that food which they have they leave no fragments ... They are people of good stature but very thin and lean I judge for want of food they are black yet I believe their hairs would be long if it was combed out but for want of combs it is matted up like a negores [sic] hair – they have all that I saw two fore teeth of their upper jaw wanting both men women and children.

Of those he saw on the water (apparently the Bardi people) Dampier wrote:

> ... they build wares [weirs] with stones cross the bays and every low water whether night or day they search those wares for what the sea hath left behind which is all that they have to depend on for a livelihood. ... They are people of good stature but very thin and lean I judge for want of food ...[20]

Williams supports Marchant and the Prestons, claiming that despite his description of the New Holland Aborigines, Dampier does not show prejudice or any animosity in his writings toward Indigenous peoples. They, and many others besides, see Dampier holding rather enlightened views toward native cultures for his time. They also find him portraying a compassionate concern for oppressed peoples. 'Prejudice was not a trait of Dampier's observational skills, as he had a more dispassionate and objective quality to his attention which distinguished him from his contemporaries', Williams wrote.[21] The reasons why a man who also penned the words 'I was well enough satisfied, knowing that the farther we went, the more Knowledge and Experience I should get, which was the main Thing that I regarded' could produce such differing accounts of the same events at New Holland remains elusive.[22]

---

19 Ibid., p. 73.

20 Ibid., pp. 149–52. Dampier's original account forms part of British Library, Sloane MS 3236.

21 Williams, 'William Dampier', p. 573.

22 William Dampier, *A New Voyage Round the World*, ed. Max Beken (London, 1998), p. 207.

The years 1691 to 1697 are crucial years for Dampier's activities when he returned home from his first voyage, prepared his diaries for publication and produced the book that made him famous. Unfortunately there is little written on the subject other than by Baer in an article entitled 'new light on the missing years'.[23] It is known, however, that he travelled with Jeoly, exhibiting him until forced to sell him as his funds ran low. Very soon after, in 1692, Jeoly died of smallpox at Oxford and Dampier is found at sea on a merchant ship and again mixing with pirates and other undesirables.[24]

In an absence of detail from this period, attention turns to the part Dampier's publisher, James Knapton, a prominent London bookseller, might have played. While there remains no concrete evidence, there are indications that Dampier allowed his works to be edited for the general readership with an eye to the contemporary societal values referred to above. A clue appears when another biographer J.A. Williamson refers to other differences between Dampier's journals and the published accounts. He notes in the case of the *Roebuck* voyage that 'it has been said that his published work was touched up by skilled editors after it left his hand ....'[25] For his part, Knapton is described as a man who had an eye for a good story, and had a good business sense. According to one commentator, O.H.K. Spate, Knapton seized on the paucity of 'voyaging literature of any note' before Dampier and in publishing his works and those subsequent he 'became a specialist in the genre'.[26] Though preceded by Exquemelin's *Buccaneers of America* (1678), published in English in 1684 and by Sharpe and Ringrose's works which appeared soon after, [27] it has been said that:

> The New Voyage began a boom in voyage narratives; Dampier's publisher, James Knapton, became a specialist in the genre ... By about 1715 the voyage narrative was well established as a distinct branch of literature, though its writers were often decidedly unliterary ... All these earlier voyages, however, except for Dampier's unhappy expedition to New Holland in 1699, were unofficial ...

---

23 Joel H. Baer, 'William Dampier at the Crossroads: New Light on the "Missing Years" 1691–1697', *International Journal of Maritime History*, 8.2 (1996): 97–118.

24 *Dampier's Voyages. Consisting of a New Voyage Round the World, A Supplement to the Voyage Round the World. Two Voyages to Campeachy, a Discourse of Winds, a Voyage to New Holland, and a Vindication, in answer to the Chimerical Relation of William Funnell*, ed. John Masefield (London, 1906), p. 3.

25 Williamson, *A Voyage to New Holland*, p. xlv.

26 O.H.K. Spate, 'The Literature of Cook's Voyages', *La Trobe Journal*, 41 (Autumn, 1988): 25–9, 25.

27 See Anna Neill, 'Buccaneer Ethnography: Nature, Culture, and Nation in the Journals of William Dampier', *Eighteenth-Century Studies*, 33.2 (2000): 165–80.

> The first official British expedition to the Pacific, however, that of George Anson in 1744–48 ... [was] published by James Knapton's sons. *With the help of competent writers a professional narrative was produced; it was not the mere chronological outpourings, overloaded with navigational detail, of the privateers' – it was structured and it set a standard.*[28] [Author's emphasis]

Rose and Louis de Freycinet

The notion that somehow 'structured' official and professional versions of exploration diaries and depictions are more reliable can now be examined using the example of Rose and Louis de Freycinet's voyage. A junior officer on Nicolas Baudin's two-ship voyage to France Australe, Louis de Saulces de Freycinet expertly charted parts of the Australian coast. Apparently he was present when the memorial pewter plate left by Willem de Vlamingh at Dirk Hartog Island in Shark Bay in 1699 was brought onboard *Naturaliste*, then under Baudin's second-in-charge Baron J.F.E. Hamelin. It is claimed that de Freycinet would not have agreed with Hamelin's decision to send it back ashore to be re-erected together with a plate or inscription recording his own visit.[29]

During subsequent explorations along the south coast in *Géographe* Baudin encountered the British navigator Matthew Flinders RN in HMS *Investigator*. He too was exploring, attempting to fill in the gaps left by the Dutch, Dampier and Cook. Within weeks of each other they had sailed northwards up Spencer's Gulf in what is now South Australia and separately proved that a passage through to the Gulf of Carpentaria separating New Holland and New South Wales into two large islands did not exist.

After leaving him on friendly terms, the French were forced to have an extended stay at Port Jackson in New South Wales due to the need to refit and re-provision the ships and due to scurvy and illness amongst the crew. They were succoured by the British colonists. Later it became a worry to Governor King that the Baudin expedition, while scientific in ethos, may also have had strategic elements, and possibly secret designs on Van Diemen's Land (Tasmania) as a future French colony. These disparate intentions are expressed in two separate utterances. The first is that of Baudin, the expedition leader, and the other by François Péron, whom Leslie Marchant in his seminal *France Australe* described as the first 'anthropologist' in Australia.[30] Ironically, it was Baudin, who had proceeded through the ranks despite

---

28 Spate, 'The Literature of Cook's Voyages', p. 25.

29 C. Halls, 'The Story of Three Plates', *Port of Fremantle Quarterly*, Winter (1974): 22–4.

30 L.R. Marchant, *France Australe: The French Search for the Southland* (Perth, 1998), p. 115.

his 'lowly' birth to command, who reflected post-revolutionary French sentiments of *Liberté, Egalité, Fraternité* in expressing very strong, and in his own words 'no doubt impolitic', views to Governor King in Sydney thus:

> To my way of thinking, I have never been able to conceive that there was justice or even fairness on the part of Europeans in seizing, in the name of their governments, a land seen for the first time, when it was inhabited by men who have not always deserved the title of savages, or cannibals, that has been freely given them.[31]

On the other hand, and quite ironically, the 'anthropologist' Péron was to urge the overthrow of the fledgling colony in his personal letters to General Decaen, the Governor of Ile de France (Mauritius), thus:

> Once the English Colony is conquered it can be easily defended by our troops against any attack with great force, and since the colony has enough subsistence it won't starve of hunger because of enemy warships. Thus it will be strong enough to hold out against British land and sea forces.[32]

Further he advised Decaen that the colony 'should be destroyed as soon as possible. To-day we could destroy it easily; we shall not be able to do so in 25 years time'. In a postscript recommending that he would prove 'of valuable assistance in such an operation' Péron also advised that 'M. Freycinet, the young officer [presumably Louis, though possibly his brother Henri], has especially concerned himself with examining all the points upon the coast of the environs of Port Jackson which are favourable to the landing of troops'.[33]

After recuperating at Port Jackson, Baudin sent Hamelin and his ship home with expedition records and with some unique flora and fauna specimens – many destined for the Empress Joséphine's garden at Malmaison on the banks of the river Seine. A small locally-built schooner *Casuarina* was purchased in its stead and de Freycinet was elevated above others, including Henri, to command. His brief was to proceed in company with *Géographe* and to undertake surveys in shallow waters. Thus de Freycinet and Baudin came to complete many fine surveys, though the relationship between the two was evidently tense. Many places on these coasts now bear French names. Of the charts produced by the voyage, de Freycinet's *Carte Générale de la Nouvelle-Hollande, Carte Général de*

---

31 Letter from Baudin to Governor King dated 23 December 1802. Cited in Susan Hunt and Paul Carter, *Terre Napoléon: Australia through French Eyes 1800–1804* (Sydney, 1999), p. 5.

32 Letter from Péron to General Decaen. Cited in Hunt and Carter, *Terre Napoléon*, p. 54.

33 Letter cited in Ernest Scott, *The Life of Captain Matthew Flinders, R.N.* (Sydney, 1914), pp. 315–37.

la Terre Napoléon à la Nouvelle-Hollande and *Carte de la Baie des Chiens-Marins* (Shark Bay) are especially notable.

Baudin died at Mauritius on their voyage home. On the return of his expedition to France, responsibility for the production of an official account to be entitled *Voyage de découvertes aux terres australes* first fell to Péron. As an expert hydrographer, and a key part of the entire Baudin voyage in his own right, Louis de Freycinet was required to have a role in finishing the report as this letter to his brother Henri indicates:

> ... the minister of the Navy requested that I cooperate in the writing of the naval and geographic section of the voyages of discovery. I received an extremely flattering letter ... I must talk to you now of Péron, my dear Freycinet ... Required as he was ... with recording the historical part of the voyage, he has already made significant progress ... There we are dear Freycinet and this is where we are up to with our work on our expedition.[34]

While it was to be expected that de Freycinet would have made a significant contribution to the work being completed by Péron, in the context of the question 'who do you trust?' the following words – which were almost penned as an afterthought – are damning in the extreme.

> Another thing, *the name of our infamous commander will not appear in this work. His memory and his name will never corrupt this glorious work which was achieved in spite of him and his perversity* which surfaced at every step. [Author's emphasis]

Péron died from tuberculosis in 1810, aged 35, leaving Louis de Freycinet to complete the second volume. With clear evidence of his bias and negative intent present in the earlier letter to his brother, the fact that Baudin is totally written out of the work comes as no surprise. Here is a case of the official report being entirely unreliable in one crucial aspect at least, that of the role of its leader. It is only in recent years that Baudin's legacy has been re-assessed and that he has begun to take his rightful place as the leader of an extremely productive voyage.[35]

---

[34] Cited in a letter from Louis Claude de Saulces de Freycinet to his brother Henri de Saulces de Freycinet dated 30 October 1806. The original letter, in the possession of the Woodside Valley Foundation in Western Australia, appeared in the Western Australian Museum's *Journeys of Enlightenment* exhibition adjacent to de Freycinet's charts and a replica of Rose de Freycinet's distinctive tent used at Shark Bay.

[35] See Jacqueline Bonnemains, Elliott Forsyth and Bernard Smith (eds), *Baudin in Australian Waters: The Artwork of the French Voyage of Discovery to the Southern Lands 1800–1804* (Melbourne, 1988), and the chapter by Jean Fornasiero and John West-Sooby in this volume.

When his full report was completed in 1816, the newly-married Louis de Freycinet proposed another voyage and was successful in obtaining the support of the Institute of France and King Louis XVIII. He then took command of the corvette *Uranie* with a complement of 120 men and 23 officers, including artists, a priest, a civilian artist Jacques Etienne Victor Arago and a stowaway, the 22-year-old Madame Rose de Saulces de Freycinet (née Pinon). Perhaps with prior knowledge of Anne and Matthew Flinders' enforced separation,[36] de Freycinet's preparations for smuggling her on board began almost as soon as he was appointed commander of the *Uranie*. It also appears that he surreptitiously extended the accommodation on board the ship to suit.[37]

Notwithstanding the illegal nature of the subterfuge, in Rose de Freycinet's account of her presence on board there exists an anthropological element that with the passing of time transcends most other results of the *Uranie* voyage. Though Rose de Freycinet was not the first woman to circumnavigate the globe, she is the first to have provided a written account. Sadly, Jeanne Baré, who was on Bougainville's round the world expedition of 1766–69, as assistant to the naturalist and physician Philibert Commerson, kept no records. As a result she is remembered only through a plant species named after her, the shipboard journals and the accounts of others. Jeanne Baré was not completely ignored by the officials, however. In one note granting her a pension for life, for example, she is deservedly described by the office of the Ministère de la Marine in the context of her contribution to natural science and her circumnavigation as an 'extraordinary woman'.[38] De Freycinet was also preceded by Marie Louise Victoire Giradin. Disguised as a man, she sailed on board the *Recherche*, one of d'Entrecasteaux's exploration ships, but died towards the end of the voyage.

Thus Rose de Freycinet was to become the first European woman to provide an account of her circumnavigation of the globe. This account appears in the form of private letters to her friend Caroline Baronne de Nanteuil. As such her record arguably far surpasses that by the next known circumnavigatrix in its importance as a primary source, that of Gabrielle-Anne Cisterne de Courtiras, Countess de Saint-Mars. She circumnavigated the world between the years 1843–51. A keen observer of customs and society, de Courtiras' account was published in Paris magazines in 1855, and the next year as the *Journal d'une Parisienne* written

---

36 De Freycinet was on board when Baudin and Flinders met at Encounter Bay and was aware of his subsequent incarceration on Ile de France (Mauritius). See Anthony J. Brown, *Ill-starred Captains: Flinders and Baudin* (Adelaide, 2000).

37 Marc Serge Rivière, *A Woman of Courage: The Journal of Rose de Freycinet on Her Voyage Around the World 1817–1820* (Canberra, 2003), p. xviii.

38 John Dunmore, *Monsieur Baret: First Woman Around the World. 1766–68* (Auckland, 2002), pp. 185–6. Quoting the office of the Ministère de la Marine.

by Alexandre Dumas.[39] Reproduced in French, German and Belgian editions it appeared in English in 1944 as *The Journal of Madame Giovanni*. Though the events, people and places described are historically correct, according to Honore Forster of the National Library of Australia, Dumas added fictional detail – vastly reducing the import of the work. The third to produce an account of her circumnavigation was a Viennese woman, Ida Pfeifer, who eventually published five books dealing with travels that commenced in the year 1844, one entitled *A Woman's Journey Round the World*. Her work received great acclaim and it attracted the attention of noted scientists and geographers.[40]

Like her two predecessors, who produced no record, yet in contrast to her two immediate successors, who published their works, Rose de Freycinet, appears to have had no grand design in respect of publication, either during or after her voyage. Delivered to Caroline on her arrival home in 1820, the letters remained in the archives of the Nanteuil family for nearly a century until 1910, when they were given to Rose de Freycinet's great-nephew Baron Henri de Freycinet. A decade or so later when scholar, Charles Duplomb wrote requesting permission to publish the 'diary' (which was then located in the Limousin region) de Freycinet wrote:

> I plan to go to the Limousin for a stay of a few weeks and shall be in a position to note everything that relates to the shipwreck *Uranie*. The journal in question concerns too many people to be published, and its extracts will have to be rigorously vetted, for its author had no thought of ever publishing it. It is written in the form of letters to an intimate friend, Caroline de Barillon, to whom, I think she gave it on her return to France. I received it a few years ago from my cousin the Baron de Retours, née de Nanteuil, and a descendant of the Barillon family.[41]

Thus the Duplomb manuscript was edited by Baron de Freycinet and 515 copies were published in 1926 with illustrations by the artists Jacques Arago and another expedition artist J. Alphonse Pellion.[42] Notwithstanding their edited nature, through her letters Rose de Freycinet has provided a new and a quite unique lens on exploration voyages. These descriptions included – apart from

39 See Honore Forster, 'Voyaging through Strange Seas: Four Women Travellers in the Pacific', *National Library of Australia News*, 10.4 (January, 2000): 3–6.

40 The work is reproduced in Forster, 'Voyaging through Strange Seas', pp. 4–5.

41 Letter dated 6 April 1923 from Henri de Freycinet to Charles Duplomb. See Rivière, *A Woman of Courage*, pp. vii–viii.

42 Rose de Freycinet, *Campagne de l'"Uranie" (1817–1820); journal ... d'après le manuscrit original accompagné de notes / par Charles Duplomb* (Paris, 1927) and Rose de Freycinet, *Journal de Madame Rose de Saulces de Freycinet, d'après le manuscript original accompagné de notes* (Paris, 1927).

detailed comment on clothing and other personal matters – what one influential commentator characterized as 'a keen observing eye for customs'.[43]

The first English-language work to be translated from the Duplomb transcript was produced in 1962 by the Australian scholar Dame Marnie Bassett entitled *Realms and Islands: The World Voyage of Rose de Freycinet in the Corvette Uranie 1817–1820*.[44] This was followed in 1996 by a work by the then Mauritius-based Professor Marc Serge Rivière entitled *A Woman of Courage: The Journal of Rose de Freycinet on her Voyage Around the World 1817–1820*.[45]

Despite the importance of these two works, they are translations of an edited source and as a result the unabridged and unedited copies of Rose de Freycinet's original letters still require detailed study. The originals are yet to be found, however. Notwithstanding, an original of Charles Duplomb's transcript was recently purchased by the Woodside Valley Foundation and there is hope that this will eventually provide scholars with the opportunity for study. Those few pages opened for the Western Australian Museum's *Voyages of Grand Discovery* exhibit show that Duplomb's transcription appears in blue ink with pencil corrections by Baron de Freycinet, and in red ink, corrections and deletions by Duplomb himself. These, and the materials beneath them, still remain legible.[46]

An example of the need to examine the Duplomb transcripts is again evident in examining Bassett's and Rivière's individual selections for their two works on Rose de Freycinet. They regularly chose different letters on which to comment, and, as expected, given their different backgrounds, there are also instances where their translations of the same letters or phrases differ. One important example is the recovery of the de Vlamingh plate in September 1818. The day after the ship arrived in Shark Bay a boat was despatched to Inscription Point on Dirk Hartog Island in order to recover the memorial. In recovering the relic, contrary to his former Captain Hamelin's wishes, de Freycinet used a time-honoured logic, referring to its exposed location, the possibility of damage or its recovery by less well-intentioned others.

> Believing that such a rare plate might again be swallowed up by the sands, or else run the risk of being taken away and destroyed by some careless sailor, I felt that its correct place was in one of these great scientific depositories which offer to the historian such

[43] John Dunmore, *French Explorers in the Pacific* (Oxford, 1969), p. 70.

[44] Marnie Bassett, *Realms and Islands: The World Voyage of Rose de Freycinet in the Corvette Uranie 1817–1820* (London, 1962).

[45] See Rivière, *A Woman of Courage*.

[46] Though the transcript was provided to the Western Australian Museum for the purposes of exhibition, it was agreed that, for conservation purposes (keeping exposure to light to an absolute minimum) only a few pages would be shown and that it would not be copied.

> rich and precious documents. I planned, therefore, to place it in the collections of the Académie Royale des Inscriptions et Belles-Lettres de L'Institut de France.[47]

Rose de Freycinet records the removal of the plate thus:

> 13 September 1818 … Louis sent a boat to Dirck Hartog's Island to remove an inscription left behind by the Dutch to mark their landing around 1600. It is a precious object which we plan to take back to Paris.[48]

While the inference here is that she had some hand in the decision to recover the plates, Bassett's translation reads 'Louis sent a boat to Dirk Hartich's to take away an inscription left by the Dutch who landed here about 1600. It is something precious to take back to Paris.'[49] Further, the artist and author Monica Oppen recently reported on her examination of a photocopy of Rose's letters in the Mitchell Library describing it as 'annotated by Bassett, marking where Duplomb deviated from the original'.[50]

While some would consider de Freycinet's act in removing the de Vlamingh plate 'sound and reasonable',[51] others, including the artist Jacques Arago, did not agree. At the time, he noted that 'M. Fabre had been directed to go in search of a plate of lead, left on this land by its first discoverer. In fact, the plate has been carried away … I abstain from all reflections on the circumstances.'[52] Arago appears to have shared Hamelin's sentiments nearly 20 years earlier when he stated in his journal that it would have been a 'crime' and a 'sacrilege' to remove the plate.[53] Further, it is possible that Freycinet also removed Hamelin's plate for it still remains to be found. Certainly the absence of the two plates created some concern just a year or so later when the Australian-born explorer Phillip Parker King (son of the governor with whom Baudin was earlier in contact) landed at Cape Inscription and 'with eager steps' hurried up to Hamelin's two posts that he could see were still standing. Expecting to view the two plates either on the

47 Louis de Freycinet, *Voyage autour du monde exécuté sur les corvettes de S. M. l'Uranie et la Physicienne pendant les années 1817, 1818, 1819 et 1820* (Paris, 1924), p. 449.

48 Quoted by Riviére, *A Woman of Courage*, p. 50.

49 Bassett, *Realms and Islands*, p. 85.

50 Personal correspondence between Monica Oppen to Michael McCarthy, 20 January 2009.

51 Dunmore, *French Explorers*, p. 73.

52 Jacques Arago, *Narrative of a Voyage Round the World, in the* Uranie *and* Physicienne *corvettes commanded by Captain Freycinet, during the years 1817, 1818, 1819, and 1820* (London, 1823). Footnote to letter LVI.

53 Frank Horner, *The French Reconnaissance: Baudin in Australia 1801–1803* (Carlton, 1987), pp. 174–5.

posts or, at worst, lying detached in the sand at the foot of the posts, his reaction, perhaps reflecting commonly-held perceptions of Indigenous propensities, is recorded by noted maritime historian Marsden Hordern thus:

> On reaching the top, however, they found the plates gone. King 'mortified', and unwilling to accept that a civilized European could have committed such an outrage, concluded that the Aboriginals had removed them.[54]

While awaiting the return of the crew carrying the plate, Freycinet moved *L'Uranie* across to Baie de Dampier (now Broadhurst Bight at Cape Peron) and established camp there. The observatory was duly set up and with water supplies very low, Freycinet arranged for the ship's water tanks to be replenished with fresh water using an 'alembic', or apparatus for distilling sea water.

The Artwork of the Voyage

The original version of Alphonse Pellion's *Baie des Chiens-Marins: Observatoire de l'Uranie* provides what appears to be a detailed record of the camp at Shark Bay. Visible are the distilling apparatus, soldiers, and military tents, arms, and Louis de Freycinet at work on his observations at a table outside Rose's very distinctive tent, with Rose and a boy seated nearby. While the official version of Pellion's work retains the tent, Rose and the boy (who was apparently taken on board at Mauritius) are omitted (see Figure 9.1). The same editing process occurred with one of Jacques Arago's depictions of a large osprey or sea eagle's nest showing him lounging unceremoniously to the right of the nest (as viewed) with a large sailor's tent left (see Figure 9.2). Apparently in choosing what was to appear in the published account, de Freycinet objected to this informality and in Arago's own words the image was altered thus:

> You would have laughed at the singularity of my accoutrement. A straw hat, with a flopped brim and terminating in a point like Robinson Crusoe's, shaded my head: a soldier's musket and bayonet gave me the appearance of a buccaneer ….[55]

Arago also records the land and its people in very negative terms and in one meeting when concerned at a developing impasse and 'the most immovable stillness' of an Elder to whom the others clearly deferred wrote:

---

[54] Marsden Hordern, *King of the Australian Coast: The Work of Phillip Parker King in the Mermaid and Bathurst 1817–1822* (Melbourne, 1997), p. 341.

[55] Arago, *Narrative of a Voyage*, letter LIV.

Figure 9.1 Official view of the *Uranie* camp in Shark Bay, Western Australia, against J. Alphonse Pellion's unofficial depiction of the *Uranie* camp in Shark Bay. Department of Maritime Archaeology, Western Australian Museum. Reproduced with permission

> At last I recollected that I had a pair of castanets in my pocket, and presuming that the sound might please them, by playing a sort of tune on them, I began to rattle them briskly. Judge of my pleasure: the old man rose with astonishment, and ... fell to dancing in such a grotesque manner that we were ready to die with laughing.[56]

The same disparaging approach often appears in his artwork where his depictions are positively unflattering, especially those of the Shark Bay Aborigines. The descriptions he published half a century later under the title 'Memories of a Blind Man' rival Dampier's [in]famous observations of the Aborigines at Cape Leveque for their negativity.[57] They bear no repetition here.

---

56 Ibid.

57 Jacques Arago, *Souvenirs d'un aveugle: voyage autour du monde* (Paris, 1868), pp. 197–200.

Figure 9.2 J.E.V. Arago, 'Nid d'Atour trouve a L'Ille d'irk Hatighs' (1818). Department of Maritime Archaeology, Western Australian Museum. Reproduced with permission

For Rose de Freycinet, the land and its peoples were frightening and inhospitable. She subsequently wrote to her mother advising that it was 'without a single regret' that she departed from 'that hell on earth'. Her mother in turn was to advise her to 'look at the drawings in Baudin's voyage ... and you will have a true idea of these people'.[58] Again this is of fundamental importance, for the imagery of the earlier Baudin voyage is strikingly positive.[59] While it needs be accepted that the images of the Indigenous people that were produced on the Baudin voyage were often of folk inhabiting a far more hospitable part of the land, it can be argued that the land and its people generally were viewed in the context of a post-Enlightenment *Liberté* and *Egalité* (*Fraternité* does not appear on the Baudin expedition letterhead). In comparison the image of the Shark Bay Aborigines produced on the de Freycinet post-Napoleonic voyages were not viewed in this way and are decidedly unflattering in some instances.

The Landing at Timor

The occasion of the landing at Dili provides yet another opportunity to examine the veracity of expedition artwork by the artists J. Alphonse Pellion, Arago and the engraver Pierre Antoine Marchais – whose job it was to prepare the agreed final images for print. There are three completely different views of the same scene which is described by Rose de Freycinet thus:

> We anchored four days ago. Our salute was returned ... The Portuguese Governor ... as soon as he learnt that I was on board, he sent me fruit and fresh bread together with an invitation to dine with him the next day ... I did not possess suitable attire for such formal occasions; a light muslin dress alone, together with a hat decorated with a few feathers, was all the finery I had. The minute our boat landed, from a fortress nearby there was such a loud salute from so many guns, and such a large crowd had scattered on the beach that I needed Louis' arm to reach the end of the gangway without stumbling. The Governor was present with all his staff. When I stepped ashore, he gave me his hand and invited me to share with him the shade of a huge parasol carried by a Timorese slave dressed in strange clothes. A similar parasol was held over Louis's head.[60]

The official and very dignified view of the officers disembarking to a salute of cannon without Rose appears in the published Atlas Historique. Produced

58 Quoted in Bassett, *Realms and Islands*, p. 92.

59 See Hunt and Carter, *Terre Napoléon*.

60 Rivière, *A Woman of Courage*, p. 59.

Figure 9.3 Alphonse Pellion, Depiction of Rose de Freycinet's landing at Dili, compared with the final engraving by Marchais. Department of Maritime Archaeology, Western Australian Museum. Reproduced with permission

Figure 9.4 J.E.V. Arago, 'Reception a Diely'. Department of Maritime Archaeology, Western Australian Museum. Reproduced with permission

by the engraver Marchais, it is based on J. Alphonse Pellion's original entitled *Premiére scene de la Fete donée par le Gouverneur de Diely a Madame de F.* This shows Rose landing in a white dress and bonnet with what are presumed to be the patriotic blue and red trimmings of the French Tricolour. It also bears a number of crosses in pencil, possibly by Louis himself, indicating those figures and detail that were to be discarded for the Marchais final. A large pencilled cross is seen over Rose's figure as she disembarks. (See Figure 9.3.)

For his rendition of the same scene Arago, in a work entitled *Reception a Diely*, shows Rose in a white empire line gown with blue boa feathers (Figure 9.4). In considering which of the informal images is the more reliable, i.e. which aligns better with Rose de Freycinet's own words, Arago's depiction appears to be the more correct, though there is a caveat. Though regularly negative and disparaging in his art and in his writings, Arago was also quite a humorist – mocking in tone even. In his account of the landing, for example, he also depicts an officer upended in the bows of the *Uranie* longboat as it comes to an abrupt halt on the beach. One welcoming officer also appears to be in a state of relative undress.[61] Did this really occur, or was Arago deliberately deconstructing the often-contrived sense of pomp and circumstance that accompanies such events?

61 While the perception could be in error, what looks like braces or ceremonial strapping appears dangling around his trousers.

Certainly his comments and observations are regularly contrarian in tone, suggesting there is a constant bias inherent in his productions.

Notwithstanding the questions raised above, Rose de Freycinet's frank musings on people, places, events and her reflections on her own reactions, and on occasion, even the appropriateness of her own attire in respect to local custom, allows us to view the accounts of both Rose and Arago as important social commentary. In that respect they represent as much of an anthropological resource as, if not more than, the more formal accounts of the officers and officials on board *L'Uranie* and on most other 'official' exploration ships that preceded it.

L'Uranie at Port Jackson

From Timor *L'Uranie* proceeded to Port Jackson and in mid November they dropped anchor in Neutral Bay. The changes evident since his visit in 1802 were a wonder to de Freycinet. The newcomer Arago was overwhelmed and as a result his comments are positively effusive.

Old friends came on board at Port Jackson and a two-storey residence was rented ashore to house the Freycinets and the scientific equipment. On the occasion of their visit to his residence at Parramatta, Governor Macquarie sent a regimental band to serenade the official party as they travelled up the harbour, such was the import of their visit. It was not all plain-sailing, however, for on the first night in town they were burgled and lost a lot of linen, clothing and silverware. Rose's account of this event is well worth reproducing if only to illustrate that one's preconceptions of place and its peoples often influence subsequent perceptions.

> We learnt that during the night our silver, table linen, our servants' clothing and other effects had been stolen from the ground floor of the house we occupy. You know the purpose of this colony and what sort of people are to be found here in plenty; you will therefore not be astonished at this misdeed: might one not say it is *roguery's classic shore*. It would be astonishing not to find thieves here as it would not to meet Parisians in Paris and Englishmen in London.[62] [Author's emphasis]

The Arrival at the Falkland Islands

Uranie left New South Wales in December 1819 for Tierra del Fuego, where, after suffering damage in a severe storm, it was decided to head for the calm waters

62 Quoted in Bassett, *Realms and Islands*, p. 179.

of the then abandoned French settlement of Port Louis in the Falkland Islands. Entering what is now Berkeley Sound in February 1820 *Uranie* struck the rock that now bears its name and the stricken ship was run ashore in a sheltered bay which also now bears the ship's name. Unable to refloat the ship, the French remained ashore for about eight weeks where they were occupied in securing as much as possible – including the de Vlamingh plate and most of the artwork and specimens. Here again Rose de Freycinet provides important unofficial perspectives. One is her account of her husband's near mental and physical breakdown in the period 10–18 March as the wreck, their subterfuge and the terrible conditions ashore appear to have driven him into a deep depression.

> Louis is still ill and stays in bed. He is suffering from nervous colic that I consider comes from the great fatigue he has experienced for 2 months. I stayed up watching him for a great part of the night and he hardly slept ... The calm Louis felt during the night was very well maintained during the whole day ... he is extraordinarily weak ... Louis gets better and better but he does not go out because the weather is very foggy and rough ... As it was very fine Louis decided to go out ... he is rather tired from this little excursion.[63]

Not without considerable mental and physical support from Rose, Louis de Freycinet recovered and continued preparations for their escape from the Falkland Islands. Eventually the castaways purchased an American vessel, the *Mercury*, and renamed it *Le Physicienne*. It had been carrying cannon and munitions despatched from the rebels at Buenos Aires to those at Valparaiso and as a result it sailed under their flag.[64] The Antarctic explorer James Weddell was in the Falkland Islands in the brig *Jane* at the time and he befriended Rose and Louis de Freycinet. They, in return gave him a boat that he named *The Rose* and then sailed off, arriving at Le Havre in November 1820 after a voyage of three years and two months. Weddell subsequently produced a chart of Berkeley Sound showing *Uranie* ashore. For its part the de Vlamingh plate was given to the authorities, lost and then repatriated to Australia after the Second World War. Today it takes pride of place in the Shipwreck Galleries of the Western Australian Museum.

In a search for other primary documents that might assist the Museum in locating the *Uranie* wreck, an illustration entitled *Reliquias d-la corveta Francesca la Urania* was found in the French Naval Archives. Appearing under the heading 'Naufrage de *l'Uranie en 1820*', it showed *Uranie* beached and on

---

63 Ibid, pp 210–11.

64 Ibid., p. 218.

its side and was expected to prove fundamental in locating the site.[65] It did not, however. Like the other artwork depicting the wreck – by the expedition artists Arago, Pellion and Nicolas Maurin – it actually served to delay the eventually-successful location of the site.[66] In not finding it at any of the places depicted, it soon became apparent after locating each of the artists' vantage points that all took considerable licence, exaggerating features of interest, foreshortening all that could be seen, for effect and for the purposes of putting into one frame elements of a much larger picture. This included objects of specific interest – including in one instance Rose de Freycinet's distinctive tent.

While misleading in many respects, and totally unhelpful in locating the wreck, *Reliquias d-la corveta Francesca la Urania* has an interesting, and quite important background in that it provides a little-known, yet important link in a chain of events leading to the present impasse over the Falklands Islands (Les Malouines). While anchored at Port St Salvador after the de Freycinets had left, Weddell recorded that he received a letter from the commander of what he called a 'patriot national frigate' which was then at anchor at Port Louis, not far from the *Uranie* wreck. In that letter Weddell records that he was informed that the Captain was 'commissioned by the supreme government of the United Provinces of South America' at Buenos Aires to take possession of the islands and that he also intended to take possession of the wreck.[67]

Postscript

Rose de Freycinet's story so captured the imagination of contemporary society that she and her husband were to be feted in the salons of Paris on their safe return to France after many adventures – and he was never censured by the officials, his colleagues, the French Navy or the king.[68] After writing an account of the voyages he then contracted cholera in an epidemic that swept Paris in 1832. Though she successfully nursed Louis of the same disease, Rose succumbed to it

65 French Naval archives, PH 71906.

66 Michael McCarthy, 'Rose de Freycinet and the French Exploration corvette *L'Uranie*, (1820): A Highlight of the "French Connection" with the "Great Southland"', *International Journal of Nautical Archaeology*, 34.1 (2005): 62–78.

67 See James Weddell, *A Voyage towards the South Pole, performed in the years 1822–24, containing an examination of the Antarctic Sea* (London, 1827).

68 Bassett notes that after court martial on the loss of his ship where de Freycinet was exonerated and honourably acquitted he had an audience with the King. See Bassett, *Realms and Islands*, p. 249.

soon after aged 37. Devastated and never recovering from his loss, de Freycinet died a decade later, his reputation as a mariner and explorer intact.

Reflecting on Péron's and de Freycinet's character assassination of Nicolas Baudin and on both Péron's and de Freycinet's failure to mention their former leader by name in the account of the Baudin voyage – except on the occasion of his death – it is truly ironic that de Freycinet did not suffer in like manner.[69] With Rose especially, there was plenty of opportunity for comment and dissent. For example, the French author Gabriel Lafond, who a few years later travelled along a similar route to that of *L'Uranie*, recorded that the de Freycinets were 'well remembered, even still discussed' among Dutch, Spanish and Portuguese officials with whom they came in contact.[70] He also noted the potentially-negative consequences of her presence.

> She was also an apple of discord, thrown ... to a crowd of young men whose jealousies and passions could not fail to be aroused. Always of dignified, becoming and discreet behaviours, never herself giving cause for a single derogatory comment, by her mere existence young Mme de Freycinet furnished a topic of conversation likely to disturb the good harmony, even the discipline, essential to a naval ship.[71]

Clearly, had de Freycinet found himself like Baudin with a large party of potentially uncontrollable civilian scientists on board, his 'official' legacy might not have been so positive. In his frank and often contrarian accounts Jacques Arago provides some inkling of what might have transpired when those objecting to de Freycinet's methods, his actions and his account of the voyage were freed from their allegiances as they eventually returned to civilian life. As Lafond notes in respect of Rose, they would have had a great deal to feed upon. When the differences in the official and the unofficial records are examined, in presenting a false perception of the reality, in removing even his wife from the record, it is evident that Louis de Saulces de Freycinet consciously and apparently unashamedly attempted to control contemporary and future perceptions of himself and of his achievements. That the bias and failings in his works are now being exposed is doubly deserved if only in the light of the attempted destruction of his former captain and complete opposite, Nicolas Baudin.

69 See Brown, *Ill-starred Captains*.

70 Quoted in Bassett, *Realms and Islands*, p. 46.

71 Ibid., p. 48.

Chapter 10

'My Own Slender Remarks': Global Networks of Slavery and Sociability in Mary Ann Parker's *Voyage to New South Wales* (1795)[1]

Katrina O'Loughlin

The long history of *Terra Australis* describes a space sketched in lines of speculation, ambition and multiple desire. Even at the end of the eighteenth century – after Cook's exemplary curiosity and minute mapping produced the fabled space of *Terra Australis* as a place of English settlement and incarceration – knowledge of the new continent of Australia remained provisional and speculative. As was the case in its earliest incarnations, the representation of *Terra Australis* responds to contemporary preoccupations and imaginative investments, in this case the intermeshing late eighteenth-century realities of slavery, anti-slavery, trade, transportation and colonialism. Mary Ann Parker's account of Australia shortly after European settlement reveals a network of such influences at work in her encounters with the landscape and culture of the new settlement. *The Voyage Around the World* (1795) positions Port Jackson at its narrative centre, but Parker's perception of New South Wales is shaped by a broader circulation of people, information and ideas through the Atlantic. Parker's approach to the colonial project at New South Wales is formed through anterior networks of hospitality and commerce connecting trading communities and colonial settlements along global travel routes. These

[1] Mary Ann Parker, *A Voyage Round the World, in the* Gorgon *Man of War: Captain John Parker. Performed and Written by his Widow; for the Advantage of a Numerous Family. Dedicated, by Permission to Her Royal Highness The Princess of Wales* (London, 1795). Modern editions of Parker's narrative include Gavin Frye (ed.), *A Voyage Round the World / by Mary Ann Parker* (Sydney, 1991) and Deirdre Coleman (ed.), *Maiden Voyages and Infant Colonies: Two Women's Travel Narratives of the 1790s* (Washington DC, 1998). All following page references to Parker's *Voyage* are to Coleman's edition of 1998 and will be cited in parentheses within the text.

material networks provide powerful cultural and ideological models for the *voyageuse*: the mediating values of Parker's text are the forms of politeness and global hospitality she enjoys in the West Indies and the Cape, complicated by the ethical dilemmas of enforced labour.

The *Gorgon*'s Mission and Itinerary

In 1791, Mary Ann joined her husband Captain John Parker (1749–94) on a voyage to the fledgling colony at Botany Bay. The *Gorgon* – under Parker's command – accompanied the transport vessels of the Third Fleet, carrying critical supplies and manpower for the fragile new settlement. She conveyed a characteristically disparate set of passengers to New Holland. Among the *Gorgon*'s supernumeraries were new convict superintendent David Burton (who was engaged simultaneously by Sir Joseph Banks in botanical work); the Reverend James Bain, the colony's second chaplain; the new deputy surveyor of roads Charles Grimes; and soldier-explorer Lieutenant Peter Rye, whose account of his expedition to the Peak of Teneriffe undertaken during the *Gorgon*'s outward voyage was published in 1793.[2] The *Gorgon*'s passengers also included Philip Gidley King, the newly-appointed Governor of Norfolk Island, who was returning to the colony with his new wife, and a fresh corps of soldiers assembled under Major Francis Grose to replace those marines sent out with the First Fleet.[3]

Mary Ann Parker was thus immersed in a shipboard community marked by a diverse range of interests and perspectives: she, like *Terra Australis*, was caught in a global web of investments and epistemologies. This single voyage navigated a wide variety of knowledge and projects characteristic of European colonialism: scientific, military, maritime, mercantile, administrative, logistical, spiritual, exploratory and entrepreneurial, anthropological, domestic and familial. The *Gorgon*'s passengers were a worldly set with diverse political and cultural connections, having between them knowledge of locations as diverse as Ireland,

---

[2] David Burton (d. 1792), botanist, surveyor and professional gardener; James Bain, (fl. 1789–94); Charles Grimes (1772–1858). Francis Grose (1758?–1814) was another career cosmopolitan whose military appointments took him to America, New South Wales (as lieutenant governor), Ireland and Gibraltar.

[3] Philip Gidley King (1758–1808), naval officer serving in the East Indies and America, the Channel Fleet and India. He married his new wife Ann Josepha Coombe only days before embarking on the *Gorgon*. When the couple left Port Jackson to take up King's post on Norfolk Island in October 1791, Parker lamented the loss of Mrs. King: 'whose affability had so much contributed to the pleasantry of our voyage thus far' (*Voyage*, 211).

Gibraltar, America, the East Indies and India. Parker too was a seasoned traveller: she had lived in France, Spain and Italy as a younger woman, and spoke Spanish fluently. The networked interests of this travelling community contributed to Parker's sense of the nature and significance of the *Gorgon*'s mission, and deeply influenced the composition of her published account. The broad geographical and cultural networks in which late eighteenth-century *Terra Australis* is caught are directly reflected in the *Gorgon*'s itinerary, which provides important contexts for the *Voyage*'s representations. From Spithead, the *Gorgon* sailed to Santa Cruz, Teneriffe; she restocked (without disembarking passengers) again at Porta Praya, St Iago; and observed Trinidad from sea before arriving at the Cape of Good Hope. After a 'residence of several months' at this critical trading post, and an attempt to salvage the cargo of the wrecked supply ship *Guardian*, the *Gorgon* spent seven weeks under sail to New Holland. Her stay at New South Wales was relatively brief before she embarked on a three-month return voyage via the Society Islands, New Zealand, Cape Horn, Falkland Islands and Cape Town again. Within such a voyage, *Terra Australis* emerges as just one new node in a vast networked geography of trade and empire.

On her return to England, Parker – then widowed – published a small travel book by subscription, reluctantly, and in her own account, 'solely for the benefit of her little flock' (206). Her tentative and self-deprecating *Voyage* was published into a fairly crowded market in 1795; interest in the penal experiment at Botany Bay in the first years of its colonization was high.[4] Parker's travel narrative represents, however, not only the first account of the colony written by a woman, but the first published impressions of Port Jackson by a private individual. This hastily and purpose-published account of the east coast of Australia is assembled from a range of documents – diaries, letters, official reports and retrospective narrative – and incorporates more than one traveller's perspective. As a strangely 'sociable' or discursive document in its own right, the *Voyage* provides us with a fascinating insight into the complexes of trade, information and influence

[4] Parker's evident reluctance and authorial self-effacement is perhaps unsurprising: her narrative follows a succession of accounts of the founding of the new colony, including: *An Historical Narrative of the Discovery of New Holland and New South Wales* (London, 1786); Arthur Phillip, *The Voyage of Governor Phillip to Botany Bay* (London, 1789); *An Authentic and Interessting [sic] Narrative of the Late Expedition to Botany Bay: as Performed by Commodore Phillips* (London, 1789); Watkin Tench's *Narrative of the Expedition to Botany Bay* (1789) and *Complete Account* (1793), surgeon and naturalist John White's *Journal of a Voyage to New South Wales* (London, 1790), and John Hunter's *Historical Journal of Port Jackson* (London, 1793). Other accounts of the *Gorgon*'s voyage include Paul G. Fidlon and R.J. Ryan (eds), *Journal and Letters of Lt Ralph Clark 1787–1792* (Sydney, 1981) and James Scott's *Remarks on a Passage to Botany Bay 1787–92* (Sydney, 1963).

traversed by late eighteenth-century voyages, and their contribution to imagining *Terra Australis* in its early years of European settlement.

The Voyage to Port Jackson

Mary Ann Parker sails for New South Wales with a sense of trepidation, buttressed by her conviction of the critical importance of the *Gorgon*'s mission. While reprovisioning at Cape Town on the outward journey, the *Gorgon* crossed paths with the English ships *Neptune* and *Juliana* on their return. These vessels were both convict transports engaged in the characteristically 'mixed business' of colonialism – returning to Europe via China with lucrative cargoes of tea. (The *Gorgon* too mirrored this mixed commercial mode of travelling. She arrives at Port Jackson with fresh administrators and critical supplies sourced primarily from the Cape. She returned freighted with scientific curiosities, new knowledge, live specimens and valuable gifts: 'kangaroos, opossums and every curiosity which that country produced'.[5]) From the *Neptune* and *Juliana*, those on the *Gorgon* learnt of the limited provisions and increasing despair at the colony. In Parker's report, 'we did not receive any favourable account of the place we were shortly going to visit; on the contrary we learnt ... there were only six months provisions in the Settlement, at full allowance' (196). This encounter between vessels is emblematic of the exchange of news and information, goods and personnel made at critical points across the voyage; transactions that powerfully condition the Parkers' approach toward their destination, and even their language for describing it. The returning ships orient the Parkers to a sympathetic and humanitarian view of the colony well before their arrival: 'in short,' Parker reflects, 'every circumstance served to assure us how anxiously they awaited the appearance of our happy bark; and made Captain Parker as anxious to relieve them' (196).

Parker is a sturdy and cheerful traveller across the long voyage, and enjoys the sociability of life on board the *Gorgon*. She glosses over the details of a rough month's voyage from Spithead to Santa Cruz ('little difficulties ... scarcely felt') and happily embraces her 'little sea-amusements in peace and tranquillity of mind. With my companion Mrs King, and the society of the ship, I seldom, if ever, found anything unpleasant, except the pitching of the ship' she declares robustly (198). Parker is curious about the natural and cultural novelties provided by travel, perhaps reflecting the scientific and ethnographic interests of her fellow passengers. The darkest moment of the voyage out finds the *Gorgon*

---

5 *Voyage*, p. 211.

caught in violent squalls among the trade-winds. Even here Parker downplays her fears, emphasizing instead the darkness and dismal sociality of a ship forced under deadlights: 'strong shutters wedged in to prevent a following sea from breaking into the ship' (191).

Waiting to make landfall on the coast of New South Wales, the *Gorgon* is caught again in a huge electrical storm, the ship rocked by an explosive 'ball of fire' over the prow, and struck by lightning. Parker provides a remarkably detailed account of the storm, continuing the scientific diarized approach she favours throughout the *Voyage*: she notes the ship's exact position at sea, the time and direction of the storm, the precise path of lightning from mast to mast, and an inventory of the resulting damage to the ship. This section might represent a direct incorporation of her husband's notes, a textual strategy that Parker uses several times in the compilation of her narrative. However the observer is also confident enough to offer her own conclusions about the atmospheric event: 'if I might hazard an opinion, I should think it was the effect of an earthquake', she suggests (199). Although sensitive to the feelings of others, and of an emotive disposition herself, Parker is rarely melancholy, and never superstitious. Despite their inauspicious arrival on the coast, she turns her mind forward once again to 'anticipate the satisfaction which our arrival would diffuse throughout the colony'.[6]

Parker is more deeply moved by human tragedy than the oppressions of the long sea voyage: she is horrified to come suddenly upon the wreck of the *Guardian* off Cape Town; sympathetic at the loss of several members of the *Gorgon*'s crew to illness or drowning, and attuned to the feelings of their grieving 'comrades' (198). Arriving at Port Jackson, Parker's attention is immediately captured by the other vessels at anchor in the bay: the *Supply* lying idle for want of suitable timber for her masts, and five convict transports. Parker's sympathies are fully aroused at the 'dreadful mortality' that had taken place on board the transports, resulting – in her account – in up to two thirds of the convicts perishing. Here the *Voyage* betrays a rare judgmental tone: 'Their appearance ... will be ever fresh in my memory', records Parker. 'I visited the hospital, and was surrounded by mere skeletons of men – in every bed, and on every side, lay the dying and the dead. Horrid spectacle!' (200). Parker's description of the wasted bodies invokes the notorious loss of life on slaving ships in the Atlantic trade; the subject of intense political and cultural debate in this period. Thomas Clarkson's numerous publications dedicated to abolition contributed to the heightened public discussion of the

[6] This is an effusion which is more than answered by Watkin Tench's account of his feelings on 21 September 1791: 'I will not say that we contemplated [the *Gorgon*'s] approach with mingled sensations: we hailed it with rapture and exultation' (Tim Flannery (ed.) *Watkin Tench's 1788: A Narrative of the Expedition to Botany Bay and a Complete Account of the Settlement at Port Jackson* (Melbourne, 2009), pp. 209–10).

1780s and 1790s, culminating in the British abolition of the Atlantic trade in 1807. Parker's truncated horror at the plight of the convicts takes up the terms of these metropolitan slavery debates, recalling Clarkson's depiction of the terrible mortality and conditions of transportation for African slaves. Their situation on board 'is beyond all description' he writes: 'confined in a pestilential prison and almost entirely excluded from the cheerful face of the day, it remains for the sickly survivors to linger out a miserable existence, till the voyage is finished'.[7]

Parker's earliest encounter with the Botany Bay convicts is haunted by an uneasy conjunction of transportation and slavery. In his *Essay on the Slavery and Commerce of the Human Species* Clarkson had tackled this connection directly, considering the ethics of slavery explicitly in terms of penal labour or 'that slavery, which is now adjudged to delinquents, among many of the states of Europe'.[8] Despite his provocative use of the term 'slavery' to describe both, Clarkson rejects the analogy between convict labour and slavery on twin grounds. The first is the sphere or purpose of labour: unlike the slave whose labour serves private wealth, the convict 'delinquent' is employed in public work. This is a just exchange in Clarkson's argument, 'because as the crimes they have committed are considered to have been crimes against the public', the criminal labours for the benefit of the public: 'no individual can justly receive the emoluments of their labour'.[9] Clarkson's second objection goes to the heart of questions of human subjectivity and self-ownership: unlike convict labour, he argues, slavery necessarily 'includes the idea of property ... to undergo which, the delinquent must previously change his nature and become a brute'.[10] In Clarkson's influential reasoning, slavery subjects people to the dehumanizing effects of commerce and subjugation, an objectification which is indefensible in moral and ethical terms.

Clarkson's careful distinction between slavery and transportation – a device that allows him to censure slavery while upholding the rule of law – works paradoxically to draw the practices closer together, and to reveal their historical and ideological proximities. It is precisely these dehumanizing factors

---

7 Thomas Clarkson, *A Summary View of the Slave Trade and of the Probable Consequences of Its Abolition* (London, 1787), pp. 139–40. For a list of Thomas Clarkson's publications in the period just prior to Parker's *Voyage* please consult the bibliography. From the *Voyage* one must conclude that Parker and her husband were conversant with the arguments of the abolitionist movement, if not Clarkson's writing specifically.

8 Clarkson's *Essay* was widely circulated and republished. The edition used here is *An Essay on the Slavery and Commerce of the Human Species, particularly the African ... Translated from a Latin Dissertation, which was honoured with the First Prize in the University of Cambridge. For the Year 1785 ... With Additions* (Philadelphia, PA, 1804), p. 117.

9 Clarkson, *Essay*, p. 118.

10 Ibid., p. 118.

of commerce and brutalization denounced by Clarkson that return, almost unbidden, in Parker's response to the emaciated bodies being unloaded from ships at Botany Bay:

> It makes me shudder when I reflect, that it will not be the last exhibition of this kind of human misery that will take place in this country, whilst the present method of transporting these miserable wretches is pursued; for, the more of them that die, the more it redounds to the interest of the ship-owners and masters, who are paid so much a-head by the government, for each individual, whether they arrive in the colony or not. (200)

Parker observes an economic arrangement between government and transport contractors that reduces a human cargo to chattels. Redolent of the practice of slave traders dumping entire consignments of slaves overboard to claim insurance payments, under the current arrangements for transportation convicts are in fact worth less than the price of their passage: the commercial contract exceeds the 'worth' of the human person (who is brutalized by such an exchange), and the convict is rendered a form of profit to the private individual. The public contract of crime and punishment approved by Clarkson is muddied by the powerful prerogatives of private commerce, an entanglement mirrored with devastating effect in the trading activities of the transports (whose mixed obligations were repeatedly cited as a contributing factor in the colony's starvation).[11] The potent physicality of Parker's description – 'the poor miserable objects' and 'mere skeletons of men' – picks up the currents of these contemporary slavery debates and underscores the discomforting connections between cargoes of felons and slaves already traced by the *Gorgon*'s outward voyage through the Atlantic.

However impassioned, this is not actually – or exclusively – Parker's language: the account of the wretched convicts is taken from her husband's report to Admiralty.[12] Parker defers to the authority of her husband, whether politely or shrewdly, in including such implicit criticisms of government arrangements for transportation. The narrative skirts around, without taking up directly, the controversial connections between convict transportation and the slave trade that so incensed a sister-traveller Anna Maria Falconbridge in her *Two Voyages to*

[11] For accounts of the intertwining discourses of slavery and transportation from the late seventeenth century see Gwenda Morgan and Peter Rushton, 'Print Culture, Crime and Transportation in the Criminal Atlantic', *Continuity and Change*, 22.1 (2007): 49–71, and Alan Lester, 'Colonial Networks, Australian Humanitarianism and the History Wars', *Geographical Research*, 44.3 (September 2006): 229–41.

[12] See Coleman's annotation to the text which attributes the material to Captain John Parker's report to Admiralty, 2 July 1792 in *Maiden Voyages*, pp. 200, 223.

the River Sierra Leone.[13] Not only were these two accounts by women observers of new British colonies published at the same period (1794/95), the travellers crossed paths at Spithead in January 1791. Both accounts are marked by scenes of horrified encounter between the travel-observer and the broken bodies of colonists. Reflecting on the incarcerated passengers of the *Gorgon*, Falconbridge insists that 'the sight of these unfortunate beings, and the thought of what they are to endure, have worked more forcibly on my feelings, than all the accounts I have ever heard or read of wretchedness before'.[14]

In the penal settlement at Botany Bay, Falconbridge discerns a troubling kinship between transportation and slavery, and uses this to great polemical effect in her own narrative of the failing freed slaves' settlement at Sierra Leone. Falconbridge's distressed response to the sight of the 'wretched' convicts on board the *Gorgon* – still moored on the English coast – effectively elevates their suffering above both traded slaves (with whose accounts she is familiar as a result of her husband's abolitionist work), and the emancipated slaves resettled in Africa.[15] The struggling colony established for freed slaves at Sierra Leone by British abolitionists and philanthropists is a peculiarly uncomfortable doppelganger for Botany Bay in the 1790s, having been an alternative proposed site for British penal settlement. Falconbridge's account of the appalling condition of the first wave of settlers at Sierra Leone – a group which controversially included emancipated slaves, black loyalists and white women – mirrors Parker's condemnatory account of the 'trade' in convicts, but without observing her niceties. 'It is scarcely possible that the British Government, at this advanced and enlightened age, envied and admired as it is by the universe, could be capable of exercising or countenancing such a Gothic infringement on human Liberty', declares Falconbridge sarcastically, carefully calibrating slavery and colonialism for sharpened polemic.[16]

---

13 Anna Maria Falconbridge, *Two Voyages to the River Sierra Leone During the Years 1791–2–3* (London, 1794), in Coleman (ed.) *Maiden Voyages*, pp. 45–168.

14 Ibid., p. 50.

15 Alexander Falconbridge had published his own *Account of the Slave Trade on the Coast of Africa* (London, 1788) prior to his employment by the Sierra Leone Company. His experiences as a surgeon on slave ships were also a key source of information for Thomas Clarkson.

16 Falconbridge, *Two Voyages*, p. 66. Falconbridge comes to understand slavery as a more humane and better-managed practice than European repatriation of black labourers, and her account is carefully calculated to support her financial complaints against the colony's Directors. See Deirdre Coleman's work on Falconbridge, 'Sierra Leone, Slavery and Sexual Politics: Anna Maria Falconbridge and the "Swarthy Daughter" of Late Eighteenth-Century Abolitionism', *Women's Writing*, 2.1 (1995): 3–23; and 'Introduction' to *Maiden Voyages*; also Katrina O'Loughlin, '"Thorns and Thistles": Anna Maria Falconbridge's *Two Voyages to the River Sierra Leone*', in the unpublished thesis 'The Paper Globe: Subject and Sociality in Eighteenth-Century

'Shrubs, Birds and Beasts of Botany Bay'

While Parker resiles from the explicit and critical conjunction of slavery and colonialism deployed by Falconbridge, these two accounts of British settlements in the 1790s are linked by common language and imagery, a shared composite textual construction, and their purposed publication for financial return. They read like sallies in a conversation where the participants never quite meet. Like Falconbridge at Sierra Leone, Parker perceives Botany Bay as an 'infant colony': embryonic, vulnerable, and desperately dependent on the goodwill and good management of her governors.[17] The natal state of the colony is inscribed in the physical landscape and 'pleasing novelties of Sidney Cove' (204). Under Parker's eye the new colony resembles a child's entertainment: 'the little habitations on shore, together with the canoes around us, and the uncommon manners of the natives in them were more than sufficient amusements', she writes after her first visit on shore (204). In Parker's account the landscape along the Hawkesbury displays potential for improvement through orderly development and application of convict labour – but only with husbandry and the intensive assistance of the government. The new settlement at Parramatta seems to sit precariously upon the landscape; Parker confesses she is pleasantly 'surprised to find that so great a progress had been made', although this professed advancement is rather undercut by her account of the actual improvements:

> There is a very good level road of great breadth, that runs nearly a mile in a straight direction from the landing place to the Governor's house, which is a small convenient building upon a gentle ascent ... On both sides of the road are small thatched huts, at an equal distance from one another. (206)

Watkin Tench's account of the same period deploys the expectant state of the settlement for more comic (and less flattering) effect: 'In a colony which contains only a few hundred hovels, built of twigs and mud, we feel consequential enough already to talk of a treasury, an Admiralty, a public library.'[18]

Women's Travel Narratives' (University of Melbourne, 2009). See also Emma Christopher's discussion of the connections between slavery and transportation in the 1780s and 1790s: 'A "Disgrace to the Very Colour": Perceptions of Blackness and Whiteness in the Founding of Sierra Leone and Botany Bay', *Journal of Colonialism and Colonial History*, 9.3 (Winter 2008).

17 For the currency and contested nature of the language of the 'mother country' and her 'infant colonies' see Coleman's 'Introduction' and Morgan and Rushton's 'Print Culture, Crime and Transportation', pp. 60–64.

18 Watkin Tench, in John Cobley (ed.) *Sydney Cove: 1791-1792* (Sydney: Angus and Robertson, 1965), p. 173.

A deep reliance of colony on 'the mother country', a dependence that is cultural, ideological and material, persists despite – or perhaps precisely because of – the huge geographical dislocations between these sites. Parker's account confirms the critical need for increased passages between Britain and New Holland, and expanded trade with all other major ports, to make the colony viable. The *Voyage* confirms the lack of encouragement and support for free settlers as well as convicts, and calls for the continuing provision of every necessity – seeds, tools, housing, more labour and expanded public stores – in order that the settlement be made able to bear a new community: 'Until that takes place the maintaining of the Colony of New South Wales will be a continually accumulating burthen to the mother-country', Parker attests (202). The importance to the infant colony of this hybrid civil and commercial support is underscored by Parker's inclusion of her husband's long report on the propriety of establishing 'a commercial Whale-Fishery on the coast of New Holland' (200–204). This proposal advocates luring whaling vessels away from the western coastline of America to the richer waters of Australia, simultaneously strengthening commerce and creating more regular traffic between Europe, Africa and the Pacific. Parker promotes the rich natural resources and possibilities of the new colony, but only under the careful superintendence of government, and in the context of carefully-regulated trade.

Once again, however, the language of maternity encircling the 'infant colony' in the *Voyage* is Captain Parker's not his wife's: the use of the metaphor of maternity is feminized, but unexpectedly gendered. Mary Ann herself does not take a specific or sustainedly maternal attitude to the colony, but a broader humanitarian approach to the support and relief of the settlement in familial and communal terms. In particular the authoress emphasizes the critical role of husbandry in the colony, no doubt influenced by the 'pressure of mind' she describes overshadowing her narrative, 'occasioned by the unexpected loss of him, who was indeed an indulgent husband and a tender parent' (220). Parker presents a fragile infant colony presided over by the impressive figure of Governor Phillip, whose house 'always proved a home to me':

> – he may be justly called, like the Monarch of Great Britain, 'The Father of his People;' and the Convict, who has forsaken the crimes that sent him to this country, looks up to him with reverence, and enjoys the reward of his industry in peace and thankfulness. (211)

The 'fatherly attention' of the governor to the Parkers is an expression of his sovereign influence over the entire settlement. The *Voyage* highlights the redemptive possibilities of the colony under Phillip's benevolent governance, creating a protective chain of God, King, Governor and Father around the struggling settlement. As Deirdre Coleman points out, such a strategy of representation simultaneously figures the redeemed convict 'as a grateful, docile child'.[19]

Despite the highly provisional status of the colony and the uncooperative nature of this new territory, Parker finds immense beauty in the landscape of New South Wales, often in the somewhat unconvincing terms of an English rural aesthetic. Parker's picturesque imagery of 'tufted woods and verdant lawns', and her favourite native plant (a white flowering shrub 'very much resembling our English Hawthorn'), represent an embryonic discourse of Australian landscape which emphasizes both the alien nature of *Terra Australis* and the overpowering conventions of a European aesthetic. 'We were all admiration at the natural beauties raised by the hand of Providence without expense or toil', enthuses Parker: 'The gentle ascents, the winding valleys, and the abundance of flowering shrubs, render the face of the country very delightful' (204). The *Voyage* displays Parker's genuine interest and pleasure in flora and fauna, and particularly birdlife: she personally superintends the return of a large number of live birds to England as specimens and personal pets. Her travelling menagerie testifies to the scientific and domestic contexts of collection in the New World. Characteristically, however, and almost certainly due to her own difficulties, Parker simultaneously has an eye to the commercial value of these colonial novelties.[20] She even reports the occasional epicurean luxury like feasting 'upon Oysters just taken out of the sea; – the attention of our sailors, and their care in opening them and placing them round their hats, in lieu of plates, by no means diminishing the satisfaction we had in eating them' (209). Parker's language renders this beauty alien and strangely superficial however: flowers seem sprung 'from rock itself'; only the 'face of the country very delightful' (204). Aspects of this verdant scenery are simply illusory: there exists 'plenty of grass, which grows with the greatest vigour and luxuriance', the writer observes, '[but] is not of the finest quality', serving horses and cows better than the colony's starving sheep (204).

Like Watkin Tench's contemporaneous account of the colony, Parker's descriptions reveal the extreme limits of European knowledge of the place in which they found themselves; the persistence of a *terra incognita* at the heart of *Terra Australis*. After three years of settlement, areas north or west of Pittwater remain largely unknown, and the mountainous bush impenetrable. Many of the rivers and bays even close to the settlement are unexplored or uncharted. The colony seems dangerously circumscribed by its own ignorance of the landscape

---

19 Coleman, *Maiden Voyages*, p. 30.

20 Invoking comparative value (in monetary terms) is perhaps also one of the ways the traveller is able to convey the relative or incommensurate meanings of these unfamiliar places to her readers.

just 20 to 30 miles beyond the settlement. The reality of the levels of ignorance at this period of the settlement's history are embedded in the travails of a group of Irish convicts who became known as the 'Chinese travellers'. Twenty-one convicts, including a pregnant woman, escaped the settlement shortly after their arrival, fully expecting to cross a river immediately to their north and arrive in China on foot.[21] The expectations and experiences of the 'Chinese travellers' are a poignant illustration of the very real spatial dissonance facing convicts from compact geographies like England and Ireland: they could have little conceptual framework for the isolation and expansiveness of this new land. In contrast to the more entrepreneurial cast of scientific and civilian administrators (like Parker, Tench, Grimes or Rye aboard the *Gorgon*), many convicts could literally have not imagined the space and distances they encountered. In this context of profound epistemological isolation, for convicts who arrived indentured and with little understanding of their new environment, the sense of total powerlessness compounded by illness and incarceration, must have indeed resembled slavery.

Parker's image of the penal colony at New South Wales is neither utopian nor wholly dystopian (in the typology outlined by Richard Phillips) but unformed or embryonic in nature.[22] The east coast of Australia is literally an emerging space, or a space of emergence, for new relations among people and landscape, and thus appears deeply inchoate. While highly provisional, the nascent landscape and sociality of the settlements along the Hawkesbury continue – as is common in colonial settings – to refer back to the world of English culture and politics. However, Parker's representation of the colony at New South Wales complicates the metropolitan and peripheral dynamic of Phillips' model. The *Voyage* transcribes an added dimension of global meaning, referring more widely, more diffusely, not only to domestic English politics, but to those European networks of colonial culture and discourse navigated by the *Gorgon* and her passengers. This dispersed pattern of influence and reference is partly an effect of her assembly of her *Voyage*: Parker's free inclusion of the work of her husband and other travellers produces a highly intertextual account of Botany Bay. It also reflects the economic and global ambitions underpinning the establishment of the new colony. The settlement of Botany Bay (like the proposed Whale-Fishery) was envisaged as a competitive 'node' in global trading networks: a major port of refreshment and reprovision modelled on the extant

---

21 See Collins and Governor Phillip's accounts reproduced in Cobley (ed.), *Sydney Cove 1791-1792*, pp. 141 and 160. See also Tench in Flannery (ed), *Watkin Tench's 1788*, pp. 209–11.

22 Richard Phillips 'Dystopian Space in Colonial Representations and Interventions: Sierra Leone as "The White Man's Grave"', *Geografiska Annaler*, 84.3–4 (2002): 189–200.

examples of Cape Town and Rio di Janeiro.[23] But the collaborative and reticulate nature of the *Voyage*'s representations also testify to the 'web-like spatiality' of colonial geography and discourse at this period; the hybrid and networked nature of information, imagery and experience characterizing colonial relations in the eighteenth century.[24]

'The Inhabitants of New South Wales'

The Aboriginal people of Sydney Cove and the Hawkesbury are subject to perhaps the greatest pressure from circulating and contesting discourses of slavery and romantic sociability in Parker's narrative. Like her account of the colony's convicts, Parker's descriptions of the people of New Holland capture the contingent nature of her travel discourse by drawing on the treatment of black subjects across the multiple sites of the traveller's experience. The text describes a curious contrast between Parker's carefully 'ethnographic' account of Aboriginal bodies – 'straight' and 'well made', but 'frightful' looking and 'loathsome' smelling – and her perception of their sensibility.[25] The account mingles descriptive physiognomy with personal revulsion. The women, generally a group of especial interest for women travellers, are described by Parker as 'extremely negligent of their persons, and ... filthy to a degree scarcely credible: their faces and bodies are besmeared with the fat of animals, and the salutary custom of washing seems entirely unknown to them' (209). Despite her severe judgement of personal appearance and cleanliness, Parker insists on the 'extraordinary honesty' and gentleness of these 'Inhabitants': qualities cast as 'natural goodness of heart' (210). A familiar reference to the idea of the 'noble savage', Parker's account of 'natural goodness' tentatively recognizes – or confers – critical forms of sovereignty on Aboriginal people: both in terms of their prior status as 'Inhabitants' of the land, and their manifest sensibility, a primary category of subjectivization for Parker with respect to others. Parker

---

23 Angus R. McGillivery suggests highly commercial motivations for the colony's foundation from the outset, identifying 'a strategically planted colony [designed] ... to fulfil the eighteenth-century maritime imperatives and visions of Pacific empires'. See 'Convict Settlers, Seaman's Greens, and Imperial Designs at Port Jackson: A Maritime Perspective of British Settler Agriculture', *Agricultural History*, 78.3 (2004): 261–88, 263.

24 Alan Lester argues that this 'networked' nature of colonial relations might also provide a more productive model of colonial history, undercutting simple metropolitan-periphery arrangements, and enabling 'us to think about the inherent relationality of nodal points or "centres" within an empire' ('Colonial Networks', p. 231).

25 See Falconbridge, *Two Voyages*, pp. 209–10.

insists 'I never felt the least fear when in their company', a state of comfort she attributes to the observation and preservation of certain forms of politesse in all engagements. Parker's rapport with the people of the Hawkesbury depends on a performative exchange of courtesy, however tentative or superficial the resultant friendship: 'Had I objected or shown any disgust at their appearance, it would have given them some reason to suppose that I was not what they term their damely, or friend', explains Parker: 'and would have rendered my being in their company not only unpleasant, but unsafe' (209–10).

Despite her own assertion of comfort and safety, the *Voyage* reveals the author's repeated anxiety about being 'surrounded' or enclosed by larger groups of both convicts and Aborigines. Parker responds to this discomforting dynamic by infantilizing both:

> The natives very frequently surround our vessel with their canoes. The women often held up their little ones, as if anxious to have them noticed by us. Sometimes, for the sake of amusement, I have thrown them ribbands and other trifles, which they would as frequently tye around their toes as any other part of their person. (210)

The *Voyage* places critical value on controlled and polite exchange between European and 'native': a strange form of settler diplomacy or reversed hospitality where the colonizer governs the terms of personal intimacy and exchange. The exchange of 'ribbands and other trifles' is a source of amusement to Parker (and offered in the same way to her readers) because they are sumptuary items 'misapplied', rendering the wearers naïve and childlike. This exchange of tokens and, in Parker's first example, the polite conventions of hospitality (under the sign of the damely or friend), represents an expansion and reconfiguration of the global sociability Parker champions across the voyage: a combination of sociable, humanitarian and familiar techniques. In fact Parker rather boldly concludes that Europeans in Australia may have been negligent in their duties of hospitable exchange toward Aboriginal people, arguing that 'From the treatment I invariably experienced, I am inclined to think favourably of them; and fully believe they would never injure our people, were they not first offended by them' (210).

The text's governing model of polite hospitality is illustrated in Parker's account of an unnamed Aboriginal man.[26] This man, of pleasing countenance and 'a most engaging deportment', observes the greatest courtesy in a visit with Parker: he is seated beside her at the meal, 'change[s] names with Captain Parker', and is politely curious about the author's 'travelling knife and fork' (210). Parker

---

[26] This unnamed man could be Banalong, whom Parker discusses later in the same section, but her text does not make the connection significantly clear.

is pleased to make him a gift of the cutlery, and he reciprocates on a return visit to the ship by displaying his gift and making use of it 'somewhat aukwardly' at a shared meal (210). In Parker's account, this Aboriginal man instinctively replicates and reciprocates her own polite behaviour around the significant object of the 'travelling' knife and fork (and the practice of a shared meal). In her more detailed portrait of Banalong, the protégé of Governor Phillip, Parker carefully expands the subjectivizing language of eighteenth-century feeling to include a 'native' man.[27] Having travelled to England with Phillip, Banalong visits Parker at home after the death of her husband:

> To describe the pleasure that overspread this poor fellow's countenance when my little girl presented to him the picture of her dear father, is impossible; it was then that the tear of sensibility trickled down his cheek; he immediately recognized those features which will never be obliterated from my memory, and spoke, with all the energy of Nature, of the pleasing excursion which they had made together up country. (210)

Parker commends Banalong's manifest sensibility, elicited in response to the portrait of her husband, and predicated on the shared recognition of a father's 'features'. Once again the connection between Parker and the indigenous man depends on an exchange of visits, food and significant objects, which leads to a scene of reciprocated feeling.

Like her engagement with the colony's convicts, Parker's encounter with indigenous Australians is shaped by the contemporary language of slavery and abolition.[28] Quoting 'The Negro's Complaint', William Cowper's poem of 1788, the *Voyage* defends Aboriginal subjectivity in the popular currency of race and feeling:

> Fleecy locks and black complexion
> Cannot forfeit nature's claim;
> Skins may differ, but affection
> Dwells in white and black the same.

---

27 Banalong or Bennelong (c. 1764–1813) was captured and brought to the settlement at Sydney Cove in November 1789 at the instruction of Phillip. In December 1792 he travelled with Phillip to England and was presented to George III. See Eleanor Dark's entry on Bennelong in the *Australian Dictionary of Biography*, vol. 1 (Melbourne, 1966), pp. 84–5.

28 Chapter 8, the section of her narrative which deals with indigenous groups at the colony, announces the segue in its subtitle: *Description of the inhabitants of New South Wales – their huts – their extraordinary honesty – account of Banalong – an instance of his sensibility – observations on the Slave Trade.*

The poem disclaims difference in terms of universal 'affection', echoing Parker's conviction of a 'natural goodness of heart' among Australian Aborigines. But slavery is not 'indigenous' to the early colonial history of Australia: one might expect the situation of American Indians or the Caribs of the West Indies rather than the 'Negro' to provide the terms of reference for Parker's encounter with the 'native' population of New South Wales. The currency – the very immediacy – of debates about slavery and abolition in Europe and the Atlantic at this period clearly prove infectious. The conjunction of slavery and transportation that emerges in Parker's consideration of the convicts, together with the black skins of the indigenous inhabitants seem to provide the language of abolition with special resonance. In this slippage between Aboriginal sovereignty and the institutions of slavery – a curious triangulation of the plight of convicts, indigenes and slaves – Parker characteristically disperses meaning across a complex of contemporary discourses surrounding race, slavery and sensibility. Her use of abolitionary discourse as a model for Australian Aboriginal sensibility, and perhaps even sovereignty, attests the force of the ideological freight accompanying the material networks of global trade.

The 'Travel of Ideas' and Networked Geographies of Colonialism

Structurally, New Holland is the centrepiece of the *Voyage*. However the constant reticulation of meaning I have tried to evoke – the allusive and dispersed quality of Parker's language, imagery and politics to multiple sites and debates outside the colony – means that her representations of *Terra Australis* and its people can be difficult to make sense of without reference to the *Voyage* as a whole. The *Voyage* navigates the networked geographies of late eighteenth- and early nineteenth-century colonialism identified by Alan Lester: the 'travel of ideas', or 'circuits of humanitarian discourse and practice that connected the colonies in Australia with other settler frontiers within Britain's growing empire'.[29] Parker's account of New South Wales testifies to the physical, material and ideological nature of these networks, which are marked by the exchange of goods, people, information and ideas between multiple locations across the Atlantic, Pacific and Indian Oceans. These physical and discursive 'webs' of colonialism evident in the latter part of the eighteenth century represent both an extension and a reconfiguration of the secular and scientific networks of the Enlightenment, which are also immediately evident in the *Gorgon*'s mission and personnel.

[29] See Lester, 'Colonial Networks', p. 231.

The *Gorgon*'s itinerary captures a specific dimension of this global movement of ideas, mapping the multiple locations and reference points for Parker's representations. Santa Cruz and Cape Town not only supply critical goods for the colony: these global port cities provide important commercial and ideological models for the new settlement, and foster influential patterns of sociability – between travelling Europeans, and between the colonial governing classes and their (indentured and dispossessed) 'others'.[30] The historical trading posts traversed by the *Gorgon* promote Parker's identification with a specific 'class' or set of global citizens – European rather than specifically British – and moving between the key ports of trade and empire. This elite colonial class includes figures like Governor Don Antonio di Gutierez and his family at Santa Cruz and the De Witts at the Cape.[31] Merchants too – the wealthy and powerful international traders based at key ports – have a great status in Parker's account, providing warm hospitality (as well as goods) to the needy travellers. A particular highlight of the Parkers' time in Cape Town is the company of the Colonel and Mrs Gordon ('a Swiss lady'): this couple model global and intellectual sociability through their hospitality and 'engaging attention to strangers' (195).[32]

The professional and military exchanges of late eighteenth-century colonialism are also in evidence in the *Voyage*, through figures like Colonel Burrington 'of the Bengal army', whom the Parkers meet at the Cape, convalescing 'for the re-establishment of his health'. The circulation of people and movement of ideas at this period is perhaps best figured in the career of William Dawes. A marine officer sent out with the First Fleet, Dawes returned to England with the *Gorgon* in 1792. A scientist and astronomer, friend of colonial historian Watkin Tench, and authority on Aboriginal languages, Dawes later accepted the governorship of Sierra Leone. He served in the position for three periods

---

30 In his 'Introduction' to *Forging European Identities*, Bernd Roeck describes the important function of these 'relay stations' or 'joints of communicational processes' of cultural transfer and exchange. From the medieval centres of London and Amsterdam to eighteenth-century hubs in Africa and the Caribbean, these locations function as cultural emporia where 'both material and non-material goods are concentrated, and the engines and agencies of exchange cluster' (Roeck, 'Introduction' to *Forging European Identities, 1400–1700, Volume IV Cultural Exchange in Early Modern Europe* (Cambridge, 2007), pp. 1–29, 23).

31 Despite the *Gorgon* being officially a 'foreign Ship of War' the family receives the Parkers warmly and with 'unlimited kindness' (*Voyage*, p. 186).

32 Jimmy Casas Klausen has argued that hospitality or the reception of strangers at this period represents a significant model of human sociability for Diderot, and was used by him to test the ethics of colonization. These paradigms of hospitality might be a helpful way of examining early colonial travel and cultural exchange more generally. See Klausen, 'Of Hobbes and Hospitality in Diderot's Supplement to the Voyage of Bougainville', *Polity*, 37.2 (April 2005): 167–92.

between December 1792 and 1803 (to the manifold disgust of Anna Maria Falconbridge, who considered a penal settlement poor training for a freed slaves colony). In his career Dawes literally traversed the late eighteenth-century geographies of slavery and abolition. After his service at Sierra Leone, and in connection with the Church Missionary Society and William Wilberforce, Dawes travelled through the Caribbean ministering to slaves and their children. He died in Antigua in 1836 working for the anti-slavery cause.

An even more intriguing – and uncanny – figure for the collapsing topographies of the colonial world crossed by Parker is the large shark caught and brought aboard the *Gorgon* on their return voyage through the South Atlantic. In the shark's belly the ship's company discover a convict prayer book bearing the date of its owner's transportation but not of release. In a further extraordinary coincidence, the name inscribed in the convict prayer book is that of one of the *Gorgon*'s mariners, who vigorously denies all knowledge and becomes the butt of the ship's jokes.

In details such as these the key values of Parker's preferred 'travelling' sociality are established, together with the networked nature of the late eighteenth-century world. In Parker's account, travel seems to foster a kind of international and peripatetic salon, where the exchange of hospitality and understanding of the requirements of travel provide the foundations of global sociability. Parker's ease within these elite colonial circles is secured by key exchanges of hospitality organized around courtesy, language and food: objects and significant practices like the travelling knife and fork, the convict prayer book and the oysters presented on a sailor's hat. Parker's fluency in Spanish and familiarity with forms of Spanish sociability at Teneriffe, for example (particularly the distinctive 'compliments di los manos' practised among the women), see her symbolically established as 'Interpreter General' between all parties. The *Voyage* is governed by the same strong sense of connection with other 'voyagers' and 'travellers', and the exemplary forms of sociability exhibited by the colonial classes in the West Indies and at the Cape: ships met in passing, shared experiences, linked projects and the sudden intimacies among strangers produced by long and uncertain voyages. The gifts of lemons 'and such other fruit as ... deemed most acceptable' repeatedly detailed by Parker throughout the voyage represent small but significant tokens of exchange among these citizens of the world.

In her turn Parker conceives of her narrative as 'the only return I shall ever have it in my power to make' for these kindnesses from strangers (189). For Parker, the writing and publication of her *Voyage* represents an extension and 'return' of the exemplary sociability experienced by the travellers at each 'node' of their colonial journey. Writing of course represents one of the most powerful 'networks' of information and ideas in this global cultural economy. Travel narratives, together with the exchange of letters, letters of introduction, news, stories, pamphlets and reports underpin the transmission of language and ideas over this expansive geography. Writing is perhaps the ultimate 'travelling discourse': like the book in the belly of the shark, writing has the power to create and reveal unexpected connections across a narrowing world.

Key features of Parker's perceptions and investments are similarly captured in the style and circumstances of the *Voyage*'s publication. Publication of her book by subscription presents an honourable and respectable way for Parker to draw attention to her husband's role in the colony, to make public some of his own observations, and to raise money for the support of her 'little flock'.[33] Perhaps more significantly, subscription publication also reconstitutes those discursive networks of travel and colonialism that are a feature of her account. Parker's list of 258 subscribers testifies to the middle-class, mercantile, naval and scientific contexts of her *Voyage*. (Her subscribers include Sir Joseph and Mrs Banks, Hannah More, William Dawes, and numerous naval and colonial connections in the West Indies, Bengal Artillery, Madras Infantry, Norfolk, and Officers of the Francis East-Indiaman.) The circulation of her book effectively maps those same civil, mercantile and discursive networks navigated by the *Voyage*, and Parker's subscribers represent an expanded form of hospitality like those small gifts of fruits and greens that sustain long distance travellers.

Conclusion

In her diaries, Elizabeth Macarthur describes Parker as 'a very amiable intelligent woman' with whom she spent 'many pleasant days together'.[34] Our 'little Circle has been of late quite brilliant' writes Macarthur with pleasure after the arrival of the *Gorgon* and her female passengers: 'we are constantly making little parties in boats up & down the various inlets of the Harbour, taking refreshments with us, & dining out under an awning upon some pleasant point of Land.'[35]

---

[33] In her Preface the authoress deliberately refutes the unjust and injurious report that she is 'worth a considerable amount of money' by insisting that 'nothing but the greatest distress could even have induced her to solicit beneficence in the manner she has done, for the advantage of her family'. Publication by subscription, a system through which readers subscribe an amount in advance to the author for the production of her narrative (and are credited at the front of the volume), represents an honourable way for Parker to 'solicit beneficence' in this public way.

[34] Elizabeth Macarthur, *The Journals and Letter of Elizabeth Macarthur, 1789–1798*, ed. Joy N Hughes (Glebe, 1984), quoted in Deirdre Coleman's discussion of Parker and Macarthur's relationship, *Maiden Voyages*, p. 31.

[35] Ibid., p. 31.

The positive, hospitable sociability which Parker enjoys across the Atlantic, and seems to introduce at Port Jackson with her own arrival, works in counterpoint to that other powerful network of associations informing Parker's text: the spectre of slavery which is raised in connection with both transported convicts and displaced indigenous people of the Hawkesbury. Like the wreck of the *Guardian* at Cape Town, from which Parker hurriedly turns – 'to avoid as much as possible any disagreeable reflections which might arise from the probability of our sharing the fate of the above vessel … I hastened to my companions, and … a Cape breakfast' – the uncomfortable reflections generated by circuits of enforced labour and long sea voyages are submerged in characteristic forms of colonial sociability:

> It is customary to arrange out the table as for dinner, except its being covered with all sorts of fruit; against each person is placed a knife, a plate, and a napkin; thus seated, the lady of the house makes tea and coffee at a side table, which the slaves hand round to the company. (194)

The 'Cape breakfast' to which the grateful traveller flees, and upon which she heaps 'many encomiums', literally incorporates the more troubling implications of colonial commerce in delightful hospitality. The shared meal encompasses key products and practices of colonial sociability – the exchange of visits, fruit, tea, coffee, even the mention of cutlery – and is attended discretely by the enforced labour of slaves.

For Parker, the Cape breakfast seems to occlude, or mediate in a more pleasing fashion, the attendant violence of colonial transportation and labour. She is clearly relieved to observe at this same meal that 'the slaves were treated at the Cape with the greatest humanity: and only in name bore the degrading distinction' (216). Parker's laudatory imagery of Creole culture in the Caribbean and at the Cape is doubly powerful: settler-colonialism and the 'management' of slavery in terms of universal hospitality provide a compelling model for relations among civilians at the new colony at Botany Bay, and a model of management for both indigenous people and convict labourers displaced by English law. These forms of courtesy and hospitality she enjoys across the Atlantic are practised by Parker with respect to the Aboriginal people of the Cove in authoritative and familial terms, gradually effacing the disturbing realities of displacement and indenture.

This chapter explores Parker's perceptions of the land and indigenous peoples of New Holland as they are framed by contemporary discussions of slavery, hospitality and sociability, and in the context of the writer's 'voyage around the world'. While the *Voyage* jostled for attention with a number of other early accounts of New South Wales, collectively these are marked by the absence of a definitive tradition or extant discourse for representing this strange convict settlement on the other side of the world, with its simultaneous realities of exploration and colonial governance, displaced indigenous inhabitants, sudden convict communities and civil / military accoutrements. The *Voyage* responds to this cultural lacuna with a dispersed, and highly relativized account. Parker's evident discomfort about publication for financial gain and her lack of confidence in her representative powers results in a hesitant and cautiously derivative narrative: her images are mediated repeatedly through her husband's records of their voyage and other contemporary accounts of the settlement at New South Wales.[36] Beyond this authorial embarrassment, Parker's is a strangely detached and deferred description of the colony, omitting certain key events, and consistently hinting at other images and debates that circulate around the edges of the text, but that are never fully taken up. The *Voyage*'s most abiding imagery is of idealized relations of maritime and colonial sociability: strangers brought together in need and sudden social intimacy at strategic points over the globe, creating webs of information and influence which condition travellers' perceptions and responses.

Parker's representation of Australia in the 1790s is less influenced by any cohesive or unitary discourse on the southern continent and its people than the broader global phenomena of cultural exchange identified by Roeck and distinctive 'planetary consciousness' described by Mary Louise Pratt.[37] Parker's perspectives on the colony at Botany Bay are informed not only by the (largely unstated) ideological projects of British imperialism and penal settlement, but in more minute and material ways by a wider network of European colonists, military, merchants and their families circumnavigating the world. The *Voyage*'s engagements trace contemporary debates on slavery and abolition, deliberations on race and indigenous sovereignty, and a range of discourses from humanitarianism to landscape aesthetics. The *Terra Australis* of Parker's *Voyage* is produced through a suggestive hybrid of literary, scientific, social and aesthetic discourses of travel which reflect the nascent status of the colony; capture the reticulate spatiality of colonial cultures and discourses in the late eighteenth century; and describe this traveller's particular journey into print. Perhaps too the networked quality of late eighteenth-century perceptions of *Terra Australis* is reflective of a particular moment of openness and uncertainty in the continent's European history. Annette Bremer and Alan Atkinson both suggest that the interval between January 1788 and late 1792 was the period of

36 See Coleman's 'Introduction' to *Maiden Voyages*, pp. 31–2.

37 Mary Louise Pratt, *Imperial Eyes: Travel Writing and Transculturation* (London, 1992), esp. pp. 9–10 and 29–30.

greatest contact or intermingling between the colonists, convicts and indigenes of New South Wales: a 'high-water mark' of cross-cultural relations prior to the next century's climate of 'antagonistic intimacy'.[38]

The contingent, apparently derivative style of Parker's text – those 'slender remarks' which seem repeatedly to defer to the authority of her husband, or to slip away in ambiguous reference – are not reducible to ignorance, simple modesty or authorial self-effacement. Instead the *Voyage* might reflect the early colonial project at New South Wales as a global phenomenon: a diverse web of interested parties, scientific, humanitarian and civil networks and 'voyagers' (193) of which Parker and her husband were active constituents, through which debates about slavery, governance and commerce circulate. *Terra Australis* does not emerge therefore from Parker's account as a wholly distinctive landscape, a singular location for the staging of authorial perception or subjectivity, but as just one site or node in the relational network of space, information and discourse constituting broader projects of colonial exploration and intervention in the late eighteenth century.

38 Annette Bremer, 'Domestic Disclosures: Letters and the Representation of Cross-Cultural Relations in Early Colonial New South Wales', *Frontiers: A Journal of Women's Studies*, 28.1–2 (2007): 77–95, 81, with reference to Alan Atkinson, *The Europeans in Australia: A History* (Melbourne, 1997), p. 162.

Chapter 11
Recovering the Imperial Context of the Mid-Victorian Exploration of Northern Australia, 1855–57

Norman Etherington

Augustus Gregory's expedition to Northern Australia in 1855–56 occupies a very small place in the annals of Australian exploration. If remembered at all, it figures mainly as revealing the potential of Victoria Downs as a cattle station watered by the Victoria River system. The illustrious leaseholders of that property constitute a who's who of wealth, beginning with Charles Brown Fisher, son of a founder of the South Australia Company. Subsequently Victoria Downs passed to a syndicate headed by Western Australia's Alexander Forrest and Sidney Kidman, who – in addition to being a distant relative of actress Nicole – was for a time the world's largest individual pastoral landholder. Afterwards it supplied British beef-paste maker Bovril with meat to be processed into a soup and spread for toast, before being absorbed into the Vestey's empire. Real estate giant, Hooker Corporation, briefly held the station, prior to it finally passing to the family firm headed by Robert and Janet Holmes à Court. This, however, constitutes but a chapter in the short and simple annals of the rich. To understand why the Gregory expedition failed to achieve the lasting fame accorded to Burke and Wills or Ludwig Leichhardt's fatal wanderings in the wastes of Central Australia, it is helpful to situate exploration within our conception of nationhood.

Narratives of nationalism as developed in the eighteenth and nineteenth centuries generally conform to the same simple template – regardless of which nation we are considering. Stripped to the bare essentials the story goes like this: people sharing important characteristics discover their common identity under conditions of hardship and oppression. After struggling to achieve control of their own affairs they set about achieving their national destiny. They subsequently celebrate their passage to independence and national fulfilment in their history books, museums and monuments. In settlement colonies like New Zealand, Canada, the United States and Australia, the national narrative cannot be firmly based on the idea of a shared history prior to immigration from many different lands. Instead it is founded on the notion of a national character forged through bonding with the land. Russel Ward's *Australian Legend* provides a classic instance of the genre.[1] In his account the experiences of convict transportation and getting in touch with the land on the frontiers of pastoralism, farming and mining gave Australians a sense that one man is as good as another and that mates stick by each other even in the most adverse conditions, such as on the battlefields of Gallipoli. Exploration best fits this national narrative, not when it is easy but when it is arduous, which is why failed explorers like Leichhardt, Burke and Wills loom much larger in the national consciousness than successes like John McDouall Stuart, who crossed the continent from South to North in 1861, and returned without much fuss. We tend to forget why the expeditions were mounted in the first place, because the point of the story within the Australian nationalist narrative is the character-building encounter with the harsh and difficult land.

This chapter seeks to rescue the Gregory expedition from relative obscurity by recovering its context, which was intimately connected to the economic and geopolitical ambitions of an expanding British Empire. To understand why the expedition was considered to be important enough to be the last Australian geographical venture funded by the British Parliament, it is necessary to visit the forgotten chronicles of what might have been. The 1850s marked one of the high points of what has been aptly called 'geography militant'.[2] During that decade David Livingstone became a household name in Britain, Fernand de Lesseps formulated his scheme for building a Suez Canal, Sir John Franklin died searching for a Northwest Passage to the Pacific Ocean, and the source of the Nile in Lake Victoria was identified by John Hanning Speke. For a time the exploration of the Australian continent appeared as important as any of these enterprises and attracted close attention from some of the greatest names in the history of science: Charles Darwin, Alfred Wallace, Thomas Huxley, Charles Lyell and Roderick Murchison.

When Murchison, as President of the Royal Geographical Society, in 1845 reviewed the great enterprises of exploration in progress he ranked explorations of Australia as equal in importance to Franklin's quest for the Northwest Passage in Canada, projects for a canal across the Isthmus of Panama, the Trigonometric Survey of India under the direction of Colonel Everest and the attempt to solve the mysteries of the Nile and the African interior.[3] What raised these particular

1 Russel Ward, *The Australian Legend* (Melbourne, 1958).

2 Felix Driver, *Geography Militant: Cultures of Exploration and Empire* (Oxford, 2001).

3 *Journal of the Royal Geographical Society of London* (hereafter *JRGS*), 15 (1845): xli–cxi.

projects to the top of geographical problems to be solved was their threefold character: strategic, economic and scientific. Their ramifications emerge clearly in the published transactions of the Royal Geographical Society of London (RGS), which are the main sources cited in this chapter.

The Royal Geographical Society in the Mid-Victorian Era

At one level the RGS was a private philanthropic organization devoted to the advancement of knowledge for its own sake.[4] Every year its president would summarize the achievements of geographers of the Western world, dispensing honours without respect to nationality. Alexander Humboldt of Prussia was readily acknowledged as the dean of modern geographers.[5] The tsar of Russia was congratulated for his patronage of major expeditions, as was the American army officer, John Charles Fremont, for his journey across the Rocky Mountains to the Oregon Territory, where conflicting British and American claims were already raising threats of possible war.[6] The society regretted the setbacks suffered by geographical science due to the European revolutions of 1848; 'men's minds have been too much preoccupied by social discord and political strife to attend to the more peaceful occupations of literature and science'.[7] The RGS awarded medals to foreign nationals, employed them on the expeditions the society sponsored, and embraced many of them as honorary fellows. It was customary for the annual presidential addresses to express disdain for merely commercial considerations in geographical research, as W.J. Hamilton did in 1848.

> The most correct and perfect maps are but the skeleton or groundwork of our science, taken in the higher and more extended sense in which we wish to see it cultivated. Its application to the progress and development of civilization, and of the animal and vegetable productions of the earth, to the distribution of the different races of the human family, and the various combinations which have arisen from their repeated intercourse, are subjects of the highest consequence, and for the clear understanding of which our maps and charts can only serve as the foundation-stone. No doubt the commercial intercourse of mankind is facilitated and kept up by these maps and charts; but we aim at a higher object,

[4] The Geographical Society of London, founded in 1830, became known as The Royal Geographical Society of Great Britain under the patronage of King William IV and was granted its Royal Charter under Queen Victoria in 1859.

[5] *JRGS*, 15 (1845): lxxxv–lxxxvii.

[6] *JRGS*, 15 (1845): xciv, lxviii.

[7] William. J. Hamilton, 'Presidential Address', *JRGS*, 19 (1849): xxxiii.

> viz., the improvement of man's moral culture by a more extended knowledge of the various productions of different climes, and by bringing before him on a large tabular scale the moral and physical condition of his race.[8]

Behind the outward appearance of internationalism and the selfless pursuit of knowledge, however, British geographers maintained the closest of connections with the military and naval establishments of their country.

The grand eighteenth-century maritime endeavours of James Cook, George Vancouver, William Bligh and others had been carried out by serving naval officers under Admiralty orders. Advances in navigation, meteorology and astronomy owed a great deal to sponsorship by the British naval establishment at Greenwich, eager to gain supremacy at sea during the century-long wars with France. The membership and executive of the RGS sparkled with naval and military brass. One of the prime movers in its establishment had been imperial strategist Sir John Barrow, who served as a secretary of the Admiralty for 40 years and president of the Society from 1835 to 1837.[9] Another was Captain (later Admiral) William Henry Smyth, who assumed the presidency in 1850. His first address in this capacity remarked astutely on the many ways in which war had assisted advances in geographical research.

> ... hence the numerous surveys and maps of conquests, from the exploits of Alexander and Caesar to those of Napoleon, Wellington, and our Indian campaigners. The French and English expeditions to Egypt were the opening and renewal of our knowledge of that singular country; and even in the recent march of the American army to Mexico, the scourge of humanity was not without important results and fruits to geography.[10]

Admiral F.W. Beechey, RGS president when news came of Augustus Gregory's successful return from northern Australia in 1856, had served with Lieutenant John Franklin on a voyage of Arctic discovery in 1818. Frederick Egerton, First Earl of Ellesmere – though primarily known as a politician when elected president of the RGS in 1854 – had begun his career as an officer in the Staffordshire regiment of yeomanry. Roderick Murchison himself had also trained for the

---

8 *JRGS*, 18 (1848): lxxi–lxxii.

9 For a full and entertaining account of Barrow and his use of naval officers on exploring expeditions after the Napoleonic wars, see Fergus Fleming, *Barrow's Boys: A Stirring Story of Daring, Fortitude, and Outright Lunacy* (New York, 1998).

10 *JRGS*, 20 (1850): lxx.

army and served with Wellington in the Napoleonic Wars.[11] Strategic thinking was second nature for such men.

They also shared an optimistic faith in Victorian narratives of progress, seeing a direct relationship between advances in geographic knowledge and the spread of British commerce across the globe. In 1845 Murchison rejoiced that a syndicate of Mexican, French and British investors appeared on the verge of cutting a canal across the Isthmus of Central America:

> When we reflect on the vast saving of nautical risk, time, and expense, which this short cut to the Pacific and all colonies ... whether on the Oregon or elsewhere, to which British enterprise can extend, and think of the vastly greater dangers and difficulties of doubling Cape Horn than those of rounding our Orkneys and the Fitful Head, we can have no hesitation in wishing hearty success to a company, which will thus at once clear away such obstacles, and open out such a grand field of commerce, by the execution of a project which has been a desideratum for centuries, and which never could be so well carried out as in these piping times of peace.[12]

De Lesseps was congratulated for launching the Suez Canal project, whose importance could not 'be overrated in a commercial point of view, and especially to this country, when considered in connection with its Indian possessions and colonies'.[13] Discoveries of economically important minerals always rated a prominent place in RGS presidential addresses. For example, W.J. Hamilton hailed the opening of the California gold fields in 1848 as

> ... mines of such extraordinary wealth as, taking only a moderate view of the accounts which have yet reached us, to throw completely into the shade all that we have yet heard of the wealth of the Brazils, or the still richer products of the gold-mines of Siberia, on the flanks of the Ural and the Altai, already so fully described to you by Sir. R. Murchison in his Address in 1844, and by M. Hoffman in an account of his journeys to the gold regions of Siberia.[14]

As Hamilton implied, Murchison took a special interest in gold, believing that he had acquired the ability to predict its discovery on the basis of specific geological

---

[11] Robert A. Stafford, *Scientist of Empire: Sir Roderick Murchison, Scientific Exploration and Victorian Imperialism* (Cambridge, 1989), pp. 4–5.

[12] *JRGS*, 15 (1845): lxx. This was but one of many failed schemes to construct a canal across Mexico's Isthmus of Tehuantepec, a long-time rival to Panama, whose canal finally opened only in 1914.

[13] *JRGS*, 26 (1856): ccxv.

[14] *JRGS*, 19 (1849): lxxxv.

formations. In 1839 he had published *The Silurian System*, which focused on the earliest fossil-bearing strata, to which he assigned the name Silurian. Because coal had its origins in decayed plant material, Murchison cogently argued that there was no point in looking in strata laid down before the appearance of the first plants on earth. Through further expeditions in Russia, Murchison extended his analysis to other strata, leading him to think that the distribution of Silurian rocks could be directly linked to the presence of precious metals at various points across the globe.[15]

With this general background it is easy to see why RGS focused greatest attention on projects at the intersection of scientific, strategic and economic interest. The lure of the Northwest Passage, shared with schemes for canals across Central America and Suez, held out the promise of cheaper coal-powered steam navigation to distant Asian markets. The Trigonometric Survey of India would consolidate Britain's control of the riches and tax revenues of 'the jewel in the crown'. Tracing the course of the great rivers of Africa promised to open a potentially rich continent to the penetration of British commerce and civilization. For several similar reasons, the exploration of Australia appeared to be an equally compelling project at mid-century.

The Promise of Australia

On the eve of Gregory's expedition Australia had emerged as a place of special interest, whether regarded from the strategic, economic or purely scientific point of view. Interest in developing military and naval facilities on the northern coast dated from the Congress of Vienna, which had restored the Indonesian archipelago to the Netherlands, putting an end to hopes that the wartime conquests of Stamford Raffles might have established a continuous string of British colonies from Singapore to Sydney. A few years later Captain J.J.G. Bremer was sent on the *Tamar* to select a site on the northern coast of Australia from which British merchants might commence operations with a view to breaking the Dutch monopoly of trade in South East Asia. His unfortunate initial choice was the place he named Fort Dundas on the western shore of Melville Island.[16] Its hundred first residents were plagued from the beginning by fevers, hostile Aborigines and pirates. Next the Admiralty despatched Captain James Stirling in 1827 to find a better location; he established Fort Wellington on Raffles Bay northeast of the present-day city of Darwin. This proved no healthier or

15 Stafford, *Scientist of Empire*, pp. 10–13.

16 J.K. Laughton, 'Bremer, Sir James John Gordon (1786–1850)', rev. Andrew Lambert, *Oxford Dictionary of National Biography* (Oxford University Press, 2004). Online at: http://www.oxforddnb.com/view/article/3313 [accessed: 30 August 2006].

happier for the settlers and soldiers transferred from Melville Island; in 1829 it too was abandoned. Notwithstanding these setbacks the great naval strategist John Barrow kept the dream alive, lobbying successfully for a new settlement at Port Essington in 1837. Once again Captain Bremer received the assignment of establishing a new northern outpost, which he named Port Victoria in 1838 after the newly crowned young queen of England.

A few months after Barrow's death in November 1848, 23-year-old Thomas Henry Huxley, on his first naval assignment as assistant surgeon on the HMS *Rattlesnake*, visited the port as part of Captain Owen Stanley's survey of the Great Barrier Reef, New Guinea and the Torres Strait. The future President of the Royal Society, renowned as 'Darwin's bulldog', penned a memorable portrait of the woebegone settlement.

> The only place we have visited since my last budget to you was Port Essington, a military post which has been an object of much attention for some time past in connection with the steam navigation between Sydney and India. It is about the most useless, miserable, ill-managed hole in Her Majesty's dominions. Placed fifteen miles inland on the swampy banks of an estuary out of reach of the sea breezes, it is the most insufferably hot and enervating place imaginable. The temperature of the water alongside the ship was from 88 to 90, *i.e.* about that of a moderately warm bath, so that you may fancy what it is on land. Added to this, the commandant is a litigious old fool, always at war with his officers, and endeavouring to make the place as much of a hell morally as it is physically. Little more than two years ago a detachment of sixty men came out to the settlement. At the parade on the Sunday I was there; there were just ten men present. The rest were invalided, dead, or sick. I have no hesitation in saying that half of this was the result of ill-management. The climate in itself is not particularly unhealthy. We were all glad to get away from the place.[17]

Later that same year the settlement closed for good, much to the annoyance of Roderick Murchison who kept the old dream alive. 'It was a subject of regret to many of us,' he recalled, 'that it should have been thought expedient to discontinue the occupation of Port Essington, and to abandon all intention of holding any other station along the northern coast of this vast continent.'[18]

What kept Barrow's project going was not the inherent value of any discovery yet made, but the perceived strategic importance of a way station between Sydney and India. He drew support from many quarters, including the swashbuckling

---

[17] Julian Huxley (ed.), *T.H. Huxley's Diary of the Voyage of H.M.S. Rattlesnake* (London, 1935).

[18] *JRGS*, 27 (1857): clxxvi.

adventurer, Rajah James Brooke, who had carved out an independent state of his own in Sarawak on the northwest coast of Borneo. According to Robert Reece, Brooke saw his kingdom 'as one of a string of British settlements linking Singapore with Port Essington'.[19] Murchison's 1845 presidential address laid out what he regarded as 'the urgent reasons for raising Port Essington in the scale of maritime establishments':

> The projected establishment of a continuous chain of steam communication between Great Britain and China, by the Straits of Malacca (no wild vision), suggests the possibility of a branch line of packets from this great trunk line to Sydney being also established at no distant period; and if so, the most eligible line for such branch would seem to be from Singapore through Torres Straits .... When we consider that the most direct line of communication between Singapore, Hong Kong, and the subordinate settlements of that great field of commerce on the one hand, and all our important Australian colonies on the other (Swan river alone excepted), lies through this gulf, and when we know that New Zealand, an important British settlement, forms the extreme southeastern limit of this great gulf, there is every prospect that eventually it must become a high road of intercourse, which will knit together the islands with our great Asiatic and Australian colonies; hence therefore I venture to say that a complete *exhaustive survey* of the whole of the gulf above defined would be a true service to science, and noble spur to national industry.[20]

For similar reasons the search for an overland route to Port Essington also seemed an urgent priority. Provided good arable or pastoral land could be located, a chain of settlements would serve the northern port with supplies and perhaps one day, goods for export.[21] An adventurous German, Ludwig Leichhardt, who had studied medicine and natural science in England, made the first successful land crossing to Port Essington, for which he received the RGS medal in 1847. The next year he was off again, hoping to trace the headwaters flowing to the Gulf of Carpentaria; eventually heavy rains forced him to retreat. Meanwhile, the veteran surveyor-general of New South Wales, Sir Thomas Mitchell, had set out to find a path between the Darling River in southeast Australia and the many rivers that run into the Gulf of Carpentaria. This, he wrote, was essential 'to the full development of the geographical resources of New South Wales' because

---

19 Robert H.W. Reece, 'Brooke, Sir James (1803–1868)', *Oxford Dictionary of National Biography* (Oxford University Press, 2004). Online at: http://www.oxforddnb.com/view/article/3549 [accessed: 26 August 2006].

20 *JRGS*, 15 (1845): lviii–lxi.

21 See the comments of Lord Colchester to this effect in *JRGS*, 16 (1846): lxxvi.

'at no great distance lay India and China, and still nearer, the rich islands of the Indian Archipelago'.[22] Murchison, who had no personal experience of the region, had only to look at a map to surmise that 'a noble bay which advances 500 miles into this continent, will, sooner or later, become the great line of intercourse between our Australian and Indian dominions'.[23]

Large hopes were held out for the Gulf region after a survey in 1841 by Captain John Lort Stokes, commanding the same H.M. Brig *Beagle* that had carried Charles Darwin on his epochal voyage around South America to Australia. Travelling for 50 miles up the Albert River, Stokes was so impressed with the agricultural potential of the well-watered lands stretching to the south, that he named them 'The Plains of Promise'. In the meantime, other reasons had emerged for tracing the course of Australia's rivers. Hard on the heels of the discovery of silver, copper and other minerals in South Australia came the sensational news of gold strikes at Ballarat and Bendigo north of Melbourne.[24] Murchison congratulated himself for having predicted the discovery, but marvelled at how far his expectations had been exceeded by the 'golden shower' that had descended on Victoria. Some sections of the upper classes began to fret that inflation of the gold-based currency would threaten their incomes – a worry Murchison attempted to assuage in his RGS presidential address of 1852.[25] At the same time he could not but wonder what riches had yet to be found.

> The colonies of New South Wales, Victoria, Adelaide and Perth, are flourishing and progressing so rapidly, that as geographers we are now led to speculate on the approaching solution of one of the most curious of all the problems which can interest this country – the true condition of the interior of this vast continent.[26]

So many attempts had failed. Edward John Eyre in 1840 had pushed north from Adelaide only to find his progress halted by the vast dried up salt lake that bears his name – not far from the eminence he called Mount Hopeless. Captain Charles Sturt, who had made the first descent of the Murray River in 1829, attempted to push north from that same river in 1844 on a journey that ended when water ran out in 'Sturt's Stony Desert'.[27] Eyre speculated in 1846

22 T.L. Mitchell, *Journal of an Expedition into the Interior of Tropical Australia in Search of a Route from Sydney to the Gulf of Carpentaria* (London, 1848), pp. 1–2.

23 *JRGS*, 23 (1853): cxxix.

24 *JRGS*, 17 (1847): xlvi; 22 (1852): lxxxii; 23 (1853): cxxiv; 27 (1857): clxxvi.

25 *JRGS*, 22 (1852): lxxxv.

26 *JRGS*, 23 (1853): cxxviii.

27 See Charles Sturt, 'A Condensed Account of an Exploration in the Interior of Australia', *JRGS*, 17 (1847): 85–129.

that perhaps the interior of the continent would eventually be 'found generally to be of a very low level, to consist of arid sands, alternating with many basins of dried-up salt lakes, or such as covered only by shallow salt water or mud, as is the case of Lake Torrens'. At best there might be 'many detached and even high ranges ... interspersed among these arid wastes' where 'intervals of a better, or even of a rich and fertile country might be met with'.[28] Leichhardt aimed to put this and other less pessimistic theories to the test in 1848 with an expedition aiming to traverse the interior from Queensland to Western Australia.[29] He was never heard from again.

Far from discouraging the members of the Royal Geographical Society, Leichhardt's fate itself became an ongoing source of interest. Speculation on the subject abounded in the years immediately prior to the departure of the Gregory expedition, along with the equally mysterious failure to hear anything of Sir John Franklin's Arctic expedition, which had set off two years earlier. Quite apart from the strategic, scientific and economic objectives underpinning the Society's work, the *frisson* of terror associated with such romantic mysteries helped arouse public support and government funding for the work of nineteenth-century exploration. Expeditions to find lost explorers often attracted more attention and financial support than the original ventures – as the search for Dr Livingstone was to demonstrate a few years later. Though Murchison still held out hope in his 1853 presidential address to the RGS that Franklin would be found, he feared that 'the chivalrous Leichhardt' had 'probably fallen a victim to his bold endeavour'.[30] Thus it remained the case that

> No region of the earth presents a greater geographical problem to solve than Australia, in the apparent termination in the interior of so many of her vast rivers .... The restless settler, forcing onwards wherever fine herbage leads him, will, it is true, eventually find his way to the very limits of productive lands. But what we geographers regret is, that so very much of the territory, which lies between such distant outposts and the regularly settled countries, has not yet been inserted on any map ....[31]

It was the hope that one of those rivers, the Victoria, might open a path to the solution of this mystery, while at the same time completing the strategic chain

---

28 Edward John Eyre, 'Considerations Against the Supposed Existence of a Great Sea in the Interior of Australia', *JRGS*, 16 (1846): 202.

29 While not disputing Eyre's hypothesis, Leichhardt held the opinion that his crossing might be effected by passing along scattered oases; see J.E. Tenison-Woods, *History of the Discovery and Exploration of Australia* (London, 1865), vol. 2, p. 518.

30 *JRGS*, 23 (1853): lxxv, cxxviii.

31 *JRGS*, 23 (1853): cxxx.

of imperial command posts on the route between Sydney and Calcutta that inspired the Gregory expedition.

Imperial Linkages of Personnel on the Gregory Expedition

The composition of the party Augustus Gregory led out from Brisbane in 1855 might seem on the surface to be a fortuitous combination of individuals. However, when studied from the perspective of the spider's web of imperial power and knowledge spun out to the corners of the earth by men such as John Barrow and Roderick Murchison, the party's composition made sense.

There were very good reasons for picking Augustus Gregory of Western Australia to lead the expedition, for the exploration of the northern coast was linked to the very foundation of the Swan River Colony. After completing his mission to establish Fort Wellington on Raffles Bay in 1827, Captain James Stirling had sailed down the west coast in search of a suitable site for a colony that would counter fears that the French exploring activity in those waters was a prelude to colonization.[32] After his travels by longboat up the Swan River convinced of its possibilities for a prosperous settlement, he returned to England where his vigorous lobbying bore abundant fruit, thanks in no small part to his friend George Murray's appointment as Secretary of State for Colonies. Soon he was on his way back to the Swan River as the colony's first governor, arriving in June 1829. The strategic imperial aspects of the settlement were made plain the same year with a proclamation claiming the whole of the Australian continent a British possession. John Septimus Roe, who accompanied Stirling as surveyor-general of the colony, was another naval officer who had seen action in the Napoleonic wars. The appointment of army and navy officers as surveyors-general of British colonies was routine in the 1830s as witnessed by the appointment of Captain Thomas Mitchell in New South Wales and Colonel William Light in South Australia. With the coming of peace Roe moved to New South Wales as master's mate to the surveying service then headed by Lt. Phillip Parker King. When King was ordered to continue the coastal survey of Australia begun by Matthew Flinders, Roe accompanied him on several voyages in the *Mermaid*. Roe revisited the northern coast again in 1824 under the command of Captain Bremer, at which time the first settlement was planted on Melville Island. After working with Stirling on the early exploration of the Western Australian interior, Roe expanded the surveying department with a cadet,

---

[32] The definitive biography is Pamela Staham-Drew's *James Stirling: Admiral and Founding Governor of Western Australia* (Perth, 2003).

Augustus Gregory, in 1841. The young man, whose father had been an army officer, was soon promoted to Assistant Surveyor and engaged in the further exploration of territory to the north of Perth, along with his brothers Frank and Henry. Thus, while the Gregory's themselves had never served in the military, they were directly connected by association and employment with the earliest naval expeditions to the northern coast of Australia.

Even without Roe's recommendation, Augustus Gregory and his brothers would have been known to the RGS from reports of their activities published in the society's journal in the years 1848 and 1852. The first concerned the discovery of the strategic raw material coal; the second, his own account of Governor Charles Fitz-Gerald's expedition to the significantly named Murchison River; and the third, the discoveries made under his own command in a search for new country to accompany the colony's burgeoning wool industry.[33] There was therefore nothing surprising about the appointment of Augustus as the leader of the expedition and his brother Henry as his associate. Their connections with earlier episodes in the exploration of the northern regions of Australia, their experience with the exploration of unknown terrain under harsh conditions and their proven ability to assess both minerals and soils of economic value made them obvious choices for the job.

Since the days of Cook it had been standard procedure to include both an artist and a botanist on ventures into the unknown. Gregory's artist, Thomas Baines, who had been official artist to the British forces during the Cape Colony's Eighth Frontier War in South Africa, would go on to serve with David Livingstone's Zambezi expedition and conduct important explorations in his own right. In Ferdinand Mueller Gregory found a counterpart to Cook's botanist Joseph Banks. When the young Banks had sought out instructors in the science of botany, he had turned to the continental scholars Linaeus and Solander. Close collaboration with Scandinavian and German scientists continued well into the Victorian age. Lacking an overseas empire of their own before the 1880s young Germans were to be found in British colonies working as missionaries, soldiers, settlers and, as in the case of Mueller, scientists. No one on the Gregory expedition came with a better recommendation. As a young man Sir William Jackson Hooker had known Joseph Banks, travelled to the continent to broaden his education, and went on to direct the institution Banks had made famous: the Royal Botanical Gardens at Kew. Mueller proved himself worthy of Hooker's commendation by going on to direct the Melbourne

33 *JRGS*, 18 (1848): xxxvii; A.C. Gregory, 'His Excellency Charles Fitz-Gerald's Expedition to the Murchison River' and A.C. Gregory and R.R. Madden, 'The Settlers' Expedition to the Northward from Perth under Mr. Assistant-Surveyor A.C. Gregory', *JRGS*, 22 (1852): 57–73.

Botanical Gardens. He did not find himself short of fellow German scientists in Australia. Richard Schomburgk from Prussia had, before his appointment as director of the Adelaide Botanic Gardens, accompanied his brother Robert on a journey of exploration to Guyana, which Murchison specially commended in his 1845 Presidential address to the Royal Geographical Society.[34]

Every well-equipped naval voyage of exploration needed a surgeon. It had been in this capacity that T.H. Huxley had made his youthful voyage to Northern Australia. The Gregory expedition's surgeon was Joseph Ravenscroft Elsey, the son of a Bank of England official. Although only 22 at the time of his appointment, Elsey had studied medicine at Guy's Hospital, London, before obtaining qualifications from both the Royal College of Surgeons and the Royal College of Chemistry. He acquitted himself in every aspect of his work: making a comprehensive report on the damage suffered by the expedition's ship, *Tom Tough*, when it ran aground, finding and naming new species of plants and birds, and publishing his own account of the health of the expedition for the *JRGS*. We can only guess at what he might have accomplished subsequently, had he not died in the West Indies on the last day of 1857, only a few months after his report was read to a meeting of the Royal Geographical Society. The inclusion of a geologist on the expedition was more of a novelty, reflecting the stupendous impact of Charles Lyell's *Principles of Geology* (1830) and Murchison's abiding faith that many more mineral riches awaited discovery on the Australian continent. Little is known of James Spottiswood Wilson, except that he, like Baines, possessed practical experience gained in South Africa. Closely connected to the British army and navy, possessing practical knowledge gained in the four quarters of the globe, and committed to the multifarious objectives of the Royal Geographical Society, the Gregory expedition represented the outward push of the British Empire as a whole, much more than it does the national interest of any of Australia's component colonies or the future federation.

Conclusion

While it has not been the intention of this chapter to retell the story of the expedition, it cannot conclude without reporting that Gregory's results hardly measured up to the grand hopes of his sponsors. Months were spent travelling from Brisbane to the Gulf of Carpenteria and Arnhem Land through territory already partially explored. More time was wasted freeing and then repairing the support brig, *Tom Tough*, which ran aground in the Victoria River. And while

[34] *JRGS*, 15 (1845): xlix.

new grazing lands were discovered at Victoria Downs, Gregory's forays into the interior merely confirmed that the Victoria was not the long hoped for broad watery highway to Central Australia. Thomas Baines' side trip in the *Tom Tough* to procure supplies in Dutch West Timor was delayed so long by worries about the seaworthiness of the vessel that he missed his appointed rendezvous with Gregory. Without new supplies the main body of the expedition had no choice but to make an overland dash back to Brisbane, arriving in December 1856 without having made any more significant discoveries.

Recovering the expedition's imperial context helps wrest nineteenth-century Australian exploration from the narrow nationalist narrative in which it has been so deeply embedded. The great objective was not to fill up the map of Australia for the benefit of patriotic school children but to link the flourishing southeastern colonies economically and strategically to Britain's burgeoning imperial enterprises in Asia. Likewise the expeditions of McDouall Stuart and Burke and Wills had the very practical, profit-seeking aim of finding ways to take stock overland from southern pastures to northern ports from which they might be shipped on to Asian markets. In the nationalist narratives those objectives are barely mentioned and the British explorers transmuted into heroic makers of the Australian nation. In the mid twentieth century Patrick White would rework the story of Ludwig Leichhardt's last journey into his novel, *Voss*, where the hero's quest cuts loose altogether from the search for geographical knowledge, grazing lands settlement or mineral resources, and becomes a flight into the spiritual unknown.

Amazingly, Roderick Murchison's dream of vast mines and flourishing cities on northern Plains of Promise that will link Australia's destiny to ports of Asia lives on, despite the meagre results achieved in the century and half since the British Parliament voted the funding for Gregory's expedition. But the Empire that launched the enterprise has vanished.

Chapter 12

The Wicked and the Fair: Changing Perceptions of *Terra Australis* through the Prism of the *Batavia* Shipwreck (1629)

Leigh T.I. Penman

In 1971, the Australian novelist Randolph Stow published a provocative article on the history and myth of the *Batavia* shipwreck of 1629.[1] Entitled 'The South Land of Antichrist', in this short piece Stow argued that the mythological history of European Australia contained an irreconcilable 'element of self contradiction', in which Australia appeared either as a prison, or as a paradise. The tension between these two visions, Stow argued, has pervaded nearly all attempts to engage with Australia's historical meta-narrative. Stow himself turned to the tales of early European exploration to illustrate this point. He contrasted the transcendent utopian vision of a *terra australis del spirito santo* attached to the Southland by the Spaniard Fernández de Quirós, with the dystopia of Jeronimus Cornelisz's kingdom on the Abrolhos in the wake of the *Batavia* wreck, an incident emblematic of a 'southland of Antichrist.'

Stow's choice of examples was apt. For it was in these two incidents that two contradictory European perceptions of *Terra Australis* collided with an unpredictable reality, each in turn producing its own legend. However, when investigating the mythology constructed around one of these incidents, the *Batavia* shipwreck, we can appreciate that the event, as a mirror for Australia's history, has not been limited to merely reflecting Stow's dialectic. Indeed, it has acted as a prism for European communities in Australia and elsewhere to engage with and express a rich spectrum of opinions, ideas and anxieties about the nation and the continent. Indeed, for a tale that represented to Stow the trope of the 'Southland of Antichrist', the *Batavia* wreck has proven itself both resilient and adaptable, capable

---

[1] Randolph Stow, 'The Southland of Antichrist: The History and Myth of the *Batavia* Disaster (1629)', in Anna Rutherford (ed.), *Common Wealth* (Aarhus, 1971), pp. 160–67.

of representing a period of both hell and earthly paradise during Australia's pre-European past; as well as, in some imaginings, a platform for a new future.

In this chapter, I want to examine some of the changing interpretations and utilizations of the 1629 wreck of the *Batavia* that have appeared in Europe and Australia between 1630 and the present. Although the *Batavia* was not the only Dutch ship wrecked on or off the coast of *Terra Australis* before European settlement, it has proven to be the only incident of Dutch contact with the continent that has attracted continuous attention over a span of nearly 400 years. Stow, in his evocation of the *Batavia* myth, dismissed the wreck itself as a 'futile, meaningless historical event', but nevertheless one 'out of which meaningful myths have been created'. Mapping the discursive contexts and shifting emphases of these varying myths and their meanings reveals how different narratives concerning the past may be constructed to reflect, encode, reinforce or transcend contemporary interests and anxieties about the Southern Continent.

The *Batavia*

The bloody tale of the *Batavia* wreck is well known, having been retold numerous times. In short, on 14 June 1629, the Dutch East India Company (VOC) flagship smashed into a reef in the Abrolhos islands, 70 kilometres off the coast of modern Geraldton, Western Australia. Following the wreck, in contravention of VOC protocol, captain Aerien Jacobsz and supercargo François Pelsaert departed for the settlement of Batavia (now Jakarta, Indonesia) to secure help. Upon their departure, a Frisian apothecary named Jeronimus Cornelisz, the senior-most remaining VOC representative, instituted a reign of terror among the survivors, beating, raping and murdering his way through the community with a handful of disciples in tow. After several months, Pelsaert managed to return with help. The reign of terror was put down, and Cornelisz and his conspirators were executed following a brief trial. Two of his cohort, however, Wouter Looes and Jan Pelgrom de Bye, were marooned on the west coast of Australia.

First Encounters

The *Batavia* wreck caused an instant sensation when news first reached the United Provinces in 1630. In early-modern literature, reports of shipwrecks brought with them varied expectations. Often seen as the result of God's wrath, they were either harbingers of disaster or, in literature at least, a precursor to the discovery of a utopia. The earliest readers about the events on the Abrolhos, however, knew exactly which they could expect. This much was announced in the attention-grabbing title of the first publication on the wreck, a hastily-issued news ballad called *Droevighe tijdinghe van de aldergouwelijkckste moorderij ... op 't schip Batavia* (Appalling news of the most horrible murders on the *Batavia*, 1630).[2] But if this initial reaction to news of the event was one of lament and sorrow, soon the Dutch public were seeking explanations. The inhuman nature of the events on the Abrolhos were quickly linked to yet another recent Dutch media sensation; the 1628 trial of the Haarlem painter Johannes Torrentius for heresy and blasphemy.[3] The link between Cornelisz and Torrentius was established in the public imagination by a letter – apparently authentic – written by a survivor of the Abrolhos mutiny. This letter claimed that in prosecuting his reign of terror, Cornelisz had followed the 'Torrentian' heresy.

The news spread rapidly. One tract printed during Torrentius's trial in 1628 was reissued, augmented with the Abrolhos letter.[4] In the subtitle to this pamphlet, Cornelisz was identified, not as one who acted according to 'Torrentian' philosophy, but a self-professed 'disciple' (*Discipel*) of the painter himself. The connection was again reinforced when the Torrentius letter was reprinted under the misleading title; *Wonderlijck verhael van het leven en gevoelen van Jan Symensz Torentius ... Mits-dien het leven ende ghevoelen van Jeronimus Cornelisz* (An extraordinary relation of the life and deeds of Jan Symensz Torrentius ... together with the life and deeds of Jeronimus Cornelisz, 1630).[5] This initial, dubious ascription of heretical inspiration to Cornelisz's deeds represented an attempt to render comprehensible an horrific event, scarcely conceivable in its brutality, which took place far from the known and secure boundaries, physical, moral and religious, of Europe: *Terra Australis* was a 'Southland of Antichrist', indeed.

It was not until 1647 that a relatively informative publication concerning the *Batavia* would appear. Separated from the initial shock of the incident, the *Ongeluckige Voyagie van 't Schip Batavia* (Unfortunate voyage of the *Batavia*, 1647), based on the journal of François Pelsaert, painstakingly outlined the

---

2 (Rotterdam for Cornelis Fransz, 1630). The text of this rare pamphlet is reprinted in Vibeke Roeper, *Het Schipbreuk van de Batavia, 1629* (Leiden, 1993), pp. 225–30.

3 On Torrentius see A. Bredius, *Johannes Torrentius. Schilder, 1589–1644* (The Hague, 1909); A.J. Rehorst, *Torrentius* (Rotterdam, 1939).

4 *Leyds Veer=Schuyts Praetjen ... Hier achtet syn by ghevoecht, Twee Brieven gheschreven in Batavia, vervanghende den handel van Ieronimus Cornelisz onder Coppman op't Schip Batavia, oock mede sijn sterven, geweest hebbende een Discipel van Iohannes Torrentius voorsz* (Unknown, 1630). Reprinted in Roeper, *Schipbreuk van de Batavia*, pp. 231–6.

5 (Haarlem, 1630). The pamphlet also reprinted the text of the *Leyds Veer=Schuyts Praetjen*, mentioned above.

Batavia's doom as a 'warning to those who would travel hither'. According to this pamphlet, during the outward voyage Cornelisz had fomented a mutinous conspiracy on board ship. Once on the Abrolhos, these mutinous intentions mixed with heterodox delusions, resulting in a cowardly and vicious bloodbath.[6] The narrative, which placed emphasis on the trials of the mutineers and featured several engravings, effectively transformed the tale into a celebration of Dutch hardiness. The *Batavia* incident was transformed into a microcosmic reflection of the positive Dutch mentality which underwrote its seaborne empire; a testament that order triumphs above chaos.

The first manifestation of broader international interest in the *Batavia* was equally benign, such as in the works of Melchisédech Thévenot (1620–93), who printed a heavily-abridged translation of the *Ongeluckige Voyagie* in his *Relations de divers voyages curievx* (Accounts of several curious voyages, 1663–72, repr. 1696).[7] This collection, which was widely read during the Enlightenment, reflected a popular interest not so much in France's expanding imperial pretensions, but rather in the collection and comparison of exceptional data concerning foreign lands, a cosmographical precursor to imperialist strivings of later times.[8] But it would not be until the mid-to-late seventeenth century, when the broken outlines of *Terra Australis Incognita* began to solidify into a *Terra Australis Cognita*, that real interest began to develop in the possibility of a European presence on the new continent, an interest which only grew in the eighteenth century. In the 1760s and 1770s, British and French ships regularly visited the austral continent, while the imagination of those that remained at home was fired by political debate and stories concerning the history of *Terra Australis*. In 1768, an English translation of Thevénot's *Batavia* excerpt was published in Edinburgh, making the events of Pelsaert and his crew – along with other accounts of early Dutch activity in the Southland – available to a broad English-speaking audience for the first time.[9]

[6] Although Pelsaert's journal was written after the fact as an exculpatory justification of his own actions, his account has remained essentially unchallenged, despite its glaring evidentiary problems. See, however, the thought-provoking article by Mike Sturma, 'Mutiny and Narrative: Francisco Pelsaert's Journals and the Wreck of the *Batavia*', *The Great Circle*, 24.1 (2003): 14–24, one of the few works to challenge the generally accepted contours of the *Batavia* narrative.

[7] Melchisédech Thévenot (ed.), *Relations de divers voyages curievx, qvi ont esté traduites d'Hacluyt, de Purchas, & d'autres voyageurs Anglois, Hollandois, Portugais, Allemands, Espagnols ...* (4 vols, Paris, 1663–72), vol. 4, sigs 4G1v–4G4v.

[8] On the text see Nicholas Dew, 'Reading Travels in the Culture of Curiosity: Thévenot's Collection of Voyages', *Journal of Early Modern History*, 10.1–2 (2006): 39–59.

[9] John Callander, *Terra Australis Cognita: or, Voyages to the Terra Australis, or Southern Hemisphere, during the sixteenth, seventeenth, and eighteenth centuries*, vol. 2 (Edinburgh, 1768), pp. 335–50.

As history demonstrates, it was the British who ultimately won the race for the Southland. But the general awareness of prior Dutch exploits on the continent, including the *Batavia* incident, meant that during the early years of exploration and settlement a cautious eye would always be cast over the legal implications of potential Dutch territorial claims. When James Cook finally claimed the colony of New South Wales for Britain in August 1770, he explicitly limited the claim to territory not already belonging to the Dutch by reason of priority of discovery.[10] Subsequently, in 1788, the western boundary of New South Wales was established at 135 degrees longitude. However, as rumours of French expansion plagued both Whitehall and Sydney, the western boundary of the colony gradually inched eastward in order to head off foreign interest. Following settlement of Fort Dundas in 1824, the border was cautiously advanced eastward to the 129th degree of latitude. This represented a territorial incursion into a nominally Dutch New Holland.[11]

Although the expanding territorial claims went unchallenged by the Dutch, it seems that something of a cultural guilt complex was nevertheless created by this presumptuous step into the legal unknown. The historical legacy of the *Batavia* wreck – more particularly the fate of her two maroons – played a key role in the expression of these anxieties. In the 1830s and 1840s, as settlers and missionaries penetrated the continent's interior, local newspapers periodically featured stories of 'white tribes' discovered in the deserts of unclaimed New Holland. Invariably all such tribes, whether of Aboriginals possessed of supposed 'European features', or indeed of shipwrecked whites, were linked by readers to the historical activities of the Dutch, particularly those 'left behind by Captain Pelsaert'.[12] Nonetheless,

---

10 Interestingly, however, in the original manuscript the crucial declarations have been struck out. '... on the Western side [of the continent] I can make no new discovery the honour of which belongs to the Dutch Navigators ~~and as such they may lay claim to it as their property~~ but the Eastern Coast from the Latitude of 38° South down to this place I am confident was never seen or visited by any European before [us and] ~~therefore by the same Rule belongs to great Brittan~~.' See Cook's original manuscript journal, entry of 22 August 1770, National Library of Australia. Online at: http://southseas.nla.gov.au/journals/cook/17700822.html [accessed: 14 March 2009].

11 On the gradual expansion of British territorial claims, see R. King, 'Terra Australis, New Holland and New South Wales: The Treaty of Tordesillas and Australia', *The Globe*, 47 (1998): 35–55.

12 On a supposed Dutch white tribe in northern Australia, see [Thomas Maslen], 'Discovery of a White Colony on the Northern Shore of New Holland', *Leeds Mercury*, 25 January 1834. The report was communicated in some 17 international newspapers, in Scotland, Ireland, South Africa, the United States, the Netherlands, and of course in Australia. Concerning this article, see Karen S. Cook, 'Thomas John Maslen and the "Great River of Desired Blessing" on His Map of Australia', *The Globe*, 61 (2008): 11–20; Karen S.

a perceived threat from French, Dutch, German, Swedish and even American colonial enterprises in the Pacific prompted the British to preemptively annex New Holland on 2 May 1831, scarcely two years after founding the Swan River District.[13] Yet despite this appropriation, and the lack of reaction from Dutch authorities, the Australian media contended that Dutch claims to sovereignty, were they asserted, could still be recognized.[14] The legacy of the *Batavia* had become one of cultural anxiety, manifested in questions of place and belonging. Clinging to the edges of the continent, early British settlers feared not only the threat of other nations by sea, but also from the wreck's supposed descendents, lurking in the rugged desert interior.

With longevity, however, concerns about territorial legitimacy ultimately faded, and Australia's colonial population established a new intellectual distance from which to consider the nation and its history. By the mid nineteenth century, Dutch priority in New Holland was openly discussed in several academic and popular publications, without negative consequences, real or imagined, for the colonies. Indeed, tales of early heroic Dutch explorers became well known throughout the Empire, and could even be celebrated.[15] The time was ripe for the British to exploit the *Batavia* myth, mainly in the fictional arena, for their own ends.

Mythology

Although treasure from the wreck was utilized as a plot in Henry Kingsley's adventure novel *The Mystery of the Island* (1877), the first significant employment

---

Cook, 'The Secret Agenda of Western Australian Explorer Robert Dale (1809–1853)', *The Globe*, 54 (2003): 23–34. Although Cook treats the original account of the Dutch tribe as fictitious, she supports a rather dubious position claiming that the alleged expedition which discovered these tribes was a 'cover' for a secret expedition led by Dale. For the reaction of Rosendo Salvado, Bishop of New Norcia, to these reports, see Rupert Gerritsen, *And Their Ghosts May Be Heard ...* (Fremantle, 2002).

13 On Swedish intentions see Robert J. King, 'Gustaf III's Australian Colony', *The Great Circle*, 27.2 (2005): 3–20. Concerning American expansion, see Jenny Booth, 'Emigration, Economics or Strategy? The British Government and the South Australia Act 1834', *Journal of Australian Studies*, 27.80 (2003): 157–64, 161.

14 See the anonymous article 'New Settlement Making on the Northern Coast', *The Sydney Gazette and New South Wales Advertiser*, Thursday, 28 July 1836, p. 2.

15 R.H. Major, *Early Voyages to Terra Australis, Now Called Australia* (London, 1859; repr. New York, c. 1970), pp. 59–74; J.H. Heaton, *Australian Dictionary of Dates and Men of the Time: Containing the History of Australasia from 1542 to date* (London, 1879); John Pinkerton, *Early Australian Voyages: Pelsart.* [sic] *Tasman. Dampier* (London, 1886).

of the *Batavia* story in fiction appeared in 1888.[16] In that year, William John Gordon, an English imitator of Verne, published his *The Captain-General: Being the Story of the Attempt of the Dutch to Colonize New Holland*, a 'boys' own adventure' novel which went through at least three editions before 1900. A naturalist and botanist by education, Gordon's limited fictional output was often set in the more exotic climes of the British Empire, and imbued with marked colonial preoccupations.[17] In the mould of his literary hero Verne, Gordon evidently saw his swashbuckling novel on the *Batavia* as a soapbox for voicing his own political and social critiques. In his preface to the work, he declared:

> [The story of the *Batavia*] is a story of the struggles of a nation in miniature, whose central figure is an unscrupulous adventurer making capital out of the political theories of the day, and advancing through democracy to despotism in a way not unknown to students of revolutions.[18]

Inspired by Pelsaert's *Ongeluckige Voyagie*, Gordon's narrative vividly portrayed the schemes of the 'turncoats' on board the *Batavia*, before describing with relish the shipwreck and the subsequent disintegration of Cornelisz's rule. Gordon saw in the *Batavia* incident a failed Dutch attempt to establish a colony on the 'Southland'.[19] This attempt would have succeeded, Gordon made clear in a moral coda for his young readers, but for the excesses of the 'thoroughly unscrupulous and almost incredibly inconsistent' Cornelisz, self-appointed 'Captain-General' of the title.[20]

Given that the Dutch never intended to colonize the Southland, Gordon's narrative is striking as a contrived warning not only to British readers – who faced real and imagined challenges from Prussian despotism, the demands of 'bomb-throwing anarchists', as well as the machinations of radical Marxists and their ilk following the 1848 revolutions – but also to inhabitants of Australia, where the book might have been read as an appeal for moderation in light of the 'blank canvas' of the nation, particularly from the likes of William Lane.[21] The

16 Henry Kingsley, *The Mystery of the Island* (London, 1877).

17 On this see Bradley Deane, 'Imperial Barbarians: Primitive Masculinity in Lost World Fiction', *Victorian Literature and Culture*, 36 (2008): 205–25.

18 W.J. Gordon, *The Captain General: Being the Story of the Attempt of the Dutch to Colonize New Holland* (London, 1888), p. xi.

19 Ibid. This mistaken notion sprang forth from Samuel Bennett, *The History of Australian Discovery and Colonisation* (Sydney, 1865), as Gordon revealed in his mawkish 'review' of his own novel, see: W.J. Gordon, 'A Colony that Failed', *The Leisure Hour*, vol. 38, 1889, p. 136.

20 Gordon, *The Captain-General*, p. 304.

21 On British perceptions of Lane, see Antony Taylor, '"'Heaven on Earth' was a Hell in Reality": Reflections on William Lane's "New Australia" in Britain', in Renata Summo-

'rough pirate' Cornelisz indeed 'wrecked the Dutch endeavour to colonize New Holland', but Gordon could hope only that 'the task left ... for the fortunate Englishmen' would not meet with a similar fate. Widely praised in Australia and Britain – although the colonial preoccupations of the work went unremarked – *The Captain-General* demonstrated that the *Batavia* incident could provide the perfect arena for exploring all manner of European ideas about Australia's past, present and future.[22] Shortly after its publication, the reading public throughout the Commonwealth was beset by a spate of works inspired by the wreck. The majority of these, however, were issued in Australia itself.

One of the earliest Australian authors to tackle the story was Ernest Favenc (1845–1908).[23] An explorer and member of Archibald's *Bulletin* bullpen, Favenc touched on the *Batavia* several times during the 1890s, mapping specifically Australian questions onto the tale. The first of these appeared in the failed educational magazine *The Story of Our Continent* (1892).[24] There, Favenc proposed a vigorous new historical model of Australia for the contemplation of young readers. In the absence of any 'reliable native [i.e. Aboriginal] tradition',

---

O'Connell (ed.), *Imagined Australia: Reflections around the Reciprocal Construction of Identity between Australia and Europe* (Bern, 2009), pp. 171–84. One might compare Gordon's warning to that issued in the same year by the Australian author Samuel Mullen, who, in a novel concerning the Chinese military occupation of Australia, warned his readers to take heed of history's lessons. See Samuel Mullen, *The Battle for Mordialloc, or, How we Lost Australia* (Melbourne, 1888), pp. 3–4.

22 See Phil Cormack, 'English Literacy and Anxiety about the Future: A Case-Study from the Turn of the Twentieth Century' (Report). Online at: www.unisa.edu.au/hawkeinstitute/cslplc/documents/CormackIFTE2003.pdf [accessed: 21 December 2009], pp. 10–11; 'An Interesting Page of Australia's History', *The West Australian*, 28 December 1891, p. 6. One Australian reader of Gordon's book was the Western Australian Guano contractor Florance Broadhurst. In addition to corresponding directly with Gordon about the book, his interest in the *Batavia* story would directly affect the later unfolding of the *Batavia* tale in Australia. See W.J. Gordon, 'Preface to the Third Edition', in *The Captain-General*, 3rd edn (London, 1891), p. xii. Cf. the account of H. Drake-Brockman, *Voyage to Disaster: The Life of Francisco Pelsaert – covering his Indian Report to the Dutch East India Company and the wreck of the ship Batavia in 1629 off the coast of Western Australia – together with the full text of his Journals concerning the Rescue Voyages, the mutiny on the Abrolhos Islands and the subsequent trials of the Mutineers* (Sydney and London, 1963), p. vii. On the Broadhurst family, see Michael McCarthy, 'Charles Edward Broadhurst, 1826–1905: A Remarkable Nineteenth-Century Failure'. Unpublished MA thesis, Perth, Curtin University, 1990.

23 On Favenc see Cheryl Frost, *The Last Explorer: The Life and Work of Ernest Favenc* (Townsville, 1983).

24 Ernest Favenc and Percy Spence, *The Story of Our Continent Told with Brush and Pen* (Sydney and Melbourne, 1892).

Favenc declared, Australia could be said to have only two histories: one was 'darkened by the shadow of the old penal laws', and was rightly to be shunned, while the other was 'illuminated by the heroism and fortitude of those who spent their lives in the service of discovery', and was thus to be celebrated.[25]

Favenc thus instrumentalized the *Batavia* story to replace the birth stain of Australia's penal past, with a new and heroic mythology. Such an approach, was, of course, not without its own prejudices. While Favenc praised the heroism of early seafarers, including the *Batavia* survivors, he lauded them above all for abstaining from contact with the 'primitive and sullen' Aboriginals during their adventures. Australia had thereby avoided the creation of a 'mongrel ... listless, peon population', the likes of which, he explained, had taken root under the mixed-races of Spanish and Portuguese colonial society.[26] For Favenc, it was imperative that Australia's history, and therefore its future, be imagined as heroic, white, European and, above all, as separate from indigenous traditions. In a political climate of 'White Australia', such an opinion was not entirely unexpected.[27]

In subsequent years, Favenc returned repeatedly to these same anxieties. His short stories often depicted European settlers as liminal figures in their own country. Favenc's first novel, *The Secret of the Australian Desert* (1895), was a macabre tale concerning the discovery of survivors of Ludwig Leichhardt's expeditionary party in central Australia, who had been corrupted into savagery after living too long among Aboriginal peoples.[28] The novel was undoubtedly inspired by the press coverage of 'Jungun', an albino Aboriginal who, in the 1880s and 1890s, toured various circuses, waxworks and museums throughout Australia as the so-called 'White Prince of the Red Desert'.[29] Many who saw

25 Ibid., p. 1.

26 Ibid., p. 7. According to Favenc, in South America, Iberian policies produced 'an enervated and listless upper class with just enough energy to quarrel amongst themselves for temporary possession of the power of governing and levying taxes. A middle class aping their vices; and a peon population of mongrel descent; part stockman, part shepherd, part miner and wholly thief. From such a fate has the stern and unkindly character of our continent saved her.'

27 For the early development of the 'White Australia' immigration policy, see Myra Willard, *History of the White Australia Policy to 1920* (Melbourne, 1923). See further Gwenda Tavan, *The Long, Slow Death of White Australia* (Melbourne, 2005).

28 Ernest Favenc, *Tales of the Austral Continent* (Sydney, 1894); Ernest Favenc, *The Secret of the Australian Desert* (London, 1895).

29 On Jungun see 'An Aboriginal Albino', *Argus* (Melbourne), Monday, 28 January 1890, p. 14; For later, quite informative, accounts, see 'Nor-West' [pseud.], 'The Story of Jungun: "White Prince of the Red Desert"', *The West Australian* (Perth), Monday, 15 January 1934, p. 14, which linked Jungun to the *Batavia* maroons; 'Memories of Jungun: Finder's Sister in Perth', *The West Australian* (Perth), Wednesday, 17 January 1934, p. 11; 'Leichardt

him speculated that he was a descendent of one of Leichhardt's party, or indeed the *Batavia* maroons. These audiences indicate a ready-made audience for the anxieties expressed in Favenc's works, and his philosophical conclusion; although whites were present on the Austral continent, they were by no means of it.

Favenc developed these themes with reference to the *Batavia* in his 1896 novel *Marooned on Australia*.[30] This book drew on the contemporary debates concerning federation, the 'native question' and the White Australia policy. In this novel, the two *Batavia* maroons encounter a continent teeming with issues remarkably similar to the hot topics of Favenc's time. Most remarkable was the ethnic diversity they confronted. In addition to the ubiquitous 'superstitious black savages', they see evidence of French explorers, and also encounter a strange, 'largely civilised' desert tribe, the white-skinned Quadrucos, led by the indefatigable King Quibibio.[31] After much trepidation, the Dutchmen band together with the Quadrucos, defeating by turns threats posed by the 'depraved savage Indians' from the interior, as well as recurrent invasions by 'Mongol Pirates' from the exterior.[32] Ultimately, one of the maroons falls prey to the horrors of the desolate environment and goes native, joining a party of 'depraved Indians'. The other returns to Europe following a chance rescue by a passing Dutch ship, but finds himself an outcast in the United Provinces, forever tainted by his experiences.

One of some 20 'lost world' novels published in Australia during the 'nervous' 1890s, *Marooned on Australia* – which, like its successors and imitators, was partly inspired by H. Rider Haggard's *She* and *Allan Quatermain* (both 1887) – was part of a literary fad in which contemporary anxieties about the place of white society in Australia were 'fabulated', in order that such anxieties might be 'conquered'; if not in real life, then at least on the written page.[33] For Favenc,

---

Mystery, Another Theory Advanced: Story of a "White" Black Man', *Western Mail* (Perth), Thursday, 18 January 1934, p. 14.

30 Ernest Favenc, *Marooned on Australia: Being the Narration by Diedrich Buys of his Discoveries and Exploits in Terra Australis Incognita about the year 1630* (London, Glasgow and Dunedin, 1897 [i.e. 1896]).

31 Ibid., p. 31 ff. In a factual epilogue to his story, Favenc refused to credit archaeological, agricultural or artistic variations found within Australia to Aboriginal peoples, insisting instead that 'these strange discoveries' 'differ so entirely from the ordinary rude memorials of the aborigines' that they 'point to the existence once of a colony of semi-civilized people' (Favenc, *Marooned*, pp. 222–4). This was a not untypical view of the period. See further Lynette Russell and Ian J. McNiven, *Appropriated Pasts: Indigenous Peoples and the Colonial Culture of Archaeology* (Lanham, MD, 2005), esp. pp. 134–9.

32 Favenc, *Marooned on Australia*, pp. 52–64, 77–88, 176–87.

33 On the significance of Lost Race novels in Australia, see Melissa Bellanta, 'Fabulating the Australian Desert: Australia's Lost Race Romances, 1890–1908', *Philament*, 3 (2002).

Asians had to be repelled and Aboriginals shunned, lest white society suffer the same degeneration as the Quadrucos.[34] But the experience of such threats, Favenc felt, was also the crucible in which a new, unique, and rugged Australian identity might be forged, hardier and stronger than that of the typical European.[35] As such, in Australia of the 1890s, the *Batavia* story could also be used in service of a growing rhetoric of independence during the Federation debates.

During these debates, the *Batavia* story and ideas of non-British exploration not only provided a platform for establishing identity; negotiations with the past reinforced the idea that white Australia somehow 'belonged' on the inhospitable continent, by reason of a continuous, if interrupted, historical presence. In 1894, George Collingridge, the Sydney-based geographer, issued a weighty tome cataloguing more than 200 pre-English 'discoveries' of Australia.[36] While Collingridge ultimately argued for Portuguese priority, others backed Spanish, French or even Egyptian discoverers. The contestations were not always serious. Patrick Moran, Catholic archbishop of Sydney and participant in the Federation debates, plundered Spanish sources to argue amusingly that, as Australia was first discovered by the Spaniard Fernández de Quirós in 1606, the federated nation must needs be a Catholic one.[37] In 1897, an English translation of Pelsaert's *Ongeluckige Voyagie*, contributed by the Dutch-Australian theosophist, anarchist and poet Willem Siebenhaar, was printed in the Christmas issue of Perth's *Western Mail*.[38] With a claimed distribution of 100,000 copies, this dramatic publication, according to a later historian, 'excited an interest that eventually spread across Australia and has never completely faded'.[39] In 1899, the Dutch scholar J.E. Heeres issued an encyclopaedic bi-lingual volume, *The Part Borne by the Dutch in the Discovery of Australia*, which excited some attention in the Australian

Online at: http://www.arts.usyd.edu.au/publications/philament/issue3_Critique_Bellanta.htm [accessed: 12 June 2009].

34 In this fashion, the Quadrucos, with their distinctly Mesoamerican nomenclature, might be taken as a commentary on the fates of the Portuguese and Spanish colonial societies critiqued in Favenc and Spence's earlier *The Story of Our Continent* (1892), referred to above.

35 See further Deane, 'Imperial Barbarians', p. 207 ff.

36 George Collingridge, *The Discovery of Australia. First Discovery of Australia and New Guinea: Being the Narrative of Portuguese and Spanish Discoveries in the Australasian Regions, Between the Years 1492–1606* (London, 1894).

37 Patrick Moran, *History of the Catholic Church in Australasia* (Sydney, 1896); The argument was repeated in a privately circulated pamphlet, Patrick Moran, *Discovery of Australia by de Quiros in the Year 1606* (Sydney, c. 1896).

38 'The Abrolhos Tragedy', *Western Mail* (Perth), Christmas special, 1897. On Siebenhaar, see Paul Eggert, 'The Dutch-Australian Connection: Willem Siebenhaar, D.H. Lawrence, Max Havelaar and *Kangaroo*', *Australian Literary Studies*, 21 (2003): 3–19.

39 Drake-Brockman, *Voyage to Disaster*, p. 21.

media despite its dreary academic prose. In the wake of successful Federation, the *Batavia* instantly became part of Australia's inalienable cultural legacy. In 1901, the woman traveller Vivienne May remarked in her *Travels in Western Australia* that relics of Dutch wrecks collected in the Abrolhos – many mistakenly believed to be from the *Batavia* – were now 'essential for every Australian to see'.[40]

An Interregnum

Perhaps contrary to expectations, interest in the *Batavia* tale and Australia's history of exploration declined in the decades immediately following Federation. Exiled to endnotes or relegated altogether to unread school textbooks, stories of such discoveries, no longer serving a useful cultural purpose, were quickly forgotten. Periodically, reports and short fiction concerning the *Batavia* and other wrecks were printed in Western Australian newspapers, but the interest they engendered was confined.[41] Even the tercentenary of Dirk Hartog's landing in 1916 passed without much spectacle. Australia was too busy building new and more conventional mythologies – such as those of the battlefield – to reinforce her newfound identity.

This would change in the 1920s, when the *Batavia* briefly reemerged in connection with the ever-present issue of race. On 28 March 1925, Daisy Bates (1859–1951), the so-called 'white queen of the Never Never', wrote a frantic article for the high-circulation *Sydney Morning Herald*.[42] Having spent years observing the negative impact of white culture on Aboriginal peoples, Bates was a fervent opponent of racial miscegenation. Incredibly, she ascribed the several 'depravities' she had observed during her lifetime to defective genes introduced through the 'indiscriminate raping of ... young girls' committed by the two

---

40 Vivenne May, *Travels in Western Australia* (London, 1973), p. 186.

41 An exception is Mary Gaunt's 'The First Australian Love Story' in her collection, *The Ends of the Earth* (London, 1916), which concerned a romance aboard the *Batavia*. In the 1910s, the Australian aviation pioneer Lawrence Hargrave – according to evidence literally found in his own backyard – propagated the theory that the Spanish had established a colony in Port Jackson, but such investigations experienced little popular resonance. In addition to several papers delivered to the Royal Society of New South Wales between 1909 and 1911, the evidence for this theory was collected by Hargrave in his pamphlet, *Lope de Vega* (Sydney, 1911). Hargrave's original notes, slides and drafts toward a book on the subject are located in Canberra, National Library of Australia, MS 352. Hargrave's theory was incorporated into a later 'lost race' novel concerning Dirk Hartog by the Sydney lawyer George Forbes, *Adventures in Southern Seas* (Sydney, 1920).

42 Daisy M. Bates, 'Our Aborigines: Pelsart's Dutchmen', *Sydney Morning Herald* (Sydney), Saturday, 28 March 1925, p. 16.

Batavia maroons. Filial cannibalism, incest and slovenliness were all practised, Bates argued, by Aboriginals 'possessed of blond hair and the flat, broad Dutch face', and nowhere else.

Bates's bizarre diatribes against the Dutch appeared in newspapers Australia-wide, and were reprinted in her influential *The Passing of the Aborigines* (1938).[43] Her articles exercised a paradoxical influence on contemporary discussions of Dutch and Aboriginal cohabitation. Commenting on the discovery of Pieter, the so-called 'Shark Bay Dutchman' in 1934, W.W. Edwards, a protector of Aboriginals in the Champion Bay district, argued,

> it is most absurd to try to connect [him] with those marooned [from the *Batavia*] for the reason that those aborigines in those days and right up to the eighteenth century were all more or less cannibals, and white men would have been killed and eaten in a very short time.[44]

Although Edwards repeated Bates's charges of cannibalism, such traits could not be ascribed, in his opinion, to the Dutch: their influence on Australian history, if they indeed exercised one, was entirely wholesome.

Separated from this specific context of nation building of the 1890s, by the 1920s the *Batavia* story had become fodder for fantasy and uninformed opinion. For, despite all of the speculation, the *Batavia* wreck remained something of a chimera. Much ink had been spilled concerning the ship and the fate of its passengers, but like so many narratives of the past, such words were quickly washed away on the fickle tide of popular interest. All this would change when the *Batavia* itself, as well as the remains of her murdered crew, were finally discovered in the Abrolhos islands.

New Directions

Since the mid nineteenth century it had been erroneously believed that the *Batavia* was wrecked somewhere in the southernmost latitudes of the Abrolhos. The misconception took root during British Admiralty surveys conducted by

---

[43] See Daisy M. Bates, 'My Natives and I: Chapter 12', *Western Mail* (Perth), Thursday, 9 April 1936, p. 10; Daisy M. Bates, 'My Natives and I: Life Story of D.M. Bates', *The Advertiser* (Adelaide), Wednesday, 29 January 1936, p. 23; Daisy M. Bates, *The Passing of the Aborigines: A Lifetime Spent among the Natives of Australia* (London, 1938), p. 107.

[44] W.W. Edwards, 'A Shark Bay Dutchman', *The West Australian* (Perth), Saturday, 17 February 1934, p. 5. See further E.H., 'Human Curios: A Shark Bay Dutchman?', *The West Australian* (Perth), Saturday, 3 February 1934, p. 4.

John Lort Stokes in 1842, when hydrographers discovered remains of Dutch wrecks, which they believed to stem from the *Batavia*. For this reason, the southernmost group bears the name 'Pelsaert' to this day.[45] But while the relics themselves were authentically Dutch, arriving in the Abrolhos with the *Zeewijck* in 1727, their association with the *Batavia* was not. When Siebenhaar translated the *Ongeluckige Voyagie* for Perth's *Western Mail* in 1897, it became evident to readers familiar with the geography of the Pelsaert group that the events of the wreck could not possibly have occurred there. Still, legends die hard, and misconceptions die harder. Indeed, it would not be until some 40 years later that the generally accepted conclusions concerning the resting place of the *Batavia* would be openly challenged, and the wreck site finally identified.

The first open questioning about the location of the *Batavia* wreck came as Australia was rocked by the turmoil of the Second World War. In 1940, the Perth journalist Malcolm Uren embarked on a series of expeditions to shipwreck sites in Western Australia in search of the *Batavia*. The results of his search were published in *Sailormen's Ghosts*, a remarkable volume which went through six editions between 1940 and 1945.[46] Concentrating on stories of mystery and treasure, it weaved numerous pieces of west-coast folklore together in a thoroughly irresistible mixture. It was quickly followed by Frank Goldsmith's *Treasure Lies Buried Here* (1946),[47] which likewise explored the world of maritime treasures in a series of disjointed narratives; the *Batavia* tale being retold, for example, as an excruciating stage play. Although both works were escapist in tone, they emphasized a burning question: where did the rotting hulk of the *Batavia* lie?

Following the conclusion of the Second World War, the search for the wreck site was accompanied by a variety of complementary political and cultural developments which hearkened back to the 1890s. In 1948, Australian citizenship was first introduced, replacing British subjectship. New perspectives about the nation's past emerged almost instantly. That same year, Charles Barrett published his *White Blackfellows: The Strange Adventures of Europeans who Lived Among Savages*,[48] in which the *Batavia* maroons received an exalted place as Australia's first 'white blackfellows', pioneers of a pluralistic modern Australian citizenship.[49] Similar shifts extended into the international realm. Following the experience of

---

[45] See John Lort Stokes, *Discoveries in Australia, 1837–43* (London, 1846).

[46] Malcolm Uren, *Sailormen's Ghosts* (Melbourne, 1940).

[47] Frank Goldsmith, *Treasure Lies Buried Here* (Perth, 1946).

[48] Charles Barrett, *White Blackfellows: The Strange Adventures of Europeans who Lived among Savages* (Melbourne, 1948), pp. 11–12.

[49] See further Catriona Elder, *Dreams and Nightmares of a White Australia: Representing Aboriginal Assimilation in the Mid-Twentieth Century* (Bern, 2009).

combat in the Pacific theatre, Australia's foreign policy had turned increasingly toward courting the favour of the United States above Britain, especially militarily. The perceived threat of nuclear peril and widespread contemporary belief in the dangers posed by the 'domino-effect' of international communism assisted this shift, and was duly reflected in the *Batavia* literature. While in 1947 Douglas Stewart's stage play *Shipwrecked!* could still provide an insular dramatic account of the Abrolhos terror, by the 1950s the international implications of the story were being deliberately evoked and explored.[50] It fell to the child protagonists of Melva Lear's *Dangerous Holiday: Five Young Treasure-Seekers in the Abrolhos* (1959) not only to discover a treasure trove left by Pelsaert on the Abrolhos islands, but also to foil the plot of a communist spy who, having stolen valuable intelligence from the new Woomera rocket range in South Australia, sought to escape Australia via submarine from a rendezvous point off Pelsaert Island.[51]

Such Cold War scenarios provided a literary counterpoint to authors attempting to pinpoint the wreck itself. By far the most significant of these was the Perth novelist Henrietta Drake-Brockman. Long fascinated by the *Batavia* story – as a girl she had read Gordon's *The Captain-General*, and visited the *Zeewijk* relics of the Abrolhos – in 1942, Drake-Brockman decided to research a novel on the wreck, and undertook the unprecedented move of writing to the Netherlands in order to acquire a copy of Pelsaert's original manuscript journal. While reading a translation prepared for her by schoolteacher E.D. Drok, Drake-Brockman realized that she had material enough not only for a novel, but a key to the wreck site itself. For Pelsaert's geographical notices clearly demonstrated that the *Batavia* had foundered not in the Pelsaert group of the Abrolhos, but in the northernmost Wallabi group. In 1955, she presented her initial argument in a respected Australian geographical journal.[52] The promised novel, which further popularized the Wallabi hypothesis, appeared under the title *The Wicked and the Fair* (1957).[53]

Coincidentally, particularly given Australia's increasing reliance – economic, military and cultural – on the United States, Drake-Brockman's conclusion had been anticipated by a disinterested Washington zoologist. In an article

---

50 Douglas Stewart, *Shipwreck: A Poetic Drama* (Sydney, 1947). A deluxe edition, featuring more than 20 etchings and paintings by Norman Lindsay was issued the following year. The play was reprinted in Stewart, *Four Plays* (Sydney, 1958).

51 Melva Lear, *Dangerous Holiday* (Sydney, 1959). Aspects of this story, such as the nefarious role of a suspicious bird watcher, were directly inspired by Uren's *Sailormen's Ghosts*. Some aspects of the tale, such as the notion of a functioning guest house on Pelsaert Island, were perhaps more fantastic than the extravagances of the Cold War-era plot.

52 Henrietta Drake-Brockman, 'The Wreck of the *Batavia*', *Walkabout*, 1 (1955): 33–9.

53 Henrietta Drake-Brockman, *The Wicked and the Fair* (Sydney, 1957).

on kangaroos for *National Geographic*, David H. Johnson noted that Pelsaert, the *Batavia*'s supercargo, was the first European to observe and describe the kangaroo.[54] The Tammar Wallaby that Pelsaert encountered is, as Johnson pointed out, indigenous to the northern Wallabi group of the Abrolhos islands, where the *Batavia* must therefore have been wrecked. A few years later, Drake-Brockman and Johnson were proven correct; the long-sought *Batavia* was discovered in 1963 – in the Wallabi group of the Abrolhos, as anticipated – by a local cray fisherman retrieving his traps. Coincidentally, the fisherman's name was Dave Johnson.

The excitement of the discovery was palpable. As the Western Australian museum and dive recovery teams visited the Wallabi group in order to free the *Batavia* from its coral tomb, news of the find – and of the potential hoard of treasure that was widely supposed to come with it – spread like wildfire. In 1963 Drake-Brockman rushed a confused non-fiction account of the wreck into print, along with a translation of Pelsaert's journal.[55] More populist accounts quickly followed, such as the tall tale of lost riches told in Kenneth W. Byron's *Lost Treasures of Australia and New Zealand* (1964).[56] The controversial Melbourne author Robert S. Close dished up a salacious tale of the wreck in his well-reviewed novel *The Voyage Continues* (1968).[57] By far the best of the contemporary works on the *Batavia*, however, was Hugh Edwards' *Islands of Angry Ghosts* in 1966.[58] Edwards, a young, charismatic and ruggedly handsome journalist and diver, provided a detailed and dramatic account of the wreck and the salvage. His book opened with Edwards and an archaeological team working on Beacon Island, or 'Batavia's Graveyard', as the Dutch maroons had named it, exhuming the skeleton of a victim of Cornelisz's reign of terror. With the signs of an axe blow to the skull and a wound to the shoulder blade clearly demonstrating the cruel and violent fate he had met, history had never seemed so close.

Bicentenaries

The rediscovery of the wreck, and of several of Cornelisz's hatcheted victims, ensured that the story of the *Batavia* was front and centre in the Australian consciousness during the decades bookended by the celebration of two important bicentenaries:

---

54 David H. Johnson, 'The Incredible Kangaroo', *National Geographic*, October (1955): p. 121.

55 See above, note 23.

56 Kenneth Byron, *Lost Treasures in Australia and New Zealand* (London, 1965).

57 Robert S. Close, *The Voyage Continues* (London, 1968).

58 The book was published simultaneously in Sydney, London and New York, and was one of the biggest selling dive books of the 1960s.

the Cook bicentenary in 1970 – marking the navigator's appropriation of *Terra Australis* for Britain – and the national bicentenary in 1988.

The inherent anglocentrism of the Cook celebrations inspired several visions of Australia's history which were inherently multicultural. In 1974, self-proclaimed 'new Australian' Priscilla Murdoch issued her *Duyfken and the First Discoveries of Australia*, a book which emphasized multiple Dutch and non-English encounters with the southern continent between 1606 and 1748.[59] In 1973, the young Australian director Bruce Beresford chose to capitalize on the success of his ode to international ockerism, *The Adventures of Barry McKenzie* (1972), by filming the story of another, very different Australian cultural icon; the *Batavia*, which subtly emphasized the unknown fate of the two maroons left on the mainland.[60] That same year, the West Australian poet Lee Knowles pictured one haunting legacy for these Dutchmen in her evocative 'Kalbarri', which emphasized possible connections historical and contemporary between white populations in Europe and Australia, as well as with the neglected Aboriginal population. Written from the perspective of a visitor to the remote Kalbarri region on West Australian coast – long suspected to have been the point where the two *Batavia* maroons set down – this visitor is shadowed by Aboriginal children, and is forced to ask himself, and by inference the reader, a striking question:

> Are these then their dark children whose
> shadows follow us
> down shelves of Rainbow Gorge within sound of the sea?[61]

But, much like the 1890s, with national identities at stake, the *Batavia* would naturally be forced to share the spotlight. Thus in 1977 Kenneth MacIntyre published *The Secret Discovery of Australia*, in which he doggedly argued the dubious case for Portuguese priority in Australia,[62] while in 1980, the so-called 'Gelignite Buccaneer', the Western Australian skin diver Allan Robinson,

---

59 Priscilla Murdoch, *Duyfken and the First Discoveries of Australia* (Artarmon, NSW, 1974).

60 'The Wreck of the Batavia', Bruce Beresford, dir. Edwin Morrisby, writer (Sydney, Reg Grundy Productions, 1973), 48 mins. A copy of this film is preserved in Canberra, National Film and Sound Archive, acc. no. 23285. Three clips from the film may be viewed online at: http://aso.gov.au/titles/documentaries/wreck-batavia/clip1/ [accessed: 12 February 2010].

61 Lee Knowles, 'Kalbarri' in *The Beloit Poetry Journal*, 25.3 (Spring 1975): 3–7. A further poem on the wreck of the *Batavia*, entitled 'Batavia Incident' was also printed in the same issue.

62 Kenneth MacIntyre, *The Secret Discovery of Australia: Portuguese Ventures 200 Years before Captain Cook* (Sydney, 1977).

declared in his biography that he had discovered a Phoenician galleon on Australia's northwest coast, and was privy to the locations of further Portuguese, Chinese and Egyptian wrecks.[63] Collingridge's *Discovery of Australia* was opportunistically reprinted by Souvenir Press of Sydney in 1983, and, following close on its heels, Lawrence Fitzgerald issued his *Java la Grande*, yet another attempt to demonstrate the Portuguese discovery of Australia.[64] Perhaps the most notable attempt to transcend perceived anglocentrism – real or imagined – during this period was Jill Hamilton's *The Captain Cook Myth* (1980), which set out to explode the 'cult of Cook' that had created him 'Australia's greatest hero'.[65] The eccentric cover design featured a statue of Cook, in a pose decidedly reminiscent of Adolf Hitler. The point of most of these works was to call into question received wisdom about Australia's history, in particular the anglocentric paradigm that defined it. At the same time, however, they also comprised attempts to anachronistically project a multicultural history back into Australia's past, to better serve its present.

This desire to let history reflect contemporary cultural diversity intensified further as the national bicentenary of 1988 neared. In 1982, MacIntyre's *Secret Discovery of Australia* appeared again in an abridged paperback edition, intended – and actually used – as a history textbook in Victorian secondary schools.[66] In 1985, Australia Post issued two strikingly-designed sets of stamps called *Terra Australis*, part of a five-year programme of bicentennial philatelic issues. The first series of four stamps, 'Navigators', bore values intended for domestic use and depicted early navigators like Willem Jansz, Abel Tasman, William Dampier, and Dirk Hartog, or their ships and instruments. The second series, 'Coastal Shipwrecks', depicted shipwreck artefacts – including items from the *Batavia* – and bore higher values intended for overseas use. As Dennis Altman has argued, these stamps were part of a conscious effort of promoting, both domestically and internationally, a picture of Australia's diverse ethnic and cultural past.[67]

---

63 Allan Robinson, *In Australia Treasure is Not for the Finder* (Perth, 1980). Robinson's story is told – rather more objectively – in several fine books, most notably in James A. Henderson, *Phantoms of the Tryall* (Perth, 1993). Robinson's misadventures and ghastly fate directly inspired the brash and tragic figure of Don Spargo in Robert Drewe's *Fortune* (Sydney, 1986).

64 Lawrence FitzGerald, *Java la Grande: The Portuguese Discovery of Australia, circa 1521* (Hobart, [1984]).

65 Jill Hamilton, *The Captain Cook Myth* (Sydney, 1980).

66 Kenneth MacIntyre, *The Secret Discovery of Australia. Portuguese Ventures 250 Years Before Captain Cook*, revised and abridged edn (Sydney, 1982).

67 For more on the political nature and context of philatelic issues, with some considered discussion of Australia's bicentennial stamp programme, see Dennis Altman, *Paper Ambassadors: The Politics of Stamps* (London, 1991), pp. 64–5, 145-6.

Australia's increasing role, cultural and financial, in South East Asia was broached through the prism of the *Batavia* wreck in Nicholas Hasluck's *The Bellarmine Jug* (1984).[68] This imaginative novel portrayed the *Batavia*'s voyage as Rosicrucian initiative to found a heretical Dutch colony on *Terra Australis*, the revelation of which would have direct implications for the independence movement in modern-day Indonesia. In 1989 and 1990, two young-adult novels directly addressed the wreck and its aftermath. Echoing perhaps the national mood and the general historical awareness the bicentenary had inspired, the past broke violently and involuntarily into the present in both stories, where two Australian teens found themselves confronted by the villains of the *Batavia* incident when they unwittingly invaded our modern world (Lisson's *The Devil's Own*, 1990), or the protagonists were inexplicably propelled into the maelstrom of the past (Crew's *Strange Objects*, 1989).[69]

Native Titles

But if the bicentennial era offered new opportunities to consider and engage with the past, public consciousness of Australia's history would return to distinctly nineteenth-century concerns following the landmark *Mabo* case of 1992.[70] This important decision, which effectively dissolved the foundational legal fiction of *terra nullius* – the idea that, at the time of Cook's proclamation of the colony of New South Wales, Australia was 'land belonging to no one' – awakened fresh anxieties concerning European liminality on the southern continent.[71] The fate of those two intrepid maroons, Looes and Pelgrom, would again loom large in the public imagination as a key to engaging with these issues.

The most significant work which appeared in the wake of the *Mabo* decision was Rupert Gerritsen's controversial work entitled *And Their Ghosts May Be*

68 Nicholas Hasluck, *The Bellarmine Jug* (Ringwood, 1984).

69 Deborah Lisson, *The Devil's Own* (Glebe, 1989); Gary Crew, *Strange Objects* (Sydney, 1990). Partially inspired by H.P. Lovecraft's horror story 'The Shadow Out of Time' (1935), Crew's superb novel has proven an enduring international bestseller, having gone through several editions, as well as being translated into multiple languages.

70 The decision itself is reproduced in *Mabo v Queensland No. 2* (1992) 175 Commonwealth Law Reports 1. Online at: http://www.austlii.edu.au/au/cases/cth/high_ct/175clr1.html [accessed: 10 June 2010].

71 An interesting manifestation of this anxiety may be seen in the title of a work by John Kenny, *Before the First Fleet: Europeans in Australia, 1606–1777* (Kenthurst, 1995), which appears to suggest the notion of continuous European occupation from 1606. The dust jacket of this book features the less alarmist and rather more accurate title, *Before the First Fleet: The European Discovery of Australia, 1606–1777*.

Heard. By means of a bewildering and imaginatively deployed variety of evidence, Gerritsen's speculative book claimed that at least 73 victims of Dutch shipwreck had been stranded on the west coast of Australia in the centuries before Cook's voyages, and had furthermore integrated with the local Aboriginal population, thereby creating a flexible hybrid community in which Europeans and Aboriginals lived together in harmony. This was a harmony effectively destroyed following the English 'invasion' and 'holocaust' of 1788.[72] Gerritsen's somewhat polemical work comprised an attempt to elaborate upon an alternative historical dynamic to race relations in Australia, by intermingling the trajectories of European and Aboriginal civilization into a single continuum. Although the intent of this argument was undoubtedly benevolent, in his insistence that archaeological, artistic and cultural anomalies present in certain Aboriginal communities may be due to the influence of stranded Dutch, his book could be interpreted as an attempt to appropriate, rather than celebrate, Aboriginal history.[73]

Gerritsen's thesis has nonetheless been taken up by others, such as the Perth-based VOC Historical Society. Founded in 2002 by Tom Vanderveldt, an Australian of Dutch descent, the group runs a popular website and runs a variety of public initiatives.[74] Their newsletter, the *New Holland Update*, describes the fate of the Batavia mutineers utilizing a distinctly modern political vocabulary. Looes and Pelgrom are described as Australia's 'first boat people', and the group claims that, like the Asian refugees seeking asylum today, these refugees of history have been just as maligned by a bigoted anglocentric administration.[75]

Recently, Vanderveldt and the VOCHS have collaborated with scientists in the Netherlands, gathering genetic data to contribute to an analysis of Western Australian Aboriginal DNA, the results of which might confirm the thesis of cohabitation.[76] The early results are promising: nearly all of the tested

---

72 Gerritsen, *And Their Ghosts May Be Heard*, p. 266. The second edition was also issued by the same publishers. Concerning Gerritsen's work, see Leigh T.I. Penman, 'The *Batavia* Legacy: The Wreck of the Dutch East India Company Ship *Batavia* (1629) in Australian History and Imaginary', in Renata Summo-O'Connell (ed.), *Imagined Australia: Reflections upon the Reciprocal Construction of Identity between Australia and Europe* (Bern and New York, 2009), pp. 153–70.

73 A similar approach is found in Gerritsen, *Australia and the Origins of Agriculture* (Oxford, 2008). See also the review of this work by Fiona Jane Walsh in *GeoJournal*, 74 (2009): 499–501.

74 The website of the group is at http://www.voc.iinet.net.au.

75 See for example *New Holland Update* 1/7 2007, [3].

76 Beatrice Thomas, 'Gene Tests Seek Dutch Links to WA Aboriginals', *The West Australian* (Perth), 18 December 2008, p. 60; Jessica Strutt, 'Aboriginals Seek Dutch DNA Link', *The West Australian* (Perth), 17 April 2010; see also the story featured on Channel 7 Australia's 'Sunday Night' programme entitled 'Australia's Secret History', Ross Coulthart,

indigenous males from the central west coast area possess Y chromosomes of European origin. Further research is necessary, however, to determine whether any of this genetic material is indeed Dutch, and when precisely it entered the genetic matrices.[77]

Although there are some who hope that the incontrovertible presence of Dutch DNA among native Australians will assist in proving claims to certain land areas under the stringent demands of the Native Title Act, the revelation of a shared genetic past is more commonly viewed as an aspirational aide to future reconciliation between European and Aboriginal communities.[78] William Mallard, one member of the Nhanda of Western Australia who claims descent from Dutch strandees, expressed his hopes as follows:

> I also descend from the European side. Not just the British. I descend before the British are come. I descend from a far greater race of people, who are explorers, who had culture, who had life, and in today's society, they are looked upon as leaders for human rights. And that's what I look to. And that's what I aim to.[79]

For Mallard, as well as others, incontrovertible proof of Dutch descent will engender nothing less than a paradigm shift for the reconciliation movement in Australia, setting debates concerning the maltreatment of Aboriginals by the

---

reporter, broadcast 19 April 2010. This rather slight report may be viewed online at: http://au.tv.yahoo.com/sunday-night/features/article/-/article/7070203/australias-secret-history-more-information/ [accessed: 20 November 2010]. The Dutch multiplatform broadcaster VPRO has organized an online collection of radio stories, videos and articles on the notion of Dutch Aboriginals, and the possible DNA link, as part of their multimedia programme in support of the *Beagle* television documentary series. See 'Dossier: Nederlandse Aboriginals?' Online at: http://beagle.vpro.nl/#/blog/item/3321/ [accessed: 27 September 2010].

[77] The 'very recent' arrival of European phylogenetic material in Australian Aboriginal populations is stressed in a recent article by Brian P. McEvoy et al., 'Whole-Genome Genetic Diversity in a Sample of Australians with Deep Aboriginal Ancestry', *American Journal of Human Genetics*, 87.2 (August 2010): 297–305, although the subjects of this particular research were not members of the Western Australian Nhanda tribe. See further Elmaar Veerman, 'Aboriginals met Hollands bloed. Gingen VOC-schipbreukelingen op in de lokale bevolking?', *Noorderlicht uw Hoofdleverancier*, 12 April 2010. Online at: http://beagle.vpro.nl/#/blog/item/3339/ [accessed: 24 September 2010].

[78] Sarah Darwin, 'It Tastes Like Shit, but You Can Live On It.' Blog entry for VPRO's *Beagle* project. Online at: http://beagle.vpro.nl/#/blog/item/3381/ [accessed: 27 September 2010].

[79] Interview with William Mallard from episode 29 of the television program *Beagle, in het kielzog van Darwin*, entitled 'Hollandse Aboriginals' [Dutch Aboriginals]. First broadcast on 18 April 2010. The episode (largely in Dutch) may be viewed online at: http://beagle.vpro.nl/#/widget/afleveringen/29 [accessed: 27 September 2010].

British administration in a new international context. While the *Batavia* wreck is here part of the background to a broader tapestry of events both historical and contemporary, the potential cultural and social effect of new revelations concerning a Dutch pre-history in Australia can be glimpsed. The transcendent visions of recognition seen in the legacy of Dutch shipwrecks by the contemporary Aboriginal population is a striking, and welcome, divergence from the anxieties previously attached to these incidents by European Australians, like Ernest Favenc.

This most recent iteration of the *Batavia* legacy, picturing descendents of the wrecks living unknown among Australia's Aboriginal population, has struck a chord in popular culture, resulting in portrayals both exciting and unconventional. Potential descendents of the *Batavia*'s crew feature prominently in a subplot to Judith Nunn's dynastic novel *Territory* (2003).[80] In Kirsty Eagar's recent novel for young adults, *Saltwater Vampires* (2010), undead members of the *Batavia's* crew return to terrorize the surfing community of Rocky Head.[81] Additionally, the Melbourne director David Blake has recently attempted to bring the drama of cohabitation to new audiences in his oddly-titled film project, *Batavia's Temper Blood Fury*. Although the project appears to have collapsed as of mid-2011, the plot concerns a detective investigating a series of 'revenge killings' carried out by descendents of the *Batavia* mutineers in present-day Western Australia. Although Blake has admitted that envisioning 'a present-day take on the murders' was inspired 'as much by the need to secure funding' as any other factor, that this idea has managed to attract generous support from various international sources, demonstrates a widespread interest in engaging with the potential modern consequences of the *Batavia* incident.[82]

Perspectives from Old Europe

But while the silver screen may carry the *Batavia* story around the world, it is important to note that the wreck and its brutal aftermath have also inspired interest in European societies outside Australia in the recent past. Dutch interest in the wreck, for example, has always been high. The rediscovery of the *Batavia*

---

80 Judy Nunn, *Territory* (Sydney, 2003).

81 Kirsty Eagar, *Saltwater Vampires* (Ringwood, 2010).

82 See the comments of Blake in Elizabeth Gosch, 'Vinnie Jones to Bring Batavia Tale to Life', *The Australian*, 13 December 2007. See also Shannon Harvey, 'Batavia Movie Set in Geraldton yet to Begin Filming', *Perth Now*, 5 May 2008. Online at: http://www.perthnow.com.au/news/batavia-moviepossibly-shipwrecked/story-0-1111116260457 [accessed: 29 July 2009].

in 1963 was perhaps the most important step in loosing the first wave of popular interest in the disaster in that country since the initial excitement of the 1630s. An initial account of the *Batavia*'s fate in the *Spiegel Historiale* (1967) inspired novelist Anthony von Kampen to author his novel *Het beest uit Zee* (The beast from the sea, 1971). Deriving its title from Daniel ch. 7, Kampen focused upon the machinations of the distinctly Faustian Cornelisz.[83] A similar focus was present in a later novel by Johannes Fabricius (1981), which bore an almost identical title.[84]

The general interest provoked by these works, in addition to an awareness of Australia's impending bicentennial celebrations, resulted in what is certainly the most remarkable manifestation of Dutch interest in the wreck. In 1985, the Dutch carpenter and shipbuilder Willem Voss began an immense project to reconstruct the *Batavia* in Lelystad, Netherlands, in order to provide a living example of Dutch maritime history to the world.[85] Based on authentic seventeenth-century shipbuilding plans, and constructed according to traditional techniques, the completed ship made its public debut in 1995 at the *Sail '95* festival in Amsterdam. Voss's project not only prompted a flurry of interest in the wreck in the Netherlands, it also inspired the Dutch maritime historian Vibeke Roeper to collect and edit all known primary source documents on the incident, issued in a superb scholarly volume of 1993.[86] The *Batavia*, for so long the subject of conjecture and speculation, was now more tangible and accessible than ever before, in a variety of realms.

The rebuilt *Batavia* quickly became a symbol of friendship, and of shared experience, between Australia and the Netherlands. During the bicentennial celebrations, the ship's dinghy was sent by Voss to Perth, where it was sailed to the Abrolhos Islands. In 2000, the *Batavia* itself was transported via barge from its mooring in Lelystad to Sydney, in order to serve as a tourist attraction during the Sydney Olympics, and as a decidedly unwieldy mascot for the visiting Dutch

83 Anthony von Kampen, *Het beest uit zee. De ondergang van het Compagniesschip 'Batavia'* (Bussum, 1971).

84 Johannes Fabricius, *Het beest uit de zee* (The Hague, 1980). A second edition was issued in 1986.

85 The early stages of the reconstruction project are outlined in the final chapter of G.C. Molewijk, *Pelsaerts journaal van de ongelukkige reis van het schip Batavia* (Weesp, 1989), pp. 92–109.

86 Roeper, *Het Schipbreuk van de Batavia.* See above, note 2. In addition to the aforementioned text by Molewijk, some other works inspired by Voss's project include Sybren Polet, 'De nachtmerrie van het Nieuwe Amsterdam', *In de arena* (Amsterdam, 1987); Eric Leijenaar, *De Batavia; het gruwelijke, waar gebeurde verhaal over de trotse Oostindiëvaarder* (Amsterdam, 1989); Mike van Niekerk, 'Agony of the *Batavia*', *Garuda Indonesia. In-Flight Magazine* (July 1993): 6–11, among others.

team.[87] One interested observer of all this activity was the British historian Mike Dash, who had long harboured an interest in the culture of the Low Countries. Inspired by the lurid and shocking tale of the wreck, he spent several years researching a volume that became the international bestseller *Batavia's Graveyard: The True Story of the Mad Heretic Who Led History's Bloodiest Mutiny* (2002), the most exhaustively researched account of the wreck available to date.[88] The astonishing success of this book has in turn inspired a wealth of novels, novellas and factual accounts of the wreck and its aftermath, by a variety of European authors.[89]

But to other nations the *Batavia* has represented still other imaginative possibilities, and reflected different realities. In France, Dabitch and Pendanx's graphic novel *Jeronimus* was recently published, which focused on the drama of the wreck, seemingly as a coda to Chirac-era politics.[90] In Germany, the *Batavia* has attracted a great deal of attention, inspiring films, non fiction and several novels.[91] While the appeal partially derives from Germany's insatiable appetite for all things antipodean, there are darker historical circumstances that have also influenced this reception. Indeed, several works focus on Cornelisz's despotic reign on the Abrolhos as an analogy for Nazi rule, such as in Gabriele Hoffmann's *Die Schiffbrüchigen* (The shipwrecked, 2004).[92] It is perhaps most apparent, however, in Rachel van Kooij's superb novel for young adults, *Der Kajütenjunge des Apothekers* (The apothecary's cabin boy, 2006), a gripping examination of conscience, complicity and culpability in the case of one boy who participated in Cornelisz's tyranny.[93] Such is the German fascination with the *Batavia* that, in 2001, the Heyne Verlag commissioned a novel by Australian

---

87 For details of the Olympic visit, see the commemorative pamphlet *Batavia: Magnificent Ship, Amazing Journey* (Sydney, 2000), pp. 4–6.

88 Mike Dash, *Batavia's Graveyard* (London, 2002). The title of the French translation of this book, *L'archipel des hérétiques: la terrifiante histoire des naufragés du Batavia* (Paris, 2002), is somewhat more evocative than the English original.

89 Two recent such works, which I became aware of only shortly before this chapter was submitted for print, include Greta van der Rol, *Die a Dry Death* (Dragon International Independent Arts.com, 2010), which concerns the rich drama of the wreck and mutiny, and Rosemary Hayes, *The Blue-Eyed Aborigine* (London, 2010), which concerns the survival of Looes and Pelgrom on the western Australian mainland.

90 Christophe Dabitch and Jean-Denis Pendanx, *Jeronimus* (3 vols, Paris, 2008–10). German and Dutch translations have also appeared.

91 Martin Paparowski's documentary *Meuterei auf der 'Batavia'* (Mutiny on the *Batavia*) was the first film in the *Höllenfahrten* ('Hell trips') series produced for Zweites Deutsches Fernsehen (ZDF) in 1998.

92 Gabrielle Hoffmann, *Die Schiffbrüchigen* (Hamburg, 2004).

93 Rachel van Kooij, *Der Kajütenjunge des Apothekers* (Vienna, 2006).

Colin Falconer about the wreck. Originally entitled 'The Kingdom', Falconer's book appeared in Germany as *Zorn der Meere* (Fury of the seas, 2001).[94] This fast and furious potboiler was narrated, in a twist that slots neatly into Randolph Stow's dialectic, by none other than Satan himself.

In these works, the wreck of the *Batavia* functions as a mirror for local issues and themes. As in Australia, the emphases of these narratives have shifted over time, subject to the ends to which they have been employed. But while it is not their overt intention, I contend that these European works nevertheless convey a peculiar conception of *Terra Australis*, one almost nineteenth-century in its preoccupations. It is one in which Europeans manifestly do not, and cannot, belong; a distant, barren and corrupting environment, far from the strictures of European decorum, where humanity's darkest desires and impulses simmer just below reason and rationality, waiting to escape.

Terra Australis through the Prism of the *Batavia* Wreck

The notion of *Terra Australis* as a land of unknown excesses and danger has thus transcended place. It has appeared and reappeared in literature and other works on the *Batavia*, issued in Australia and elsewhere, dating back to the very first news of the wreck in the United Provinces in 1630. Then, the reports of the incident were rationalized with reference to the heresy of Torrentius. The locations – physical, mental and spiritual – of the wreck were deliberately shifted in the popular imagination beyond the boundaries of the known and acceptable world. This sense of the southern continent as a distant and corrupting place would appear again in Ernest Favenc's works concerning the *Batavia* of the 1890s. These portrayed *Terra Australis* as a place where the European – physically confined to the edge of the continent, trapped between the black 'savages' within and the Yellow Peril without – must forge a new identity in order to transcend the brutality of their environment and the 'degeneracy' that accompanied it. The ghosts of these ideas can be seen in some later manifestations of the *Batavia* myth, particularly Daisy Bates's rambling and ill-considered condemnation of Dutch castaways Looes and Pelgrom in the 1920s.

However, as the European society in Australia became ever more comfortable with the strange land upon which they found themselves, the purposes to which

[94] Colin Falconer, *Zorn der Meere* (Stuttgart, 2001), a translation into Romanian followed in 2002. Another German-language novel which concerns the *Batavia* is Silke Kerne, *Batavia 1629* (Munich, 2000); a sort of diffuse travelogue of Western Australia and Indonesia is provided in Wolf-Ulrich Cropp, *Die Batavia war ihr Schicksal. Seeabenteuer eines Ostindienfahrers* (Bielefeld, 1997).

the stories told about its history began to change. Following the Second World War, within a new multicultural Australia, the shipwreck was one of the most memorable and remarkable signposts to the nation's historically diverse past. This is nowhere more apparent than during the bicentenaries of the 1970s and 1980s, in which attempts were made to move away from the spectre of anglocentrism, and promote the ethnic and cultural diversity of the nation's history. The 'lost white tribes' which terrorized the imagination of Australians during the 1830s and 1840s would reappear, albeit with a very different significance, following the rejection of *terra nullius* in 1992. Suddenly, after taking just over 200 years for Europeans to become comfortable with their position on the continent and the idea of being 'Australian', the very legality of this European presence was seemingly challenged by native title. It can hardly be surprising, therefore, that several parties, among them Gerritsen or the VOCHS, have since this time begun to look for a heretofore neglected history of European and Aboriginal cooperation and cohabitation. The offspring of native Australians and Dutch victims of marooning and shipwreck, real or imagined, have since become an important feature of several negotiations with Australia's past, and perhaps also for mapping out a potentially harmonious future.

Although in his landmark essay on the subject Randolph Stow called the *Batavia* incident a 'futile, meaningless historical event', he nevertheless recognized that it was one 'out of which meaningful myths have been created'. Throughout the history of its interpretation, the wreck of the *Batavia* has functioned as a mirror which has reflected back to its audience a surprisingly diverse and durable set of myths and historical realities. In a very real sense, such myths and realities have been intimately concerned with mapping out the contours and topography of *Terra Australis*. This particular *Terra Australis*, however, is not that of the physical world, which explorers and cartographers of the past pursued so intently, but rather one of the imagination.

Bibliography

Albertus Magnus, *De natura loci*, Paul Hossfield (ed.), *Opera Omnia*, vol. 2 (Aschendorf: Monasterii Westfalorum, 1980).

Alfonse, Jean, *Les voyages auantureux du capitaine Ian Alfonce, saintongeois* (Poitiers: 1559).

Alfonse, Jean, *La cartographie, avec l'espère et* régime du soleil et du nord, par Jean Fonteneau dit Alfonse de Saintonge, capitaine pilote de François 1er. Paris: Bibliothèque Nationale de France, MS français 676; Georges Musset (ed.) (Paris: E. Leroux, 1904).

Alighieri, Dante, *La Commedia secondo l'antica vulgata*, Giorgio Petrocchi (ed.) (4 vols, Milan: Mondadori, 1966–67).

Alighieri, Dante, *The Divine Comedy, vol. 1: Inferno*, Charles S. Singleton (trans.) (2 vols, London: Routledge and Kegan Paul, 1970).

Altman, Dennis, *Paper Ambassadors: The Politics of Stamps* (London: Angus & Robertson, 1991).

An Authentic and Interessting [sic] Narrative of the Late Expedition to Botany Bay, as Performed by Commodore Phillips ... Written by an officer just returned ... (London: Printed by W. Bailey, 1789).

An historical narrative of the discovery of New Holland and New South Wales. Containing an account of the inhabitants, soil, animals, and other Productions of those Countries, and including a particular Description of Botany Bay (London: Printed for John Stockdale, 1786).

'An Interesting Page of Australia's History', *The West Australian*, 28 December 1891, p. 6.

Ando, Clifford, *Imperial Ideology and Provincial Loyalty in the Roman Empire* (Berkeley, CA: University of California Press, 2000).

Anthiaume, Albert-Marie-Ferdinand, *Cartes marines* (2 vols, Paris: Dumont, 1916).

Anthiaume, Albert-Marie-Ferdinand, 'Un pilote et cartographe havrais au XVIème siècle: Guillaume le Testu', *Bulletin de géographie historique et descriptive* (1911): 135–202; repr. *Acta Cartographica*, 19 (1974): 10–77.

Arago, J., *Narrative of a Voyage Round the World, in the* Uranie *and* Physicienne *corvettes, commanded by Captain Freycinet, during the years 1817, 1818, 1819, and 1820; on a scientific expedition undertaken by order of the French government. In a series of letters to a friend, by J. Arago, Draftsman to the Expedition. With twenty-six engravings. To which is prefixed, The Report made to the Academy of Sciences on the General Results of the Expedition* (London: Treuttel and Wurtz, 1823), printed in facsimile by N. Israel (Amsterdam: Da Capo Press, 1971).

Arago, J., *Souvenirs d'un aveugle: voyage autour du monde* (Paris, 1868).

Aristotle, *Meteorologica*. Online at: www.greektexts.com/library/Aristotle/Meteorology/eng/486.html [accessed: 20 November 2010].

Atkinson, Alan, *The Europeans in Australia: A History* (Melbourne: Oxford University Press, 1997).

Augustine of Hippo, *De Civitate Dei*, B. Dombart and A. Kalb (eds), Corpus Christianorum, Series Latina (2 vols, Turnhout: Brepols, 1955).

Bacon, Roger, *The 'Opus Maius' of Roger Bacon*, John Henry Bridges (ed.) (2 vols, Oxford: Clarendon Press, 1897).

Baer, J.H., 'William Dampier at the Crossroads: New Light on the "Missing Years" 1691–1697', *International Journal of Maritime History*, 8.2 (1996): 97–118.

Bailey, Martin, 'The Discovery of the Lost *Mappamundi* Panel: Hereford's Map in a Medieval Altarpiece?', in P.D.A. Harvey (ed.), *The Hereford World Map: Medieval World Maps and Their Context* (London: British Library, 2006), pp. 79–93.

Bakewell, Peter, 'Spanish America: Empire and Its Outcome', in John H. Elliott (ed.), *The Hispanic World: Civilization and Empire: Europe and the Americas: Past and Present* (London: Thames and Hudson, 1991), pp. 65–84.

Barber, Peter, 'The Christian Knight, the Most Christian King and the Rulers of Darkness', *The Map Collector*, 52 (1990): 8–13.

Barber, Peter, 'Malaspina and George III, Brambila, and Watling: Three Discovered Drawings of Sydney and Parramatta by Fernando Brambila', *Australian Journal of Art*, 11 (1993): 31–55.

Barnes, G., 'Curiosity, Wonder, and William Dampier's Painted Prince', *Journal for Early Modern Cultural Studies*, 6.1 (2006): 31–50.

Barrett, Charles, *White Blackfellows: The Strange Adventures of Europeans who Lived among Savages* (Melbourne: Hallcraft, 1948).

Barton, Tamsy, 'Augustus and Capricorn: Astrological Polyvalency and Imperial Rhetoric', *Journal of Roman Studies*, 85 (1995): 33–51.

Bassett, M., *Realms and Islands: The World Voyage of Rose de Freycinet in the Corvette Uranie 1817–1820* (London: Oxford University Press, 1962).

Batavia: Magnificent Ship, Amazing Journey (Sydney: Australian National Maritime Museum, 2000).

Bates, Daisy M., *The Passing of the Aborigines: A Lifetime Spent among the Natives of Australia* (London, 1938).

Bellanta, Melissa, 'Fabulating the Australian Desert: Australia's Lost Race Romances, 1890–1908', *Philament*, 3 (2002). Online at: http://www.arts.usyd.edu.au/publications/philament/issue3_Critique_Bellanta.htm [accessed: 12 June 2009].

Bennett, Samuel, *The History of Australian Discovery and Colonisation* (Sydney: Hanson and Bennett, 1865).

Beresford, Bruce, 'The Wreck of the Batavia', Edwin Morrisby, writer (Sydney: Reg Grundy Productions, 1973). 48 mins. [Canberra, National Film and Sound Archive, acc. no. 23285].

Berger, John, *Ways of Seeing* (London: BBC and Penguin, 1972).

Berggren, J. Lennart and Alexander Jones, *Ptolemy's Geography: An Annotated Translation of the Theoretical Chapters* (Princeton, NJ and Oxford: Princeton University Press, 2000).

Best, Michael (ed.), 'The Staging of a Morality Play (2)', *Internet Shakespeare Editions*, University of Victoria. Online at: http://internetshakespeare.uvic.ca/Library/SLT/stage/early%20stages/castle2.html [site updated November 2010].

Bonnemains, J., E. Forsyth and B. Smith (eds), *Baudin in Australian Waters: The Artwork of the French Voyage of Discovery to the Southern Lands 1800–1804* (Melbourne: Oxford University Press, 1988).

Bonner, Elizabeth, 'Did the French Discover Australia? The First French Scientific Voyage of Discovery, 1503–1505', in David W. Lovell (ed.), *Revolution, Politics, and Society: Elements in the Making of Modern France* (Canberra: Australian Defence Force Academy, 1994), pp. 40–48.

Booth, Jenny, 'Emigration, Economics or Strategy? The British Government and the South Australia Act 1834', *Journal of Australian Studies*, 27.80 (2003): 157–64.

Bosch-Barrett, Manuel, *Doña Isabel Barreto: Adelantada de las islas Salomón* (Barcelona: Juventud, 1943).

Bouzrara, Nancy and Conley, Tom, 'Cartography and Literature in Early Modern France', in David Woodward (ed.), *The History of Cartography*, vol. 3, part 1 (Chicago, IL: Chicago University Press, 2007), pp. 427–37.

Bredius, A., *Johannes Torrentius. Schilder, 1589–1644* (The Hague: Martinus Nijhoff, 1909).

Bremer, Annette, 'Domestic Disclosures: Letters and the Representation of Cross-Cultural Relations in Early Colonial New South Wales', *Frontiers: A Journal of Womens Studies*, 28.1–2 (2007): 77–95.

Brome, Richard, *The Antipodes*, in Anthony Parr (ed.), *Three Renaissance Travel Plays* (Manchester: Manchester University Press, 1995).

Brosses, Charles de, *Histoire des Navigations aux Terres Australes contenant ce que l'on sçait des mœurs & productions des Contrées découvertes jusqu'à ce jour: & où il est traité de l'utilité d'y faire de plus amples découvertes, & des moyens d'y former un établissement* (Paris: Durand, 1856; Amsterdam and New York: Nico Israel and Da Capo Press, 1967).

Brown, Anthony J., *Ill-starred Captains: Flinders and Baudin* (Adelaide: Crawford House, 2000).

Brown, Anthony J. and Christine Cornell, 'Legacy of the Encounter: British and French Place Names of the South Australian Coast', *South Australian Geographical Journal*, 99 (2000): 10–15.

Brown, Anthony J. and Gillian Dooley (eds), *Matthew Flinders Private Journal* (Adelaide: Friends of the State Library of South Australia, 2005).

Bulman, J.C. and J.M. Nosworthy (eds), *Timon* (Oxford: Oxford University Press, Malone Society Reprints, 1980).

Burningham, Nick, 'Wilhelm Janszoon and the Dutch Exploration of Australia' (published c. 2007). Online at: http://www.voc.iinet.net.au/nickb.html [accessed: 2 November 2011].

Byron, Kenneth, *Lost Treasures in Australia and New Zealand* (London: Angus & Robertson, 1965).

Callander, John, *Terra Australis Cognita, or, Voyages to the Terra Australis, or Southern Hemisphere, during the sixteenth, seventeenth, and eighteenth centuries* (3 vols, Edinburgh: A. Donaldson, 1768).

Calvert, A.F., *The Discovery of Australia* (London: George Philip and Son, 1893).

Cameron, Alan, 'Paganism and Literature in Late Fourth-Century Rome', in *Christianisme et formes littéraires de l'antiquité tardive en Occident* (Geneva: Foundation Hardt, 1977), pp. 1–30.

Cameron, Ian (Pseud. Donald Gordon Payne), *Lodestone and Evening Star: The Epic Voyages of Discovery 1493 B.C.–1896 A.D.* (New York: Dutton, 1966).

Camino, Mercedes Maroto, 'A Waxen World: Early Modern Women and Geographical (Un)Awareness', *Parergon*, 16.2 (1999): 101–32.

Camino, Mercedes Maroto, *Producing the Pacific: Maps and Narratives of Spanish Exploration (1567–1606)*, Portada Hispánica, 18 (Amsterdam and New York: Rodopi, 2005).

Camino, Mercedes Maroto, *Exploring the Explorers: Spaniards in Oceania 1519–1794* (Manchester and New York: Manchester University Press, 2008).

Capella, Martianus, *De nuptiis Mercurii et Philologiae*, James Willis (ed.) (Leipzig: Teubner, 1983).

Cary, M. and E.H. Warmington, *The Ancient Explorers*, revised edn (Harmondsworth: Penguin, 1963).

Case, Arthur E., 'The Geography and Chronology of *Gulliver's Travels*', in Arthur E. Case, *Four Essays on* Gulliver's Travels (Gloucester, MA: Peter Smith, 1958), pp. 50–68.

Casey, Dawn, 'Being Collected'. Lecture given at the University of Sydney on 28 August 2006, reproduced in full at: http://sydney.edu.au/museums/whatson/webcasts/lecture_beingcollected.shtml [accessed: 30 October 2011].

Cassius Dio Cocceianus, *Dio's Roman History*: in nine volumes, with an English translation by Earnest Cary; from the basis of the version of Herbert Baldwin Foster. Loeb classical library; 176 (Cambridge, Mass: Harvard University Press; London : Heinemann, 1925).

Certeau, Michel de, *The Writing of History*, Tom Conley (trans.) (New York: Columbia University Press, 1988).

Christopher, Emma, 'A "Disgrace to the Very Colour": Perceptions of Blackness and Whiteness in the Founding of Sierra Leone and Botany Bay', *Journal of Colonialism and Colonial History*, 9.3 (2008).

Cicero, *De re publica*, K. Ziegler (ed.) (Leipzig: Teubner, 1964).

Claeys, Gregory (ed.), *Utopias of the British Enlightenment* (Cambridge: Cambridge University Press, 1994).

Clarkson, Thomas, *An essay on the slavery and commerce of the human species, particularly the African, translated from a Latin Dissertation, which was honoured with the first prize in the University of Cambridge, for the year 1785* (London: T. Cadell and J. Phillips, 1786).

Clarkson, Thomas, *A Summary View of the Slave Trade and of the Probable Consequences of Its Abolition* (London: J. Phillips, 1787).

Clarkson, Thomas, *An essay on the impolicy of the African slave trade* (London: J. Phillips, 1788).

Clarkson, Thomas, *An essay on the comparative efficiency of regulation or abolition, as applied to the slave trade. Shewing that the latter only can remove the evils to be found in that commerce* (London: J. Phillips, 1789).

Clarkson, Thomas, *The substance of the evidence of sundry persons on the slave-trade* (London: J. Phillips, 1789).

Clarkson, Thomas, *An Essay on the Slavery and Commerce of the Human Species, particularly the African ... Translated from a Latin Dissertation, which was honoured with the First Prize in the University of Cambridge. For the Year 1785 ... With Additions* (Philadelphia, PA: Nathan Wiley, 1804).

Clarkson, Thomas, *History of the Rise, Progress and Accomplishment of the Abolition of the African Slave Trade by British Parliament* (London: Longman, Hurst, Rees, and Orme, 1808).

Clifford, James, 'Introduction: Partial Truths', in James Clifford and George E. Marcus (eds), *Writing Culture: The Poetics and Politics of Ethnography* (Berkeley, CA: University of California Press, 1986), pp. 1–26.
Clifford, James, *The Predicament of Culture: Twentieth-century Ethnography, Literature, and Art* (Cambridge, MA and London: Harvard University Press, 1988).
Close, Robert S., *The Voyage Continues* (London: Michael Joseph, 1968).
Cobley, John, *Sydney Cove, 1791–1792* (Sydney: Angus & Robertson, 1965).
Cockayne, Aston, *The Obstinate Lady*, Catherine M. Shaw (ed.) (New York: Garland, 1986).
Coleman, Deirdre, 'Sierra Leone, Slavery and Sexual Politics: Anna Maria Falconbridge and the "Swarthy Daughter" of Late Eighteenth-century Abolitionism', *Women's Writing*, 2.1 (1995): 3–23.
Coleman, Deirdre (ed.), *Maiden Voyages and Infant Colonies: Two Women's Travel Narratives of the 1790s* (Washington DC: Leicester University Press, 1998).
Coleman, Deirdre, *Romantic Colonization and British Anti-Slavery* (Cambridge: Cambridge University Press, 2005).
Collingridge, George, *The First Discovery of Australia and New Guinea. Being the Narrative of Portuguese and Spanish Discoveries in the Australasian Regions, Between the Years 1492–1606* (London: 1894, repr. Sydney: Souvenir, 1986).
Collingridge, George, *The Discovery of Australia* (Sydney: Hayes Bros., 1895).
Conley, Tom, 'Pierre Boaistuau's Cosmographic Stage: Theater, Text and Map', *Renaissance Drama*, 23 (1992): 59–86.
Conley, Tom, *The Self-Made Map: Cartographic Writing in Early Modern France* (Minneapolis, MN: University of Minnesota Press, 1996).
Cook, Karen S., 'The Secret Agenda of Western Australian Explorer Robert Dale (1809–1853)', *The Globe*, 54 (2003): 23–34.
Cook, Karen S., 'Thomas John Maslen and the "Great River of Desired Blessing" on His Map of Australia', *The Globe*, 61 (2008): 11–20.
Cooper, H.M., *The Unknown Coast (A Supplement)* (Adelaide: [s.n.], 1955).
Cormack, Phil, 'English Literacy and Anxiety about the Future: A Case-Study from the Turn of the Twentieth Century' (Report). Online at: http://www.unisa.edu.au/cslplc/publications/Cormack%20IFTE%20July%2003.pdf [accessed: 21 December 2009].
Cornell, Christine (trans.), *The Journal of Post Captain Nicolas Baudin Commander-in-Chief of the Corvettes* Géographe *and* Naturaliste (Adelaide: The Friends of the State Library of South Australia, 2004 [1974]).
Cosgrove, Denis, *Social Formation and Symbolic Landscape* (Madison, WI: University of Wisconsin Press, 1998).

Cowper, William, 'The Negro's Complaint', *The Gentleman's Magazine*, December 1793 [written 1788].

Crawford, Ian M., 'Aboriginal Cultures in Western Australia', in C.T. Stannage (ed.), *A New History of Western Australia* (Perth: University of Western Australia Press, 1981), pp. 3–34.

Crawford, Michael H., *Roman Republican Coinage* (2 vols, Cambridge: Cambridge University Press, 1974).

Crew, Gary, *Strange Objects* (Sydney: Hodder Headline, 1990).

Cropp, Wolf-Ulrich, *Die Batavia war ihr Schicksal. Seeabenteuer eines Ostindienfahrers* (Bielefeld: Delius Klasing, 1997).

Cutter, Donald, 'Introduction', in Andrew David, Felipe Fernández-Armesto, Carlos Novi and Glyndwr Williams (eds), *The Malaspina Expedition, 1789–1794: Journal of the Voyage by Alejandro Malaspina*, vol. 1 (London and Madrid: Hakluyt Society and Museo Naval, 2001–2004), pp. xxix–lxxvii.

d'Ailly, Pierre, *Ymago Mundi*, Edmond Buron (ed.) (3 vols, Paris: Maisonneuve Frères, 1930).

Dabitch, Christophe and Jean-Denis Pendanx, *Jeronimus* (3 vols, Paris: Futuropolis, 2008–10).

Dalrymple, Alexander, *An Account of the discoveries made in the South Pacifick Ocean previous to 1764* (London, 1767).

Dalrymple, Alexander, *An Historical Collection of the Several Voyages and Discoveries in the South Pacific Ocean* (2 vols, London, 1770–71).

Dampier, William, *A New Voyage Round the World* (London, 1697); ed. Max Beken (London: Hummingbird Press, 1998).

Dampier, William, *Dampier's Voyages*, John Masefield (ed.) (2 vols, London, 1906).

Darwin, Sarah, 'It Tastes Like Shit, but You Can Live On It.' Blog entry for VPRO's *Beagle* project. Online at: http://beagle.vpro.nl/#/blog/item/3381/ [accessed: 27 September 2010].

Dash, Mike, *Batavia's Graveyard* (London: Verso, 2002).

David, Andrew, 'The Voyage of Alejandro Malaspina 1789–1794', *Annual General Meeting of The Hakluyt Society 7 July 1999* (London: The Hakluyt Society, 2000), pp. 3–24.

Deane, Bradley, 'Imperial Barbarians: Primitive Masculinity in Lost World Fiction', *Victorian Literature and Culture*, 36 (2008): 205–25.

Desanges, Jehan, 'Les relations de l'Empire romain avec l'Afrique nilotique et érythréenne, d'Auguste à Probus', in Hildegarde Temporini (ed.), *Aufstieg und Niedergang der Römischen Welt*, vol. 2.10.1 (Berlin and New York: de Gruyter, 1988), pp. 3–43.

Destombes, Marcel, *Mappemondes A.D. 1200–1500* (Amsterdam: Israel, 1964).

Dew, Nicholas, 'Reading Travels in the Culture of Curiosity: Thévenot's Collection of Voyages', *Journal of Early Modern History*, 10.1–2 (2006): 39–59.

Drake-Brockman, Henrietta, 'The Wreck of the *Batavia*', *Walkabout*, 1 (1955): 33–9.

Drake-Brockman, Henrietta, *The Wicked and the Fair* (Sydney: Angus & Robertson, 1957).

Drake-Brockman, Henrietta, *Voyage to Disaster. The Life of Francisco Pelsaert – covering his Indian Report to the Dutch East India Company and the wreck of the ship Batavia in 1629 off the coast of Western Australia – together with the full text of his Journals concerning the Rescue Voyages, the mutiny on the Abrolhos Islands and the subsequent trials of the Mutineers* (Sydney: Angus & Robertson, 1963).

Drewe, Robert, *Fortune* (Sydney: Pan Books, 1986).

Driver, Felix, *Geography Militant: Cultures of Exploration and Empire* (Oxford: Blackwell, 2001).

Dunmore, J., *French Explorers in the Pacific* (Oxford: Clarendon Press, 1969).

Dunmore, J., *Monsieur Baret: First Woman Around the World. 1766–68* (Auckland: Heritage Press, 2002).

Duplomb, C., *Campagne de l'"Uranie" (1817–1820); journal ... d'après le manuscrit original accompangé de notes / par Charles Duplomb*, copy NLA (Paris: Societe d'Editions, 1927).

Duyker, Edward, *François Péron, an Impetuous Life* (Carlton: Miegunyah Press, 2006).

Eagar, Kirsty, *Saltwater Vampires* (Camberwell, Vic: Penguin, 2010).

Eccleston, Gregory C., 'The Neglect of Baudin's Manuscript Charts of the Victorian Coastline', *The Globe*, 66 (2010): 27–58.

Edwards, Hugh, *Islands of Angry Ghosts* (New York: Henry Holt, 1966).

Eggert, Paul, 'The Dutch-Australian Connection: Willem Siebenhaar, D.H. Lawrence, Max Havelaar and *Kangaroo*', *Australian Literary Studies*, 21 (2003): 3–19.

Elder, Catriona, *Dreams and Nightmares of a White Australia: Representing Aboriginal Assimilation in the Mid-Twentieth Century* (Bern: Peter Lang, 2009).

Elliott, John H., *The Old World and the New 1492–1650* (Cambridge: Cambridge University Press, 1970).

Étienne de Rouen, *Le Dragon Normand et autres poèmes d'Étienne de Rouen*, Henri Omont (ed.) (Rouen: Société de l'histoire de Normandie, 1884).

Evans, James and J. Lennart Berggren, *Geminos's Introduction to the Phenomena: A Translation and Study of a Hellenistic Survey of Astronomy* (Princeton, NJ: Princeton University Press, 2006).

Fabian, Johannes, *Time and the Other: How Anthropology Makes Its Object* (New York: Columbia University Press, 1983).

Fabricius, Johannes, *Het beest uit de zee* (The Hague: Leopold, 1980).

Falconbridge, Alexander, *An Account of the Slave Trade on the Coast of Africa* (London: J. Phillips, 1788).

Falconbridge, Anna Maria, *Two voyages to Sierra Leone, during the years 1791–2–3, in a series of letters, By Anna Maria Falconbridge. To which is added, a letter from the author, to Henry Thornton, Esq. M.P. and chairman of the Court of Directors of the Sierra Leone Company* (London: Printed for the author, and sold by different booksellers throughout the Kingdom, 1794).

Falconer, Colin, *Zorn der Meere* (Stuttgart: Heyne, 2001).

Fausett, David, *Writing the New World: Imaginary Voyages and Utopias of the Great Southern Land* (Syracuse, NY: Syracuse University Press, 1993).

Fausett, David, *Images of the Antipodes in the Eighteenth Century: A Study in Stereotyping* (Amsterdam: Rodopi, 1994).

Favenc, Ernest, *Tales of the Austral Continent* (Sydney: The Bulletin, 1894).

Favenc, Ernest, *The Secret of the Australian Desert* (London: Blackie, 1895).

Favenc, Ernest, *Marooned on Australia. Being the Narration by Diedrich Buys of his Discoveries and Exploits in Terra Australis Incognita about the year 1630* (London, Glasgow and Dunedin: Blackie & Son, 1897 [i.e. 1896]).

Favenc, Ernest and Percy Spence, *The Story of Our Continent Told with Brush and Pen* (Sydney and Melbourne: Robert Thompson, 1892).

Ferguson, John Alexander, *Bibliography of Australia: 1784–1830* (Canberra: National Library of Australia, 1975).

Fernández de Enciso, Juan, *Suma de geographia* (Sevilla, 1519).

Fernández de Navarrete, Martín, *Colección de los viages y descubrimientos que hicieron por mar los españoles* (5 vols, Buenos Aires: Editorial Guarania, 1946).

Fernández-Armesto, Felipe, 'Review of Gavin Menzies, *1421: The Year China Discovered the World*', *Literary Review*, Dec/Jan (2002/2003): 21–2.

Fidlon, Paul G. and R.J. Ryan (eds), *Journal and Letters of Lt Ralph Clark 1787–1792* (Sydney: Australian Documents Library in association with the Library of Australian History, 1981).

Finlay, Robert, 'How Not to (Re)Write World History: Gavin Menzies and the Chinese Discovery of America', *Journal of World History*, 15 (2004): 229–42.

Fischer, Joseph and F.R. von Wieser, *The Oldest Map with the Name America of the Year 1507 and the Carta Marina of the Year 1516 by Martin Waldseemüller* (Innsbruck, 1903).

FitzGerald, Lawrence, *Java La Grande: The Portuguese Discovery of Australia* (Hobart: The Publishers Pty. Ltd., 1984).

Flamant, Jacques, *Macrobe et le néo-Platonisme latin, à la fin du IVe siècle* (Leiden: Brill, 1977).

Flannery, Tim (ed.), *Watkin Tench's 1788: A Narrative of the Expedition to Botany Bay and a Complete Account of the Settlement at Port Jackson* (Melbourne: Text Publishing, 1996, repr. 2009).

Fleming, Fergus, *Barrow's Boys: A Stirring Story of Daring, Fortitude, and Outright Lunacy* (New York: Granta, 1998).

Flinders, Matthew, *A Voyage to Terra Australis; Undertaken for the Purpose of Completing the Discovery of that Vast Country, and Prosecuted in the Years 1801, 1802 and 1803, in His Majesty's Ship, the* Investigator (London: G. & W. Nicol, 1814).

Foigny, Gabriel de, *La Terre australe connue* (1676), Pierre Ronzeaud (ed.) (Paris: STFM, 1990)*uë*.

Foigny, Gabriel de, *The Southern Land, Known*, David Faussett (ed. and trans.) (Syracuse, NY: Syracuse University Press, 1993).

Forbes, George, *Adventures in Southern Seas* (Sydney: Macmillan, 1920).

Fornasiero, Jean, Peter Monteath and John West-Sooby, 'Hydrographic Reputations: Matthew Flinders, Nicolas Baudin and Australia's Unknown Coast', *Hydrographic Journal*, 121 (July 2006): 17–23.

Fornasiero, Jean and John West-Sooby, 'Baudin's Books', *Australian Journal of French Studies*, 39.2 (2002): 215–49.

Forster, H., 'Voyaging through Strange Seas: Four Women Travellers in the Pacific', *National Library of Australia News*, 10.4 (January, 2000): 3–6.

Foucault, Michel, 'The Subject and Power', in Hubert Dreyfus and Paul Rabinow (eds), *Michel Foucault: Beyond Structuralism and Hermeneutics* (Chicago, IL: University of Chicago Press, 1982), pp. 208–26.

Freycinet, L. de, Letter to the Minister of Marine, dated 29 August 1811, Archives Nationales de France, BB4 996.

Freycinet, L. de, *Voyage Autour du Monde, entrepris par ordre du Roi ...: Historique*, vol. 1 (Paris, 1827).

Freycinet, L. de, *Voyage autour du monde exécuté sur les corvettes de S. M. l'Uranie et la Physicienne pendant les années 1817, 1818, 1819 et 1820* (Paris, 1924).

Freycinet, Rose de, *Journal de Madame Rose de Saulces de Freycinet, d'après le manuscript original accompagné de notes* (Paris, 1927).

Frost, Cheryl, *The Last Explorer: The Life and Work of Ernest Favenc* (Townsville: Foundation for Australian Literary Studies, 1983).

Frye, Gavin (ed.), *A Voyage Round the World by Mary Ann Parker* (Sydney: Hordern House, 1991).

Gaunt, Mary, 'The First Australian Love Story', in Mary Gaunt, *The Ends of the Earth* (London, 1916)

Gautier Dalché, Patrick, 'L'œuvre géographique du cardinal Fillastre', *Archives d'histoire doctrinale et littéraire du Moyen Âge*, 59 (1992): 319–83.

Gautier Dalché, Patrick, 'Le paradis aux antipodes? Une *Distinctio divisionis terre et paradisi delitiarum* (XIVe siècle)', in Dominique Barthélemy and Jean-Marie Martin (eds), *'Liber largitorius': Études d'histoire médiévale offertes à Pierre Toubert par ses élèves* (Geneva: Libraire Droz, 2003), pp. 615–37.

Gautier Dalché, Patrick, *La Géographie de Ptolémée en Occident (IVe–XVIe siècle)* (Turnhout: Brepols, 2009).

Geminos, *Introduction aux phénomènes*, Germaine Aujac (ed.) (Paris: Les Belles Lettres, 1975).

George, A.S., *William Dampier in New Holland: Australia's First Natural Historian* (Melbourne: Bloomings Books, 1999).

Gerald of Wales, *Itinerarium Kambriae*, James F. Dimock (ed.), Rolls Series, 21 (London: Longman, 1868).

Gerritsen, Rupert, *And Their Ghosts May Be Heard* ... (Fremantle: Fremantle Arts Centre, 1994, repr. 2002).

Gerritsen, Rupert, *Australia and the Origins of Agriculture* (Oxford: Archaeopress, 2008).

Gervase of Tilbury, *Otia imperialia*, S.E. Banks and J.W. Binns (eds and trans) (Oxford: Clarendon Press, 2002).

Ghim, Walter, *Vita Mercatoris*, Hans-Heinrich Geske (ed.), *Duisburger Forschungen*, 6 (1962): 246–72.

Goldie, Douglas Mark, *The Idea of the Antipodes: Place, People and Voices* (New York: Routledge, 2010).

Goldsmith, Frank, *Treasure Lies Buried Here* (Perth: C.H. Pitman, 1946).

Gordon, W.J., *The Captain-General: Being the Story of the Attempt of the Dutch to Colonize New Holland* (London: Frederick Warne, 1888).

Gordon, W.J., 'A Colony that Failed', *The Leisure Hour*, vol. 38, 1889, p. 136.

Gordon, W.J., 'Preface to the Third Edition', in *The Captain-General*, 3rd edn (London: Frederick Warne, 1891), p. xii.

Gosch, Elizabeth, 'Vinnie Jones to Bring Batavia Tale to Life', *The Australian*, 13 December 2007.

Graubard, Mark and John Parker (eds), *New Tidings of the Land of Brazil* (Minneapolis, MN: University of Minnesota Press, 1957).

Graves, Robert, *The Isles of Unwisdom* (London: Cassell, 1950).
Grynaeus, Simon and Johann Huttich, *Novus Orbis Regionum ac Insularum Veteribus Incognitarum* (Paris and Basle, 1532).
Gurval, Robert Alan, *Actium and Augustus* (Ann Arbor, MI: University of Michigan Press, 1995).
Hall, Joseph, *Mundus Alter et Idem Siue Terra Australis ante hac semper incognita longis itineribus peregrini Academici nuperrime lustrata* (Frankfurt, 1605).
Hall, Joseph, *Another World and Yet the Same: Bishop Joseph Hall's* Mundus Alter et Idem, John Millar Wands (ed. and trans.) (New Haven, CT: Yale University Press, 1981).
Halls, C., 'A Story of Three Plates', *Port of Fremantle Quarterly*, Winter (1974): 22–4.
Hamilton, Jill, *The Captain Cook Myth* (Sydney: Angus & Robertson, 1980).
Hamy, E.T., 'Francisque et Andre d'Albaigne: cosmographes lucquois au service de la France', *Bulletin de Géographie Historique et Descriptive* (Paris, 1894): 405–33.
Hamy, E.T., 'Nouveaux documents sur les frères d'Albaigne et sur le projet de voyage et de découvertes présenté à la cour de France', *Bulletin de Géographie Historique et Descriptive* (Paris, 1899): 101–10.
Hamy, E.T., 'Documents relatifs à un projet d'expéditions lointaines présentés à la cour de France en 1570', *Bulletin de Géographie Historique et Descriptive* (Paris, 1903): 266–73.
Hargrave, Lawrence, *Lope de Vega* (Sydney: Privately Printed, 1911).
Harvey, P.D.A., *Mappa Mundi: The Hereford World Map* (Toronto: Toronto University Press, 1966).
Harvey, P.D.A. (ed.), *The Hereford World Map: Medieval World Maps and Their Context* (London: British Library, 2006).
Harvey, Shannon, 'Batavia Movie Set in Geraldton yet to Begin Filming', *Perth Now*, 5 May 2008. Online at: http://www.perthnow.com.au/news/batavia-movie possibly-shipwrecked/story-0-1111116260457 [accessed: 29 July 2009].
Hasluck, Nicholas, *The Bellarmine Jug* (Ringwood, Victoria: Penguin Books, 1984).
Hayes, Rosemary, *The Blue-Eyed Aborigine* (Melbourne: Frances Lincoln, 2010).
Heaton, J.H., *Australian Dictionary of Dates and Men of the Time. Containing the History of Australasia from 1542 to date* (London: S.W. Silver & Co., 1879).
Henderson, James A., *Phantoms of the Tryall* (Perth: St George, 1993).
Hervé, Roger, 'Australia in French Geographical Documents of the Renaissance', J.M. Forsyth (trans.), *Royal Australian Historical Society's Journal and Proceedings*, 41 (1955): 23–38.

Hervé, Roger, *Découverte fortuite de l'Australie et de la Nouvelle Zélande par des navigateurs portugais et espagnols entre 1521 et 1528* (Paris: Bibliothèque Nationale, 1982).

Hervé, Roger, *Chance Discovery of Australia and New Zealand by Portuguese and Spanish Navigators between 1521 and 1528*, John Dunmore (trans.) (Palmerston North: Dunmore Press, 1983).

Hiatt, Alfred, 'The Map of Macrobius before 1100', *Imago Mundi*, 59 (2007): 149–76.

Hiatt, Alfred, *Terra Incognita: Mapping the Antipodes before 1600* (London and Chicago, IL: British Library and University of Chicago Press, 2008).

Higueras Rodríguez, María Dolores, *Catálogo crítico de los documentos de la expedición Malaspina (1789–1794)* (3 vols, Madrid: Museo Naval, 1985–94).

Hilder, Brett, *The Voyage of Torres: The Discovery of the Southern Coastline of New Guinea and Torres Strait by Captain Luis Báez de Torres in 1606* (St Lucia: University of Queensland Press, 1980).

Hoffmann, Gabrielle, *Die Schiffbrüchigen* (Hamburg: Europa, 2004).

Hordern, Marsden, *King of the Australian Coast: The Work of Phillip Parker King in the Mermaid and Bathurst 1817–1822* (Melbourne: Melbourne University Press, 1997).

Hordern, Peregrine and Nicholas Purcell, *The Corrupting Sea: A Study of Mediterranean History* (Oxford: Blackwell, 2000).

Horner, Frank, *The French Reconnaissance: Baudin in Australia 1801–1803* (Carlton: Melbourne University Press, 1987).

Horner, Frank, 'The Baudin Expedition to Australia 1800–1804', in J. Bonnemains, E. Forsyth and B. Smith (eds), *Baudin in Australian Waters: The Artwork of the French Voyage of Discovery to the Southern Lands 1800–1804* (Melbourne: Oxford University Press, 1988), pp. 1–15.

Hughes, Joy N. (ed.), *The Journals and Letter of Elizabeth Macarthur, 1789–1798* (Glebe: Historic Houses Trust of New South Wales, 1984).

Hunt, S. and P. Carter, *Terre Napoléon: Australia through French Eyes. 1800–1804* (Sydney: Historic Houses Trust of New South Wales in association with Hordern House, 1999).

Hunter, John, *An Historical Journal of the transactions at Port Jackson and Norfolk Island, with the discoveries which have been made in New South Wales ... since the publication of Phillip's voyage ...* (London: Printed for John Stockdale, 1793).

Huxley, Julian (ed.), *T.H. Huxley's Diary of the Voyage of H.M.S. Rattlesnake* (London: Chatto and Windus, 1935).

Ingleton, Geoffrey, *Matthew Flinders, Navigator and Chartmaker* (Guildford: Genesis Publications and Hedley Australia, 1986).

Isidore of Seville, *Etymologiarum sive originum libri xx*, W.M. Lindsay (ed.) (2 vols, Oxford: Clarendon Press, 1911).

Iturbe, Juan de, 'Sumario Breve', in Celsus Kelly O.F.M. (ed.), *La Austrialia del Espíritu Santo: The Journal of Fray Martín de Munilla O.F.M. and Other Documents Relating to The Voyage of Pedro Fernández de Quirós to the South Sea (1605–1606) and the Franciscan Missionary Plan (1617–1627)*, The Hakluyt Society, 2 (Cambridge: Cambridge University Press, 1966), pp. 273–93.

Jameson, Shelagh, 'The Chronology of the Campaigns of Aelius Gallus and Caius Petronius', *Journal of Roman Studies*, 58 (1968): 71–84.

Johnson, David H., 'The Incredible Kangaroo', *National Geographic*, October 1955.

Johnston, Sandra F. (ed. and trans.), *The Castle of Perseverance*. Online at: http://homes.chass.utoronto.ca/~ajohnsto/cascomp.html [Site updated November 2010].

Jones, Brian W., *The Emperor Domitian* (London and New York: Routledge, 1992).

Journal of the Royal Geographical Society.

Judge, E.A., 'Veni. Vidi. Vici, and the Inscription of Cornelius Gallus', in E.A. Judge and James R. Harrison (eds), *The First Christians in the Roman World: Augustan and New Testament Essays* (Tübingen: Mohr Siebeck, 2008), pp. 72–84.

Kampen, Anthony von, *Het beest uit zee. De ondergang van het Compagniesschip 'Batavia'* (Bussum: C. de Boer, Jr, 1971).

Kelly, Celsus, O.F.M. and Gerard Bushell O.F.M. (eds), *Austrialia Franciscana*, vol. 6 (6 vols, Madrid: Franciscan Historial Studies and Archivo Iberoamericano, 1973).

Kenny, John, *Before the First Fleet: Europeans in Australia, 1606–1777* (Kenthurst: Kangaroo Press, 1995).

Kerne, Silke, *Batavia 1629* (Munich: Ehrenwirth, 2000).

King, R., 'Terra Australis, New Holland and New South Wales: The Treaty of Tordesillas and Australia', *The Globe*, 47 (1998): 35–55.

King, R., 'Gustaf III's Australian Colony', *The Great Circle*, 27.2 (2005): 3–20.

King, R., 'The Jagiellonian Globe: A Key to the Puzzle of Jave la Grande', *The Globe*, 62 (2009): 1–50.

Kingsley, Henry, *The Mystery of the Island* (London, 1877).

Klausen, Jimmy Casas, 'Of Hobbes and Hospitality in Diderot's Supplement to the Voyage of Bougainville', *Polity*, 37.2 (April 2005): 167–92.

Kline, Naomi Reed, *Maps of Medieval Thought: The Hereford Paradigm* (Woodbridge: Boydell Press, 2001).

Knowles, Lee, 'Batavia Incident' and 'Kalbarri', *The Beloit Poetry Journal*, 25.3 (Spring 1975): 3–7.

Kooij, Rachel van, *Der Kajütenjunge des Apothekers* (Vienna: Jungbrünnen, 2006).

Lactantius, *Opera omnia*, Samuel Brandt and Georg Laubmann (eds) (2 vols, Vienna: Bibliopola Academiae Litterarum Caesareae Vindobonensis, 1890).

Langdon, Robert, *The Lost Caravel* (Sydney: Pacific Publications, 1975).

Langdon, Robert, *The Lost Caravel Re-explored* (Canberra: Brolga, 1988).

Las Casas, Bartolomé de, *La Destruction des Indes* (1552), traduction de Jacques de Miggrode (1579), A. Milhou and J.-P. Duviols (eds) (Paris: Editions Chandeigne, 1995).

Law, R.C.C., 'The Garamantes and Trans-Saharan Enterprise in Classical Times', *The Journal of African History*, 8.2 (1967): 181–200.

Leadbetter, Bill, 'Imperial Ideology and the Christians in the Late Third Century', in M. Dillon and L. Garland (eds), *Religion in the Ancient World* (Amsterdam: Hakkert, 1995), pp. 535–45.

Lear, Melva, *Dangerous Holiday* (Sydney: Angus & Robertson, 1959).

Leijenaar, Eric, *De Batavia; het gruwelijke, waar gebeurde verhaal over de trotse Oostindiëvaarder* (Amsterdam: Omega, 1989).

Leoni, Sylviane and René Ouellet (eds), *Mythes et géographies des mers du Sud* (Dijon: Editions Universitaires de Dijon, 2006).

Leschenault, Théodore, *Extrait de la relation de l'expédition de découverte commandée par le Cit.*[n] *Baudin Capitaine de Vaisseau*, manuscript held by the Archives nationales de France, Série Marine, 5JJ 56.

Lester, Alan, 'Colonial Networks, Australian Humanitarianism and the History Wars', *Geographical Research*, 44.3 (September 2006): 229–41.

Lestringant, Frank, *Mapping the Renaissance World: The Geographical Imagination in the Age of Discovery*, David Fausett (trans.) (Berkeley, CA: University of California Press, 1994).

Lestringant, Frank and Monique Pelletier, 'Maps and Descriptions of the World in Sixteenth-Century France', in David Woodward (ed.), *The History of Cartography*, vol. 3.2 (Chicago, IL: University of Chicago Press, 2007), pp. 1463–79.

Lévêque de Pontharouart, Jacques, *Paulmier de Gonneville: son voyage imaginaire* (Beauval en Caux: J. Lévêque de Pontharouart, 2000).

Leyds Veer-Schuyts Praetjen ... Hier achtet syn by ghevoecht, Twee Brieven gheschreven in Batavia, vervanghende den handel van Ieronimus Cornelisz onder Coppman op't Schip Batavia, oock mede sijn sterven, geweest hebbende een Discipel van Iohannes Torrentius voorsz (1630).

Linschoten, Jan Huyghen van, *The Voyages of John Huyghen van Linschoten to the East Indies*, Hakluyt Society, 1st series, 71 (2 vols, New York: Burt Franklin, 1970; orig. edn 1884).

Lipscombe, Trevor, 'Two Continents or One? The Baudin Expedition's Unacknowledged Achievements on the Coast of Victoria', *Victorian Historical Journal*, 78.1 (May 2007): 23–41.

Lisson, Deborah, *The Devil's Own* (Glebe: Walter McVitty, 1989).

Lollis, Cesare de (ed.), *Raccolta di documenti e studi pubblicati dalla R. Commissione Colombiana* (Rome: Ministero della pubblica istruzione, 1894).

Lucan, *De bello civili*, D.R. Shackleton Bailey (ed.), 2nd edn (Stuttgart: Teubner, 1997).

Lucian, 'A True Story II', in *Opera* (8 vols, London: Heinemann, 1913–67), vol. 1.

Lucian, 'The Ship or the Wishes', in *Opera* (8 vols, London: Heinemann, 1913–67), vol. 6.

Lull, Ramon, *Declaratio Raimundi*, in Theodor Pindl-Büchel (ed.), *Raimundi Lulli Opera Latina 76–81*, Corpus Christianorum Continuatio Mediaevalis, 79 (Turnhout: Brepols, 1989).

Lutz, Cora E. (ed.), *Iohannis Scotti Annotationes in Marcianum* (Cambridge, MA: Medieval Academy of America, 1939).

McCall, Daniel F., 'Herodotus on the Garamantes: A Problem in Protohistory', *History in Africa*, 26 (1999): 197–217.

McCarthy, M., 'Charles Edward Broadhurst, 1826–1905: A Remarkable Nineteenth-Century Failure'. Unpublished MA thesis, Perth, Curtin University, 1990.

McCarthy, M., 'HM Ship *Roebuck* (1690–1701): Global Maritime Heritage?', *International Journal of Nautical Archaeology*, 33.1 (2004): 54–66.

McCarthy, M., 'Rose de Freycinet and the French Exploration corvette *L'Uranie*, (1820): A Highlight of the "French Connection" with the "Great Southland"', *International Journal of Nautical Archaeology*, 34.1 (2005): 62–78.

McCarthy, M., 'Rose de Freycinet and the *Uranie*', in M. Stanbury and V. Northey (eds), *Journeys of Enlightenment: French Exploration of Terres Australes* (Perth: Western Australian Museum, 2008), pp. 10–13.

McEvoy, Brian P. et al., 'Whole-Genome Genetic Diversity in a Sample of Australians with Deep Aboriginal Ancestry', *American Journal of Human Genetics*, 87.2 (August 2010): 297–305.

McGillivery, Angus R. 'Convict Settlers, Seaman's Greens, and Imperial Designs at Port Jackson: A Maritime Perspective of British Settler Agriculture', *Agricultural History*, 78.3 (2004): 261–88.

McIntyre, Kenneth G., *The Secret Discovery of Australia: Portuguese Ventures 200 Years before Captain Cook* (Medindie, South Australia and London: Souvenir Press, 1977; revised and abridged edn Sydney: Picador, 1982).

McKiggan, Ian, 'The Portuguese Expedition to Bass Strait in AD 1522', *Journal of Australian Studies*, 1 (1977): 2–32.

Macrobius, *Commentarii in Somnium Scipionis*, J. Willis (ed.) (Leipzig: Teubner, 1963).

Major, R.H., *Early Voyages to Terra Australis, Now Called Australia* (London: Hakluyt Society, 1859; repr. New York, c. 1970).

Major, R.H., 'Further Facts in the History of the Early Discovery of Australia', *Archaeologia*, 44 (1873): 233–41.

Malaspina, Alejandro, *The Malaspina Expedition, 1789–1794: Journal of the Voyage by Alejandro Malaspina*, Andrew David, Felipez Fernández-Armesto, Carlos Novi and Glyndwr Williams (eds), vol. 1 (3 vols, London and Madrid: Hakluyt Society and Museo Naval, 2001–2004).

Malinowski, Bronislaw, *Argonauts of the Western Pacific: An Account of Native Enterprise and Adventure in the Archipelagoes of Melanesian New Guinea* (London: Routledge, 1922).

Manegold of Lautenbach, *Liber contra Wolfelmum*, Wilfried Hartmann (ed.) (Weimar: Böhlaus Nachfolger, 1972).

Manilius, *Astronomica*, G.P. Goold (ed.) (Stuttgart: Teubner, 1998).

Map, Walter, *De nugis curialium*, M.R. James, rev. C.N.L. Brooke and R.A.B. Mynors (ed. and trans.) (Oxford: Clarendon Press, 1983).

Marchant, L.R., *France Australe: The French Search for the Southland* (Perth: Artlook Books, 1982; new edn, Scott Four Colour Print, c. 1998).

Marchant, L.R., *An Island unto Itself: William Dampier and New Holland* (Perth: Hesperian Press, 1988).

Marden, H., *King of the Australian Coast: The Work of Phillip Parker King in the Mermaid and Bathurst 1817–1822* (Melbourne: Miegunyah Press, 1997).

Martin, Roger (ed.), *Mémoire sur les établissements anglais à la Nouvelle Hollande, à la Terre de Diémen et dans les archipels du grand Océan Pacifique*, *Revue de l'Institut Napoléon*, no. 176 (1998).

Masefield, J. (ed.), *Dampier's Voyages. Consisting of a New Voyage Round the World, A Supplement to the Voyage Round the World. Two Voyages to Campeachy, a Discourse of Winds, a Voyage to New Holland, and a Vindication, in answer to the Chimerical Relation of William Funnell* (London: E. Grant Richards, 1906).

Mattern, Susan P., *Rome and the Enemy: Imperial Strategy in the Principate* (Berkeley, CA: University of California Press, 1999).

Mattiacci, Silvia, *I carmi e frammenti di Tiberiano* (Florence: Olschki, 1990).

May, Vivienne, *Travels in Western Australia* (London: Macmillan, 1973).
Menzies, Gavin, *1421: The Year China Discovered the World* (London: Bantam, 2002).
Menzies, Gavin, *1434: The Year a Magnificent Chinese Fleet Sailed to Italy and Ignited the Renaissance* (London: HarperCollins, 2008).
Mette, Hans, *Sphairopoiia: Untersuchungen zur Kosmologie des Krates von Pergamon* (Munich: Beck'sche, 1936).
Milanesi, Marica, 'Il commento al "Dittamondo" di Guglielmo Capello (1435–37)', in Marco Bertozzi (ed.), *Alla corte degli Estensi: Filosofia, arte e cultura a Ferrara nei secoli XV e XVI* (Ferrara: Università degli Studi, 1994), pp. 365–88.
Mitchell, Thomas L., *Journal of an Expedition into the Interior of Tropical Australia in Search of a Route from Sydney to the Gulf of Carpentaria* (London: Longman, Brown, Green and Longmans, 1848).
Mitchell, W.J.T., *Picture Theory: Essays on Verbal and Visual Representation* (London and Chicago, IL: University of Chicago Press, 1994).
Molewijk, G.C., *Pelsaerts journaal van de ongelukkige reis van het schip Batavia* (Weesp: Heureka, 1989).
Moran, Patrick, *History of the Catholic Church in Australasia* (Sydney: Privately Printed, 1896).
Moran, Patrick, *Discovery of Australia by de Quiros in the Year 1606* (Sydney: Privately Printed, c. 1896).
Moretti, Gabriella, *Gli antipodi: avventure letterarie di un mito scientifico* (Parma: Pratiche Editrice, 1994).
Morgan, Gwenda and Peter Rushton, 'Print Culture, Crime and Transportation in the Criminal Atlantic', *Continuity and Change*, 22.1 (2007): 49–71.
Mullen, Samuel, *The Battle for Mordialloc, or, How we Lost Australia* (Melbourne: Mullen, 1888).
Munilla, Fray Martín de, 'La Austrialia del Espíritu Santo: The Journal of Fray Martín de Munilla O.F.M. and Other Documents Relating to The Voyage of Pedro Fernández de Quirós to the South Sea (1605–1606) and the Franciscan Missionary Plan (1617–1627)', in Celsus Kelly O.F.M. (ed.), *La Austrialia del Espíritu Santo*, The Hakluyt Society, vol. 1 (Cambridge: Cambridge University Press, 1966), pp. 134–270.
Murdoch, Priscilla, *Duyfken and the First Discoveries of Australia* (Artarmon, NSW: Antipodean Publishers, 1974).
Neill, A., 'Buccaneer Ethnography: Nature, Culture, and Nation in the Journals of William Dampier', *Eighteenth-Century Studies*, 33.2 (2000): 165–80.
Neville, Henry, *The Isle of Pines, or, A late Discovery of a fourth Island near Terra Australis, Incognita by Henry Cornelius Van Sloetten* (London, 1668).

'New Settlement Making on the Northern Coast', *The Sydney Gazette and New South Wales Advertiser*, Thursday, 28 July 1836, p. 2.

Niekerk, Mike van, 'Agony of the *Batavia*', *Garuda Indonesia. In-Flight Magazine* (July 1993): 6–11.

Nunn, Judy, *Territory* (Sydney: Arrow, 2003).

O'Loughlin, Katrina, '"Thorns and Thistles": Anna Maria Falconbridge's *Two Voyages to the River Sierra Leone*', in unpublished thesis 'The Paper Globe: Subject and Sociality in Eighteenth-Century Women's Travel Narratives', University of Melbourne, 2009.

Padilla Suárez, Ignacio, *Las Antípodas y el siglo*, Alastair Reid (trans.) (London: Scribner, 2004).

Pagden, Anthony, *Lords of All the World: Ideologies of Empire in Spain, Britain and France c.1500–c.1800* (New Haven, CT and London: Yale University Press, 1995).

Panegyricus Messallae, in Hermann Tränkle (ed.), *Appendix Tibulliana* (Berlin: de Gruyter, 1990).

Paparowski, Martin, dir. *Meuterei auf der 'Batavia'* (Mutiny on the *Batavia*), 42 mins. Zweites Deutsches Fernsehen (ZDF), Germany, 1998.

Parker, Mary Ann, *A Voyage Round the World, in the* Gorgon *Man of War: Captain John Parker. Performed and Written by his Widow; for the Advantage of a Numerous Family. Dedicated, by Permission to Her Royal Highness The Princess of Wales* (London: Printed by John Nichols, 1795).

Paulmier, Jean, L'abbé, *Mémoires touchant l'établissement d'une mission chrestienne dans le troisième Monde, Autrement appellé, La Terre Australe, Meridionale, Antartique, & Inconnuë. Presentez à Nostre S. Pere le Pape Alexandre VII. Par un Ecclesiastique Originaire de cette mesme Terre* (Paris, 1664).

Paulmier, Jean, L'abbé, *Mémoires touchant l'établissement d'une mission chrestienne dans le troisième Monde, Autrement appellé, La Terre Australe, Meridionale, Antartique, & Inconnuë. Presentez à Nostre S. Pere le Pape Alexandre VII. Par un Ecclesiastique Originaire de cette mesme Terre*, established, presented and annotated by Margaret Sankey (Paris: Honoré Champion, 2006).

Penman, Leigh T.I., 'The *Batavia* Legacy: The Wreck of the Dutch East India Company Ship *Batavia* (1629) in Australian History and Imaginary', in Renata Summo-O'Connell (ed.), *Imagined Australia: Reflections upon the Reciprocal Construction of Identity between Australia and Europe* (Bern and New York: Peter Lang, 2009), pp. 153–70.

Péron, François and Louis Freycinet, *Voyage de découvertes aux Terres Australes exécuté par ordre de Sa Majesté l'Empereur et Roi, sur les corvettes le* Géographe, *le* Naturaliste*; et la goélette le* Casuarina*, pendant les années 1800, 1801, 1802, 1803, et 1804* (Paris: 1807–1816), comprising the following volumes:

Historique, vol 1, by François Péron, 1807; *Historique*, vol. 2, by F. Péron, continued by L. Freycinet, 1816; *Atlas*, 1st part, by Lesueur and Petit, 1807; *Atlas*, 2nd part, by L. Freycinet, 1811; *Atlas, Navigation et géographie*, by L. Freycinet, 1812 (sic: appeared December 1814); *Navigation et géographie*, by L. Freycinet, 1815. (A second edition of the historical narrative, revised and enlarged, was published in 1824, also with an atlas.)

Phillip, Arthur, *The Voyage of Governor Phillip to Botany Bay* (London: John Stockdale, 1789).

Phillips, Richard, 'Dystopian Space in Colonial Representations and Interventions: Sierra Leone as "The White Man's Grave"', *Geografiska Annaler*, 84.3–4 (2002): 189–200.

Pigafetta, Antonio, *The First Voyage Round the World by Magellan*, Lord Stanley of Alderley (ed.), Hakluyt Society, 1st series, 52 (New York: Burt Franklin, 1874).

Pigafetta, Antonio, *Magellan's Voyage: A Narrative Account of the First Circumnavigation*, R.A. Skelton (trans. and ed.) (2 vols, New Haven, CT and London: Yale University Press, 1969).

Pinkerton, John, *Early Australian Voyages. Pelsart.* [sic] *Tasman. Dampier* (London: Cassell and Company, 1886).

Plato, *Timaeus, Critias, Cleitophon, Menexenus, Epistles*, R.G. Bury (trans.) (Cambridge, MA: Harvard University Press, 1929).

Plato, *The Republic*, Paul Shorey (trans.) (London: Heinemann, 1935).

Plato, *Phaedo*, C.J. Rowe (ed.) (Cambridge: Cambridge University Press, 1993).

Playford, Philip, *Carpet of Silver: The Wreck of the Zuytdorp* (Nedlands: University of Western Australia Press, 1999).

Pliny the Elder, *Naturalis historiae libri xxxvii*, C. Mayhoff (ed.) (6 vols, Leipzig: Teubner, 1892–1909).

Pohlsander, H.A., 'Victory: The Story of a Statue', *Historia*, 18 (1969): 588–97.

Polet, Sybren, 'De nachtmerrie van het Nieuwe Amsterdam', *In de arena* (Amsterdam: De Bezige Bij, 1987).

Polo, Marco, *Le divisament dou monde*, Gabriella Ronchi (ed.) (Milan: Mondadori, 1982).

Popelinière, Henri-Lancelot Voisin, sieur de la, *L'Amiral de France, et par occasion, de celuy des autres notions, tant vieiles [sic] que nouvelles* (Paris: T. Périer, 1584).

Popelinière, Henri-Lancelot Voisin, sieur de la, *Les trois mondes*: *Les Trois Mondes de La Popelinière*, Anne-Marie Beaulieu (ed.) (Geneva: Droz, 1997).

Pratt, Mary Louise, *Imperial Eyes: Travel Writing and Transculturation* (London and New York: Routledge, 1992).

Preston, D. and M. Preston, *A Pirate of Exquisite Mind: Explorer, Naturalist and Buccaneer: The Life of William Dampier* (New York: Walker and Co., 2004).

Ptolemy, Claudius, *Ptolemy's Geography: An Annotated Translation of the Theoretical Chapters*, J. Lennart Berggren and Alexander Jones (trans) (Princeton, NJ: Princeton University Press, 2000).

Quirós, Pedro Fernández, *Descubrimiento de las regiones austriales*, Roberto Ferrando (ed.), *Crónicas de América*, vol. 25 (Madrid: Historia 16, 1986).

Quirós, Pedro Fernández, 'Viajes de Quirós', in Justo Zaragoza (ed.), *Historia del descubrimiento de las regiones austriales hecho por el general Pedro Fernández de Quirós* (3 vols, Madrid: Manuel G. Hernández, 1876. Facsimile in 1 vol.; Madrid: Dove, 2000), pp. 115–308.

Rainaud, Armand, *Le continent australe, hypothèses et* découvertes (Paris: Colin, 1893; repr. Amsterdam: Meridian, 1965).

Recueil de plusieurs Relations tant anciennes que modernes, Voyages, Navigations, et Mémoires, c'est a dire, La cinquième partie de L'Univers dépeinte dans toutes les Mappes-monde, sous le tiltre de Terres Australes Incogneuës pour Monsieur du Chesne Conseiller du Roy, Historiographe de France, en Advocat en tous les Conseils de sa Majesté 1659 [BnF, NAF 7454], transcribed in Abbé Paulmier, *Mémoires*, Margaret Sankey (ed.), pp. 317–34.

Rehorst, A.J., *Torrentius* (Rotterdam: W.L. & J. Brusse NV, 1939).

Requeste presentée a Philippes III Roy d'Espagne, par Pierre Ferdinand de Quir, Portugais, pour tascher d'obliger sa Majesté catholique a envoyer des Colonies dans les Terres Australes, contenant une sommaire description d'une partie de ces Terres, descouvertes par Ledit de Quir et ceux de sa flotte. Traduite par Monsieur de Flacourt, l'un des directeurs de la Compagnie françoise d'orient, et Commandant de la Colonie françoise de Madagascar [BnF, Ms NAF, 7454], transcribed in Abbé Paulmier, *Mémoires*, Margaret Sankey (ed.), pp. 289–302.

Review of 'Péron, *Voyage de découvertes aux Terres Australes*', *Quarterly Review*, 4.7 (August 1810) : 42–60.

Richardson, W.A.R., 'Toponymy and the History of Cartography', *Journal of the Royal Australian Historical Society*, 78, Parts 1 and 2 (June 1992): 125–9.

Richardson, W.A.R., 'Mercator's Southern Continent: Its Origins, Influence and Gradual Demise', *Terrae Incognitae*, 25 (1993): 67–98; repr. in Tony Ballantyne (ed.), *Science, Empire and the European Exploration of the Pacific* (Aldershot: Ashgate Variorum, 2004), pp. 12–42.

Richardson, W.A.R., 'East and Southeast Asia: Cartographers' Attempts to Reconcile the Maps of Ptolemy and Martellus with Marco Polo's *Travels* and with Portuguese Charts', *Terrae Incognitae*, 32 (2000): 1–22.

Richardson, W.A.R., 'South America on Maps before Columbus? Martellus' "Dragon's Tail" Peninsula', *Imago Mundi*, 55 (2003): 25–37.

Richardson, W.A.R., *Was Australia Charted before 1606? The* Jave la Grande *Inscriptions* (Canberra: National Library of Australia, 2006; repr. 2008).

Rivas, Michèle, 'Un navigateur-naturaliste d'origine poitevine célèbre en Australie, méconnu dans sa patrie: Nicolas Baudin (1754–1803)', *Revue historique du Centre-ouest*, 5 (2006): 77–8.
Rivière, M.S., *A Woman of Courage: The Journal of Rose de Freycinet on Her Voyage Around the World 1817–1820* (Canberra: National Library of Australia, 1996).
Robinson, Allan, *In Australia Treasure is Not for the Finder* (Perth: Vanguard, 1980).
Robson, John, 'Somebody Knows What!: The Cartographic Results of the Visits by Vancouver and Malaspina to New Zealand in the 1790s', *New Zealand Map Society Journal*, 15 (2002): 35–47.
Roeck, Bernd, 'Introduction' to *Forging European Identities, 1400–1700, Volume IV Cultural Exchange in Early Modern Europe* (Cambridge: Cambridge University Press, 2007), pp. 1–29.
Roeper, Vibeke, *Het Schipbreuk van de Batavia, 1629* (Leiden: Zutphen, 1993).
Rol, Greta van der, *Die a Dry Death* (Dragon International Independent Arts.com, 2010)
Romer, F.E., *Pomponius Mela's Description of the World* (Ann Arbor, MI: University of Michigan Press, 1998).
Romm, James S., *The Edges of the Earth in Ancient Thought* (Princeton, NJ: Princeton University Press, 1992).
Ross, John, 'The Framing and Printing of the Motte Editions of *Gulliver's Travels*', *Bibliographical Society of Australia and New Zealand Bulletin*, 20 (1996): 5–19.
Rougé, Jean, *Expositio Totius Mundi et Gentium: introduction, texte critique, traduction, notes et commentaire*, Sources Chrétiennes, 124 (Paris: du Cerf, 1966).
Russell, Lynette and Ian J. McNiven, *Appropriated Pasts: Indigenous Peoples and the Colonial Culture of Archaeology* (Lanham, MD: Alta Mira, 2005).
Rye, Peter, *An excursion to the Peak of Teneriffe, in 1791; being the substance of a letter to Joseph Jekyll ... from Lieutenant Rye of the Royal Navy* (London: Printed for R. Faulder, 1793).
Said, Edward, *Orientalism* (London and Harmondsworth: Penguin, 1978).
Sankey, Margaret, 'Est ou Ouest: le mythe des terres australes en France aux XVIIe et XVIIIe siècles', in Issur, Kumari and Vinesh Hookoomsing (eds), *L'Océan Indien dans les littératures francophones: pays réels, pays rêvés, pays révélés* (Paris and Réduit: Editions Karthala et Presses de l'Université de Maurice, 2001), pp. 13–26.
Sankey, Margaret, 'L'Abbé Paulmier méconnu: le mythe et l'histoire des *Terres australes* en France au dix-septième et dix-huitième siècles', *Studies*

in Eighteenth-Century French Fiction in Honour of Angus Martin, Wallace Kirsop (ed.), *Australian Journal of French Studies*, 38.1 (2001): 54–68.

Sankey, Margaret, 'Madagascar et la terre de Gonneville: tribulations d'un mythe des origines', *Ports et voyages dans le sud-ouest de l'océan Indien, XVII–XXe siècles, Revue Historique des Mascareignes*, 5 (2004): 147–56.

Scott, Ernest, *Terre Napoléon: A History of French Exploration and Projects in Australia* (London: Methuen & Co., 1910).

Scott, Ernest, *The Life of Captain Matthew Flinders, RN* (Sydney: Angus & Robertson, 1914).

Scott, James, *Remarks on a Passage to Botany Bay 1787–92: A First Fleet Journal* (Sydney: Public Library of New South Wales in association with Angus & Robertson, 1963).

Seneca, *Epistulae morales*, L.D. Reynolds (ed.) (2 vols, Oxford: Clarendon Press, 1965).

Seriman, Zaccaria, *Viaggi di Enrico Wanton ai terre incognite australi ed ai Regni delle scimie e de' cinocefali* (2 vols, Bern, 1764).

Seymour, M.C. (ed.), *Mandeville's Travels* (Oxford: Clarendon Press, 1967).

Shakespeare, William, *Othello*, Norman Sanders (ed.) (Cambridge: Cambridge University Press, 1984).

Shakespeare, William, *The Riverside Shakespeare*, G. Blakemore Evans et al., 2nd edn (Boston, MA: Houghton Mifflin, 1997).

Shirley, James, *James Shirley's The Humorous Courtier*, Marvin Morillo (ed.) (New York: Garland, 1979).

Shirley, Rodney W., *The Mapping of the World: Early Printed World Maps 1472–1700* (London: Holland Press, 1983; 2nd edn Riverside: Early World Press, 2001).

Sidebotham, Steven E., 'Ports of the Red Sea and the Arabia-India Trade', in Vimala Begley and Richard Daniel de Puma (eds), *Rome and India: The Ancient Sea Trade* (Madison, WI: University of Wisconsin Press, 1991), pp. 12–38.

Siebenhaar, Willem, 'The Abrolhos Tragedy', *Western Mail* (Perth), Christmas special, 1897.

Sigmond, J.P. and L.H. Zuiderbaan, *Dutch Discoveries of Australia: Shipwrecks, Treasures and Early Voyages off the West Coast* (Adelaide: Rigby, 1978).

Smeeks, Hendrik, *The Mighty Kingdom of Krinke Kesmes (1708)*, Robert H. Leek (trans.) (Amsterdam: Rodopi, 1995).

Smith, Bernard, *The Antipodean Manifesto: Essays in Art and History* (Oxford: Oxford University Press, 1976).

Smith, Bernard, *Imagining the Pacific: In the Wake of the Cook Voyages* (New Haven, CT and London: Yale University Press, 1992).

Spate, O.H.K., 'The Literature of Cook's Voyages', *La Trobe Journal*, 41 (1988): 25–9.
Stafford, Robert A., *Scientist of Empire: Sir Roderick Murchison, Scientific Exploration and Victorian Imperialism* (Cambridge: Cambridge University Press, 1989).
Staham-Drew, Pamela, *James Stirling: Admiral and Founding Governor of Western Australia* (Perth: University of Western Australia Press, 2003).
Stahl, William Harris, *Commentary on the Dream of Scipio by Macrobius: Translated with an Introduction and Notes* (New York: Columbia University Press, 1952).
Stallard, Avan Judd, 'Origins of the Idea of the Antipodes: Errors, Assumptions, and a Bare Few Facts', *Terrae Incognitae*, 42 (2010): 34–52.
Stannage, C.T. (ed.), *A New History of Western Australia* (Perth: University of Western Australia Press, 1981).
Stevenson, E.L., *Terrestrial and Celestial Globes* (New Haven, CT: Yale University Press, 1921).
Stevenson, Seth W., *A Dictionary of Roman Coins* (London: B.A. Seaby, 1884; repr. 1964).
Stewart, Douglas, *Shipwreck: A Poetic Drama* (Sydney: Shepherd Press, 1947).
Stewart, Douglas, *Four Plays* (Sydney: Angus & Robertson, 1958).
Stokes, John Lort, *Discoveries in Australia, 1837–43* (London: T. and W. Boon, 1846).
Stow, Randolph, 'The Southland of Antichrist: The History and Myth of the *Batavia* Disaster (1629)', in Anna Rutherford (ed.), *Common Wealth* (Aarhus: Akademisk Boghandel, 1971), pp. 160–67.
Strabo, *Geography*, H.L. Jones (trans.) (Cambridge, MA: Harvard University Press, 1917).
Strutt, Jessica, 'Aboriginals Seek Dutch DNA Link', *The West Australian* (Perth), 17 April 2010.
Sturma, Mike, 'Mutiny and Narrative: Francisco Pelsaert's Journals and the Wreck of the *Batavia*', *The Great Circle*, 24.1 (2003): 14–24.
Swift, Jonathan, *Gulliver's Travels*, Harold Williams (ed.) (London: First Edition Club, 1926).
Swift, Jonathan, *Gulliver's Travels*, Herbert Davis (ed.) (Oxford: Blackwell, 1959).
Tangl, Michael (ed.), *Die Briefe des heiligen Bonifatius und Lullus* (Berlin: Weidmannsche Buchhandlung, 1916).
Tavan, Gwenda, *The Long, Slow Death of White Australia* (Melbourne: Scribe, 2005).

Taylor, Antony, '"'Heaven on Earth' Was a Hell in Reality": Reflections on William Lane's "New Australia" in Britain', in Renata Summo-O'Connell (ed.), *Imagined Australia: Reflections around the Reciprocal Construction of Identity between Australia and Europe* (Bern: Peter Lang, 2009), pp. 171–84.

Tcherkézoff, Serge, *Tahiti-1768: Jeunes filles en pleurs* (Papeete: Au Vent Des Isles, 2004).

Tench, Watkin, *A Narrative of the Expedition to Botany Bay; with an account of New South Wales, its productions, inhabitants, &c. To which is subjoined, a list of the civil and military establishments at Port Jackson. By Captain Watkin Tench, Of The Marines* (London: J. DeBrett, 1789).

Tench, Watkin, *A Complete Account of the Settlement at Port Jackson, in New South Wales, including an accurate description of the situation of the colony; of the natives; and of its natural productions: taken on the spot by Captain Watkin Tench of the Marines* (London: Sold by G. Nicol and J. Sewell, 1793).

Tenison-Woods, J.E., *History of the Discovery and Exploration of Australia* (2 vols, London: Sampson Low, Son, and Marston, 1865).

Thévenot, Melchisédech (ed.), *Relations de divers voyages curievx, qvi ont esté traduites d'Hacluyt, de Purchas, & d'autres voyageurs Anglois, Hollandois, Portugais, Allemands, Espagnols …* (4 vols, Paris: Iacqves Langlois, 1663–72).

Thomas, Beatrice, 'Gene Tests Seek Dutch Links to WA Aboriginals', *The West Australian* (Perth), Thursday, 18 December 2008, p. 60.

Thomas, Richard S., 'Virgil's Georgics and the Art of Reference', *Harvard Studies in Classical Philology*, 90 (1986): 171–98.

Thorndike, Lynn, *The Sphere of Sacrobosco and Its Commentators* (Chicago, IL: Chicago University Press, 1949).

Toulouse, Sarah, 'Marine Cartography and Navigation in Renaissance France', in David Woodward (ed.), *The History of Cartography*, vol. 3.2 (Chicago, IL: University of Chicago Press, 2007), pp. 1550–68.

Trickett, Peter, *Beyond Capricorn: How Portuguese Adventurers Secretly Discovered and Mapped Australia and New Zealand 250 Years before Captain Cook* (Adelaide: East Street Publications, 2007).

Trickett, Peter, *Para Além de Capricórnio* (Lisboa: Caderno, 2008).

Trustees of the British Museum, *Coins of the Roman Republic in the British Museum* (London: British Museum Press, 1970).

Tyler, Stephen, 'Post-modern Ethnography: From Document of the Occult to Occult Document', in James Clifford and George E. Marcus (eds), *Writing Culture: The Poetics and Politics of Ethnography* (Berkeley and Los Angeles, CA and London: University of California Press, 1986), pp. 122–40.

Tyssot de Patot, *Voyages et avantures de Jaques Massé* (Bourdeaux (La Haye): Jaques L'Aveugle, 1710).

Updegraff, Robert T., 'The Blemmyes I: The Rise of the Blemmyes and the Roman Withdrawal from Nubia under Diocletian (with additional remarks by L. Török)', in Hildegarde Temporini (ed.), *Aufstieg und Niedergang der Römischen Welt*, vol. 2.10.1 (Berlin and New York: de Gruyter, 1988), pp. 44–106.

Uren, Malcolm, *Sailormen's Ghosts: The Abrolhos Islands in Three Hundred Years of Romance, History and Adventure* (Melbourne: Robertson & Mullen, 1940).

Vairasse, Denis, *Histoire des Sévarambes, Peuples qui habitent une partie du troisième continent, communément appelé la terre australe. Contenant un compte exact du Gouvernement, des Mœurs, de la Religion, du langage de cette Nation, jusques aujourd'huy inconnuë des Peuples de l'Europe.* Traduit de l'anglois, 4 tomes (2 vols, Paris: Claude Barbin, 1677–79; repr. Geneva: Slatkine Reprints, 1979).

Van Duzer, Chet, 'Cartographic Invention: The Southern Continent on Vatican MS Urb. Lat. 274, fols 73v–74r (c. 1530)', *Imago Mundi*, 59 (2007): 193–222.

Veerman, Elmaar, 'Aboriginals met Hollands bloed. Gingen VOC-schipbreukelingen op in de lokale bevolking?', *Noorderlicht uw Hoofdleverancier*, 12 April 2010. Online at: http://beagle.vpro.nl/#/blog/item/3339/ [accessed: 24 September 2010].

Vega, Lope de, *El Nuevo Mundo descubierto por Cristóbal Colón*, J. Lemartinel and Charles Minguet (eds) (Lille: Presses universitaires de Lille, 1980).

Veiras, Denis, *L'Histoire des Sévarambes* (2 vols, Paris, 1677).

Veiras, Denis, *The History of the Sevarites or Sevarambi: A Nation inhabiting part of the third Continent, Commonly called, Terrae Australes Incognitae* (2 vols, London, 1679).

Verger, Jacques, 'Patterns', in Hilde de Ridder-Symoens (ed.), *A History of the University in Europe*, vol. 1 (Cambridge: Cambridge University Press, 1992), pp. 35–74.

Virgil, *Aeneid*, in R.A.B. Mynors (ed.), *Opera* (Oxford: Clarendon Press, 1969).

Vitoria, Francisco di, *Leçons sur les Indiens et le droit de guerre* (1532), Maurice Barbier (ed.) (Geneva: Droz, 1966).

Vries, Susanna de, *Desert Queen: The Many Lives and Loves of Daisy Bates* (Sydney: HarperCollins, 2008).

Wallis, Helen, 'Java-la-Grande: The First Sight of Australia?', in Helen Wallis (ed.), *The Maps and Text of the Boke of Idrography presented by Jean Rotz to Henry VIII* (Oxford: Roxburghe Club, 1981), pp. 58–67.

Wallis, Helen (ed.), *The Maps and Text of the Boke of Idrography presented by Jean Rotz to Henry VIII* (Oxford: Roxburghe Club, 1981).

Wallis, Helen, 'Visions of Terra Australis in the Middle Ages and Renaissance', in William Eisler and Bernard Smith (eds), *Terra Australis: The Furthest Shore* (Sydney: International Cultural Corp. of Australia, 1988), pp. 35–8.

Walsh, Fiona Jane, 'Review of Gerritsen's *Australia and the Origins of Agriculture*', *GeoJournal*, 74 (2009): 499–501.

Walter of Châtillon, *Alexandreis*, Marvin L. Colker (ed.) (Padua: Antenore, 1978).

Ward, Russel, *The Australian Legend* (Melbourne: Oxford University Press, 1958).

Weddell, J., *A Voyage towards the South Pole, performed in the years 1822–24, containing an examination of the Antarctic Sea, &c*, 2nd edn (London: Longman, Rees, Orme, Brown and Green, 1827).

Weinstock, Stefan, *Divus Julius* (Oxford: Oxford University Press, 1971).

Werner, Louis, 'Libya's Forgotten Desert Kingdom', *Saudi Aramco World*, May/June (2004): 8–13.

Westrem, Scott D., 'Against Gog and Magog', in Sylvia Tomasch and Sealy Gilles (eds), *Text and Territory: Geographical Imagination in the European Middle Ages* (Philadelphia, PA: University of Pennsylvania Press, 1998), pp. 54–75.

White, John, *Journal of a Voyage to New South Wales* (London: J. Debrett, 1790).

Willard, Myra, *History of the White Australia Policy to 1920* (Melbourne: Melbourne University Press, 1923).

Williams, G.C., 'William Dampier: Science, Exploration and Literary Influence Including his Hydrographic Treatise of 1699', *Proceedings of the California Academy of Sciences*, series 4, 59.14 (2008): 533–663.

Williams, Glyndwr, *The Expansion of Europe in the Eighteenth Century: Overseas Rivalry, Discovery and Exploitation* (London: Blandford, 1966).

Williamson, J.A. (ed.), *William Dampier: A Voyage to New Holland* (London: The Argonaut Press, 1939).

Young, Gary K., *Rome's Eastern Trade: International Commerce and Imperial Policy 31 BC–AD 305* (London and New York: Routledge, 2001).

Zanker, Paul, *The Power of Images in the Age of Augustus*, Alan Shapiro (trans.) (Ann Arbor, MI: University of Michigan Press, 1988).

Index

For Product Safety Concerns and Information please contact our EU representative GPSR@taylorandfrancis.com
Taylor & Francis Verlag GmbH, Kaufingerstraße 24, 80331 München, Germany

www.ingramcontent.com/pod-product-compliance
Ingram Content Group UK Ltd.
Pitfield, Milton Keynes, MK11 3LW, UK
UKHW020019200926
475702UK00014B/1965
9781138110977